Fences
for
Home
and
Garden

FENCES

FOR HOME AND GARDEN

A Complete Guide to Selecting and Installing
Wood, Masonry, Metal, and Living Fences

• •

JEFF BENEKE

Illustrations by Melanie Powell

Storey Publishing

*The mission of Storey Publishing is to serve our customers by
publishing practical information that encourages
personal independence in harmony with the environment.*

Edited by Nancy W. Ringer and Sarah Guare Slattery
Cover design by Ian O'Neill
Book design by Kent Lew
Text production by Jennifer Jepson Smith
Cover photography by © Catriona Tudor Erler, back;
　© Dietrich Leppert/Shutterstock.com, front (b.l.);
　© Hamish Gray/Shutterstock.com, front (t.); © Roger
　Foley, front (m.l.); © Scott Barrow/Getty, front (b.r.)
Interior photography credits appear on page 252
Illustrations by Melanie Powell

Text © 2005, 2025 by Jeff Beneke
Previously published under the title *The Fence Bible*
　(Storey, 2005)

• •

Storey books may be purchased in bulk for business,
educational, or promotional use. Special editions or
book excerpts can also be created to specification.
For details, please contact your local bookseller or the
Hachette Book Group Special Markets Department at
special.markets@hbgusa.com.

• •

Storey Publishing
210 MASS MoCA Way
North Adams, MA 01247
storey.com

Storey Publishing is an imprint of Workman
Publishing, a division of Hachette Book Group, Inc.,
1290 Avenue of the Americas, New York, NY 10104.
The Storey Publishing name and logo are registered
trademarks of Hachette Book Group, Inc.

ISBNs: 978-1-63586-916-3 (paperback);
978-1-63586-917-0 (ebook)

Printed in China by Shenzhen Reliance Printing Co.
Ltd. on paper from responsible sources
10 9 8 7 6 5 4 3 2 1

REL

Library of Congress Cataloging-in-Publication Data
　on file

For Molly, Erica, and Kathryn.
My inspiration. My hopes. My heroes.

■　■　■

Contents

Introduction

· ·

LIKE MANY OTHER FOLKS, my first experiences in fence construction were reactive and defensive, motivated by the pressing realization that deer, woodchucks, and rabbits were as interested in the fruits of my gardening as I was. I like to think that I have learned a thing or two along the way, and one of them involves the benefits of planning rather than reacting. This book is about planning, and then building, fences.

In neighboring lots of my childhood, fences served first and foremost as home-run walls. It really did not matter how far you had to hit the ball, as long as you could say "it's over the fence." Some fences were forbidding; they enclosed neighbors we barely knew, who wished to keep it that way. When a ball sailed into one of those yards, we pursued it warily, if at all. Other fences were intended to keep transient ballplayers out of the yard, but the folks who lived within them were not nearly so alien. They were instead friends, good neighbors, and we understood that their fences were not antisocial statements.

Fences can mean many things; even the word itself is full of symbolic intensity and variety. In his Pulitzer Prize–winning play *Fences*, August Wilson manages to gather many of these references and meanings in a single work. The "fences" in this play refer to a physical structure built by the main character on his property, to the legal barriers that American society enacted to justify and condone racism and segregation, and to the emotional defenses that individuals erect in their own minds that create unintended divisions within families.

Scores of song and book titles contain the word, and when we hear someone sing or say "Don't fence me in" or "I can mend your fences" or "He's sitting on the fence," we understand the reference instinctively, even though in each case the term "fence" refers to something different. No journalist or essay writer or, alas, book author, it seems, can even approach the subject without invoking Robert Frost and his poem "Mending Wall." Ask ten people at random what they think about fences, and there is a good chance that eight of them will respond "Good fences make good neighbors." And, of those eight, perhaps three or four will preface the response with "As Robert Frost said . . ." Robert Frost, that cantankerous Californian who moved east to raise chickens and transform himself into the embodiment of the Yankee homesteader, has long been cited as the principal spokesperson for the idea that we will all get along better if and when we fence ourselves in.

There is just one thing wrong with this sentiment: It's not true. Frost, in fact, did not say or even imply that "good fences make good neighbors." In this wonderful poem, which begins "Something there is that doesn't love a wall," two men have met at the boundary dividing their property. It is "spring mending-time," and they are jointly repairing the stone wall that marks their property lines, as they do every year, which has been damaged by rabbit hunters and their dogs. The narrator of the poem, who we can reasonably assume is speaking for Frost, presses his neighbor on why the stone wall is really necessary. Yes, he thinks, if we had cows, we would want to keep them enclosed, but surely my apple trees are not going to stray into your stand of pine trees. The neighbor (not the author) can only come up with one justification: "Good fences make good neighbors." His father said it was so and, therefore, it must be so. Frost's point really seems to be just the opposite of what is popularly understood: that building fences is really more of an antisocial act than a contribution to social harmony. "Before I built a wall," he suggests,

> I'd ask to know
> What I was walling in or walling out
> And to whom I was like to give offence.

With a little effort you should be able to find a recording of Robert Frost reading his poem. Give it a good listen and then dispense with the notion that building a fence will automatically improve the climate in your neighborhood. Maybe it will, maybe it won't. Fences, by themselves, do not necessarily "say" anything. They do not form communities nor divide neighbors. Fences do not make good neighbors; neighbors do.

There is also a good deal of anecdotal evidence that good fences can make neighbors hate each other. While I was working on an early draft of this book, the press was full of reports of escalating "hedge rage" in England. In one case, a man shot and killed his neighbor after arguing over the maintenance on a 35-foot-tall cypress hedge that divided their gardens. The alleged shooter was upset that the unkempt hedge creaked in the wind and cast a long shadow over his garden. This is an extreme and tragic conclusion to a much broader problem in England. The *New York Times*, I kid you not, referred to a "fledgling

hedge-dispute empowerment movement" that has taken shape in England under a group called Hedgeline, which has a website full of complaints about troublesome hedges.

Across the Atlantic, in the Hamptons on Long Island, hedge disputes also rage. As people there build more and more houses, closer and closer together, they have also erected miles upon miles of 10- to 15-foot-tall hedges in an effort to capture some sense of privacy. Some are neatly trimmed ("country club" style) and some are not ("country" style), but either way they are having a noticeable impact on the environment. Longtime inhabitants resent the proliferation of these "green walls" and seek to maintain some sense of the open fields that graced the area not so long ago. These folks might be more inclined to embrace the real sentiments of Robert Frost than those of his poetic neighbor.

If good fences do not necessarily make good neighbors, it nevertheless remains true that good neighbors can make good fences. And they can make bad ones. This book was written to encourage the former. It differs from many of the fine books that have been published on the subject by trying to be comprehensive. Books devoted exclusively to wood fences or stone fences or living fences (such as hedges) are not particularly useful in helping you to decide which type of material to use in building your own fence. They assume that you have already made that choice. This book steps back from that conclusion with the goal of helping you decide what kind of fence to build. With the decision made, you may then turn to the appropriate section of the book for detailed instructions on how to build it. It is also possible that you may want to build more than one fence on your property, each with a unique function that is best served using different materials. Whether you want a utilitarian animal enclosure, an attractive and functional privacy screen, a decorative landscaping feature, or a bit of each, the best results will come after you have carefully considered all of the styles and materials available and assessed your needs, abilities, and budget. Good planning, good materials, and good construction make good fences. ■

Fencing, Then and Now

WHAT IS A FENCE? The question seems simple and straightforward enough, but the answer can be another story. New fences change the landscape and disrupt views and well-traveled pathways. Changes, even good ones, can create discomfort initially. True, with apologies to Sigmund Freud, sometimes a fence is just a fence. Often, however, it is much more.

Throughout history, fences have been objects of beauty, incitements to class warfare, revolutionary advances in agricultural production, and rigid obstacles to human interaction. Just as the term *fence* has its multiple meanings and a diverse symbolic resonance over the centuries, so the actual structure we commonly refer to as a fence has itself generated a complex history in the realms of agriculture, landscaping, law, and neighborly ethics.

Fences not only protect gardens but also help define them.

Some of this history can be gleaned from one of my favorite reads, the venerable *Oxford English Dictionary*. The very first definition of *fence* offered by the *OED* is as a medieval verb meaning "the action of defending." To fence, therefore, was to engage in an act of self-defense. The second definition refers to what we now know as the sport of fencing, that is, swordplay. A fence is also "that which serves as a defence." By the early sixteenth century the term was being used to describe "an enclosure or barrier . . . along the boundary of a field, park, yard, or any place which is desired to defend from intruders." Today, people who traffic in stolen goods are known as "fences." "Don't fence me in" begs not to be restricted, while "she's sitting on the fence" implies a lack of conviction. And, as we will discuss shortly, the combativeness and negativity that are entwined in the etymology of the word are reflected in the legal history of fencing.

In many respects, fences have helped to define who and what we are today. At one time, our ancestors roamed the fields, forests, and deserts in search of food and shelter. Fences offered the opportunity to keep cultivated crops free of wandering animals, or to keep wandering animals from straying too far. They defined property holdings, giving those who worked the land incentive to improve it over time. And once we adjusted to the idea of staying in one place, we learned to construct housing that could last, and we formed communities. When crop surpluses appeared, we discovered that not all of us had to work at subsistence all of the time, and thus we found some time to create music, games, art, wine, and other standards of modernity.

From traditional wood enclosures to unique uses of common materials to designs of fanciful shapes, fences offer unlimited opportunities to accessorize your property.

History

Throughout medieval Europe, village life depended on a system of open fields for pasturage. Although there were many variations in field systems from village to village and region to region, the most typical feature was the right for all to graze their animals on the common pasturage. Historians speculate that some of the specific field systems were prehistoric, meaning that they originated before written records were produced. For societies that depended for their survival on the food they produced themselves, field systems defined how people lived, worked, ate, and cooperated. Disputes over the rules of cultivation and grazing were often matters of life and death, and they tended to frame, if not completely overshadow, many of the conflicts that today we somewhat simplistically reduce to religious or political squabbles. Fences in particular tended to upset people.

Individual landholders worked long strips of arable land and meadow, which were often located far from each other. Long stretches of field were easier to plow, since they required less time turning around. Landholders produced their own crops of grain and hay, but they shared use of the common pastures. Draft animals, which were particularly needed in the hard fields of northern Europe, required substantial pasturage. Villages faced the task of ensuring this common pasturing, and also of controlling it. The animals had to be kept away from the fields under cultivation.

The entire community of landholders decided on crop rotation. When all of the crops had been harvested, the land was then made available for grazing by all of the livestock owned by the landholders.

The biggest human threat to this agricultural system was enclosure, or the act of denying access to the commons. In the later Middle Ages, especially in England, common lands gradually began passing into private hands. This movement away from common ownership and toward private property was often marked by the appearance of fences. Thus, enclosure came to be seen by many as a denial of justice, pure and simple. And no item was more successful at establishing enclosure than the fence. Many early American colonists brought the experience of enclosure and commons with them to the New World.

As early as 1632, the Virginia colony required crops to be fenced, and in 1646 added to this requirement a formal definition of what constituted a legal fence. Other colonies adopted similar policies. By mandating that crops be fenced in, the result was that large tracts of land were free to be used for pasture, which could not be fenced. Laws requiring fences spread west with the settlers.

Materials used for fences varied significantly from one region to the next. In some parts of the East, fieldstones (refuse from plowed fields) were commonly laid alongside fields to keep them out of the way (and,

ZIGZAG FENCE

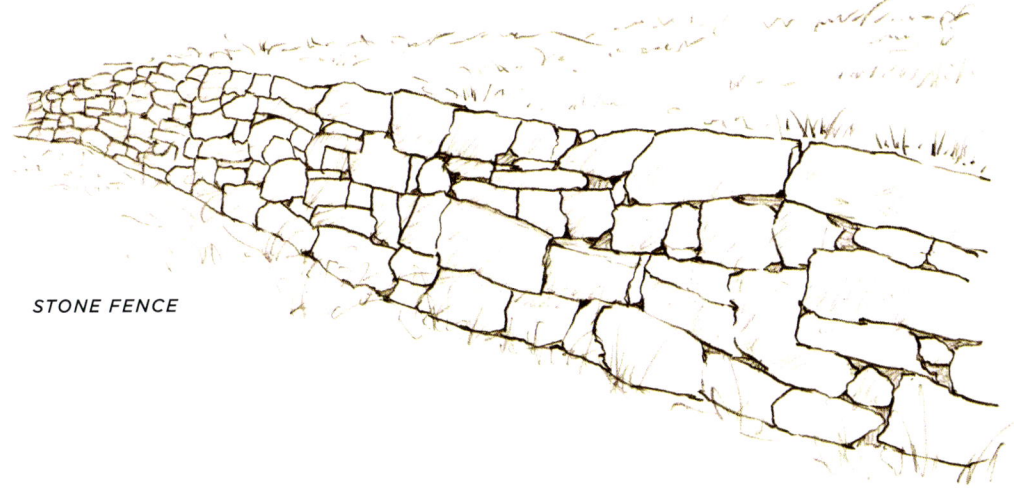

STONE FENCE

thereby, creating de facto fences). Where timber was abundant, the zigzag rail fence was widely used (it required much less backbreaking work than stones, though it was very labor-intensive). But the demand for fences severely strained timber supplies, and board and picket fences began to appear. Laid straight, they consumed less linear footage, and the use of milled boards constituted a much more efficient use of wood than the unmilled logs. As settlers moved into the treeless prairie, even those options became extreme. Wire fencing was tried, but it proved less than useful at controlling cattle and horses. Often, fences were simply constructed out of whatever was available, whether it be brushes, stumps, sod, or mud.

Since trees were abundant in the East, zigzag fences were particularly popular in that part of the country. Colonial Virginians developed a system of using eight rails, each about 10 feet long, for each section of the fence. The sections were set at about 60-degree angles, often with two vertical rails holding the horizontal rails in place on each end. This type of fence came to be known as the Virginia worm fence. Made with oak, cedar, chestnut, or other types of wood, they lasted for many years, required little upkeep, and were easy to build.

Early pole fences represented an improvement over the zigzag style in a couple of respects. They were laid straight, thus taking up less room and using less timber, and they relied on wood pins driven into bored holes for joinery, which created a strong connection. Slim, flexible twigs were also used to tie post sections together.

Good stone fences required some skill and a lot of sweat to construct. A solid foundation was a prerequisite, and subsequent layers were carefully designed to keep the fence level, with gaps between stones staggered and the weight equally distributed. Fences high enough to contain horses and cattle were particularly tough to build. When some timber was available, often a simplified pole fence could be laid over a shorter stone fence to maximize its effectiveness.

Milled lumber and metal nails helped create the board fence. Posts with flat sides would be sunk into the ground, with equally flat boards attached with nails. Board fences fit together nicely and looked neater, and once farmers and village dwellers began to appreciate their fancier creations, the era of fence design began. Rails were installed diagonally, vertical pickets were fastened to horizontal rails, cap rails were placed atop posts, and the posts themselves began to receive special, decorative attention.

POLE FENCE

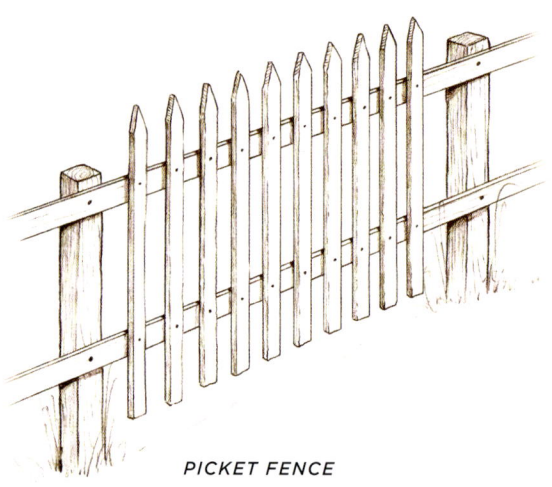

PICKET FENCE

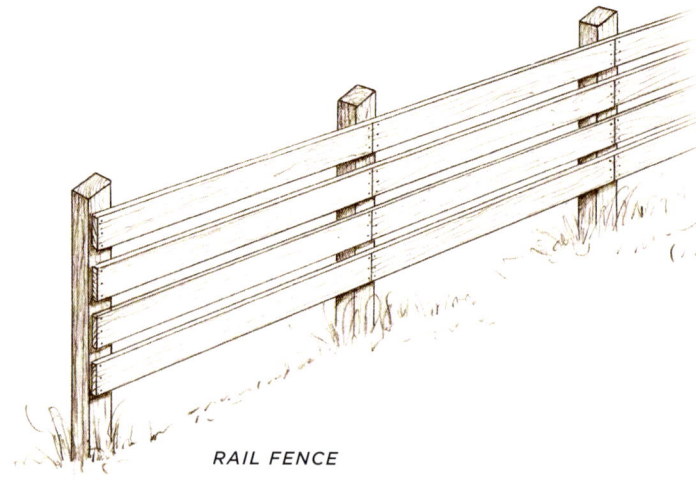

RAIL FENCE

Picket fences relied on even smaller pieces of milled lumber. Attached to horizontal rails, which themselves were tied to vertical posts, the resulting weave created a strong structure that used lumber efficiently and could effectively keep animals out of gardens. Pointed pickets added further protection against animals and people trying to climb over the top. People began painting their fences, as carpenters sought ever-new designs to increase the visual appeal of fences. Once the ornamental characteristics were appreciated, people began putting up fences as decorative features around their property. And at some point along that road, people began writing books on how to build fences.

A functional, though hardly attractive, variation on the first picket fences was a hybrid metal and wood fence. Small strips of lath were woven into two or more strands of wire. Long runs of this type of fence could be built off-site, or even mass-produced, rolled up, and transported to the fence site. This type of inexpensive "utility" fence is still available and is used for gardens and snow fencing.

Settlers east of the Mississippi River had to clear tracts of trees before they could plant crops. They built their fences from what they cut down. One style utilized the trunks of large pine trees as posts, with smaller rails inserted into mortises in the posts. Brushwood and shrubbery were laid alongside cleared fields to create impromptu fences. When a fence was not needed, the refuse was often burned.

As settlers moved beyond the timbered regions of the East into the Great Plains, they found a shortage of both trees and stones, which meant that fence building was nearly impossible. Thus the fertile fields that would later become known as the Breadbasket of the World were left uncultivated for lack of fencing, which was required to define property lines, protect crops from animals, and restrict the movement of livestock. Settlements in these regions centered around scattered wood lots and water sources.

Eventually it was discovered that what these regions had in abundance (sod) could be turned into fences and even houses. Sod cutters were developed that could cut 16-inch-wide strips to a depth of about 3 or 4 inches. With the sod removed, and recycled, the fields were ready to be plowed and planted. Mud fences were also common in some areas.

Historians maintain that the lack of adequate fencing sharply curtailed settlement in the treeless prairies of the United States. Attracted by the 1862 Homestead Act and its promise of 160 low-cost acres of land, early settlers found that they could not grow crops without fences, and they could not build fences at a reasonable price. So, many of them simply passed on through, heading for the West Coast. For some who stayed behind, the idea of fencing in animals was dismissed. Instead, the concept of common land for grazing cattle became widespread. Early settlers in west Texas began ranching with this mentality, gathering wild cattle (longhorns) that had moved north from Mexico. These new cowboys essentially roamed the open range with the cattle in search of water and pasture.

Others continued to seek cost-effective means of building fences. For a brief period (roughly 1850 to

1870), hedge fences became the rage west of the Mississippi River. Osage hedge was particularly popular. This thorn- and fruit-bearing tree grows to 40 feet. Millions of osage hedge plants were raised in nurseries in Texas, Arkansas, and other southern states and then sold to farmers to be planted for fencing. The wood from the hedge was also good for fuel, fence posts, and tool handles.

But the reasons for the success of osage hedge led to its downfall. The thorny character of the plant was the principal reason it was so successful. Cattle, horses, and animals of prey just did not like trying to breech the barrier, since it hurt and injured them. So some enterprising farmer-entrepreneurs began looking for a way to manufacture a product with similar characteristics, which would obviously be much more efficient than growing the plant. In 1873 an Illinois farmer named Joseph F. Glidden perfected a process for producing twisted wire that contained coiled barbs, and he began selling his barbed wire the next year. Other manufacturers quickly joined in, and factories sprouted almost overnight, producing what quickly became one of the defining commodities of nineteenth-century American innovation. The incipient osage-hedge industry quickly evaporated.

In its first year on the market, about 10,000 pounds of barbed wire was sold. Within only six years, that figure rose to over 80 million pounds. Barbed wire was relatively cheap, and it could be installed quickly. Posts had to be installed, which did take some time,

BARBED-WIRE FENCE

but once that chore was done, the wire could be unwound almost as fast as it was attached to the posts.

Suddenly, large tracts of public land were enclosed in barbed wire by land companies, shutting those itinerant cattlemen out of water sources. One single enclosure reported to the United States General Land Office in 1884 contained 600,000 acres, and another swallowed up 40 townships. Ambitious cattlemen installed hundreds of miles of barbed wire fencing. By 1888 more than 7 million acres of public lands were enclosed illegally, by which time the federal government had begun taking steps to curtail and reverse the practice of enclosing public lands.

Simultaneously, most privately owned rangeland on the Great Plains was fenced in. Trail driving and free grazing, staples of life on the Plains in previous decades, quickly disappeared. Owners of livestock who planted few crops favored "free grass" (that is, open grazing), and maintained that crops should be fenced in so that animals could be free to roam and graze. Farmers, however, kept few animals and wanted open fields to grow their crops. They promoted "herd law," which required that pasturelands be fenced, leaving fields under cultivation unfenced. Where these two cultures collided, conflicts arose. As the historian Walter Prescott Webb wrote in his classic 1931 study, *The Great Plains*, "It is not too much to say that in the middle and later years of the decade 1870–1880 the questions pertaining to fencing occupied more space in the public prints in the prairie and Plains states than any other issue—political, military, or economic" (p. 282).

Barbed wire helped end the era of cowboys, cattle trails, and the open range. There was resistance, in the form of fence cutting, which became widespread in Texas in the 1880s. And there were frequent violent clashes between poor homesteaders and wealthier cattlemen. But, in a very brief period of time, barbed wire ensured the predominance of farming over ranching throughout the Great Plains. Fencing also proved to be an effective deterrent to another industry, as cattle rustlers were no longer able to count on an open range to make their escape with herds of cattle belonging to others. Cattlemen increasingly turned their attention to improving the breeds of livestock and making water available.

Fence Law

WHAT IS A FENCE? I posed this question earlier and return to it now from a different angle. Not surprisingly, fences are often subject to strict legal definitions and restrictions, many of which are direct outgrowths of the struggles discussed above. Building codes, zoning ordinances, and subdivision homeowners' associations may specify whether or not you can build a fence at all, what style of fence you can build, how high the fence can be, and how far it must be kept from the property line or from the street.

If you want to build a fence, the first place to begin your legal research is with the local building department. Your building department may not actually call itself The Local Building Department, however, and in some areas it may not be so much a department as an individual with a small office and irregular hours. If you do not know whom to call, search online for the website of your local government body. For my own township, the relevant phone number is listed under "Zoning and Building," while the larger nearby city boasts a listing for "Building Department." In other areas, you will need to find the Planning Department or the city attorney's office.

Once you find an appropriate official, be prepared to explain the type of fence you want to build, how long and high you want to build it, and how close you want to place it to the neighbors' property, to sidewalks, or to the street. The official should be able to tell you right away whether or not you will need a building permit and what limitations may be placed on your plans.

In reality, fences are constructed all the time that do not meet the letter of the law. They may be a few inches too high or too close to the street, or they may have been built without a proper permit. When the affected parties have no complaints, a type of "don't ask, don't tell" policy often prevails. But keep in mind that even if this informality is the norm in your community, all it takes is one person (perhaps a new neighbor) to challenge the legality of your fence, and that could prove to be an expensive and emotionally draining experience.

Building Permits

If you do need a building permit, find out what information you need to submit, how much it will cost, how long you are likely to have to wait, and at what stage or stages you will need to have the fence inspected. If you plan to take your time building the fence, you should also ask how long the permit will remain valid; permits are typically valid for no more than one year.

Obtaining a building permit can be a quick, painless affair, or it can be a protracted hassle. If your proposed project has to be reviewed by an architectural review board or some similar body, you may first have to present detailed drawings of the fence, including information on color, type of material, and design features such as the shape of the finials that will rest on top of the posts. You may be required to revise your plans, and you could find yourself waiting for many months before obtaining a decision. Should you find yourself in that circumstance (waiting, waiting, waiting), resist the urge to get a head start on your fence, as such cheating may well incite a negative backlash.

Zoning Ordinances

Most towns have a book of zoning ordinances, which is often as thick as (or thicker than) the local phone book. Most zoning matters relate to delineating residential, commercial, and industrial areas. While property owners often scream to the heavens about the alleged infringement on property rights that zoning imposes, it is because of zoning laws that someone cannot build a gas station next to your house or a strip club next to your school.

Zoning laws frequently address fences. Common restrictions include placing a 3-foot height limit on fences facing the street and a 6-foot limit on side and backyard fences. They will likely define the required setback, if any. Often, you cannot build a fence, deck, driveway, or addition right up to the edge of your property line. Instead, you will need to set it back several feet from the property line.

When a Fence Is Not a Fence

If you run into a code or zoning problem with a fence project, it is worthwhile finding out exactly what the relevant board considers to be a "fence." You may find that a fence is defined as something that is built, and not something that is planted. If you are prohibited from building a fence high enough to suit your needs, you may nevertheless be perfectly within your rights to plant a hedgerow or some trees that can exceed that height. Such a "living fence" (see Chapter 6) may not afford the level of security you desire, but it can be an effective privacy screen. And neighbors might be much less offended by some shrubbery than by a solid fence.

Or you may want to ponder the example of Thomas White, who served as President George W. Bush's Army secretary until his sudden resignation in April 2003. White was the vice chairman of the Enron Corporation until his Pentagon appointment in 2001. In the months prior to Enron's bankruptcy, White sold his 405,710 shares of Enron stock for $14 million. He used the handsome profit to buy a $6.5 million chunk of beachfront property in Naples, Florida, and proceeded to build a 15,145-square-foot house, at a cost of $4 million. But White wanted to build a 6-foot-9-inch-high wall around his property, while the local building code had a 3-foot height limit on such structures. White's initial request was rejected by the city council, which feared that the proposed wall would make the property look like a fortress. So White revised his request, proposing instead to build a 3-foot-11-inch wall topped with 18 inches of wrought iron, and he was granted a variance. Evidently dissatisfied with the compromise, however, the Army secretary then decided to sell the house for a cool $13.9 million.

There are a few lessons to be learned from Secretary White's travails. The first is that prudent timing in the sale of stock options can net you a nice piece of property. Second, local boards can and will say no to your requests if they think it will adversely affect the look of the community. And, third, if you want to build a fence higher than allowed, offer to use a more transparent material on top of the fence, such as metal or lattice, and you may get your way.

A high, solid stucco wall [LEFT] may not be allowed by local code, but you might be able to achieve the same effect by building a shorter fence and topping it with lattice [RIGHT] or vines.

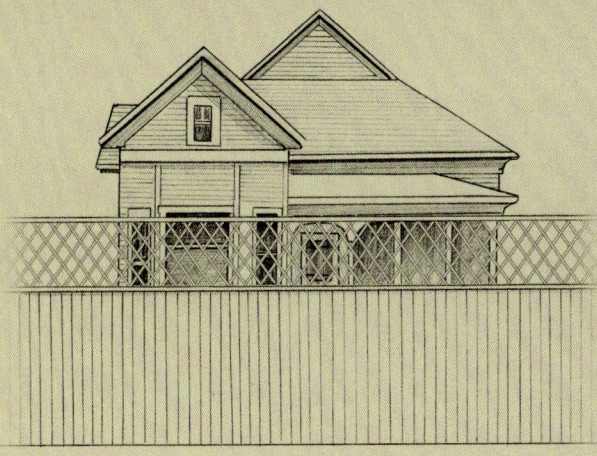

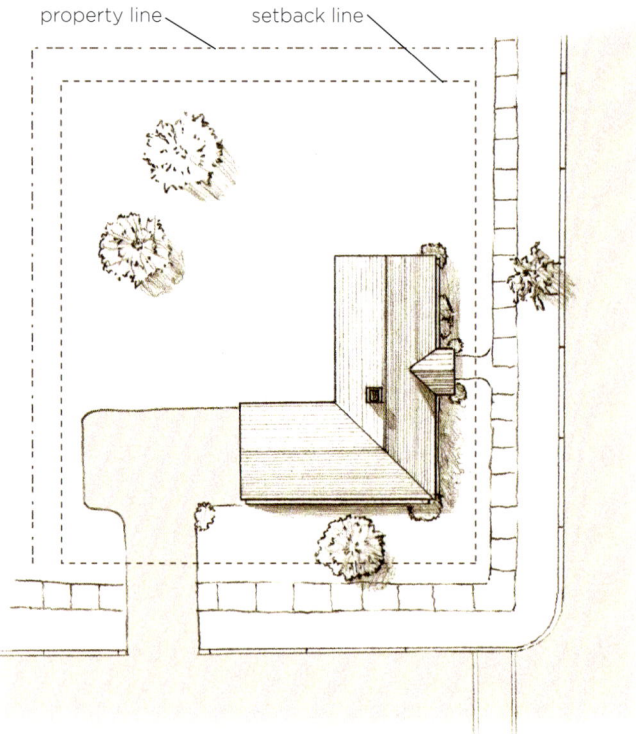

property line · setback line

Even though you own the property out to the property line, you may not be allowed to add new construction beyond the setback line.

plans. Perhaps it is because the zoning ordinance was devised to preserve the architectural character of the neighborhood and, by implication, property values. Or a safety issue may be involved, such as keeping fences on corner lots low enough not to restrict traffic visibility. Challenging these types of restrictions can be very difficult, if not fruitless.

On the other hand, say that you want to build a privacy fence to block an unpleasant view, but that the view in question is uphill from the intended fence location. In this case, you could argue that a height limit of 6 feet restricts you from building an effective fence, while an 8- or 10-foot fence would satisfy your needs just fine. If there were no other issues involved, and particularly if your neighbors had no objections, you might be able to obtain a variance quite easily.

But there are no guarantees. When the rap artist Eminem, whose stage name derives from the initials of his legal name, Marshall Mathers, became a sudden sensation, he found himself with a fencing versus zoning dilemma. Local youth started littering his Sterling Heights, Michigan, yard with empty M&M candy wrappers. Mathers requested that the city grant him a variance to build a 12-foot-high fence to limit the mischief, but his request was denied. The performer responded by selling the house and moving into an upscale, gated community. Most of us cannot afford to take such steps, but if you can convince the zoning board that your project would not adversely affect safety or property values, and that the ordinance

Variances

If you want to build a fence 20 feet from the street, but the zoning law says it must be at least 30 feet from the street, you can apply for a variance. A variance is simply permission to build a structure even though it violates the existing requirements. Assessing variance requests is a major duty of zoning boards, and you should not hesitate to apply for one if you feel you have a legitimate argument.

If you apply for a variance, be prepared to do some research and present a good case. Begin by finding out why you are being prohibited from building to your

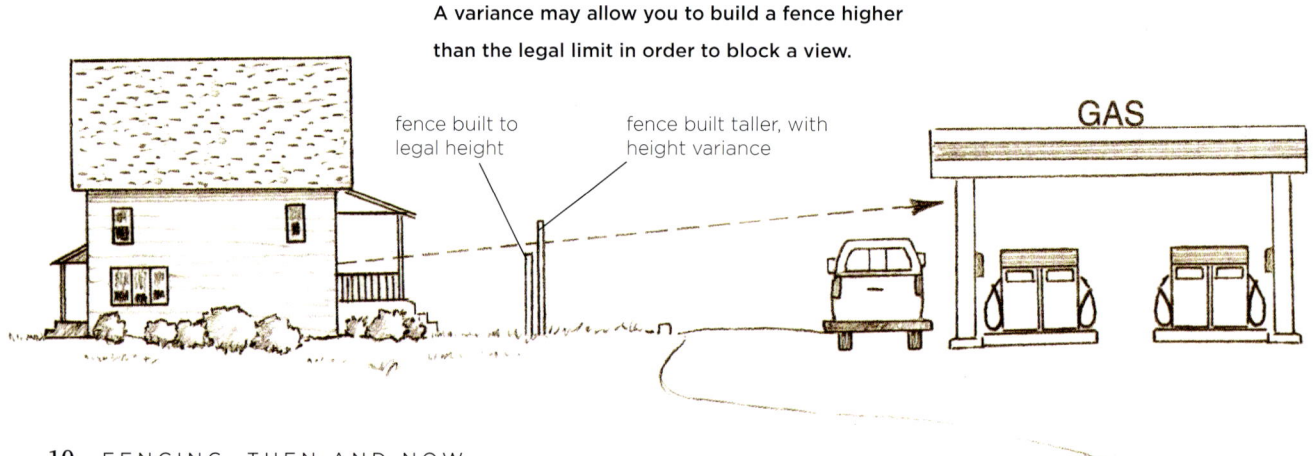

A variance may allow you to build a fence higher than the legal limit in order to block a view.

fence built to legal height

fence built taller, with height variance

GAS

places an undue burden (financial or otherwise) on your plans, you may have better success with your variance request than did Eminem.

Easements

Easements can also throw a monkey wrench into fence-building plans. "Easement" is a shorter name for "property use agreement," and it defines the rights of parties other than the property owner to use the property for specific purposes. A common type of easement is the right of a utility company to drive their trucks onto or through your property to tend to repairs. Sometimes adjoining properties have "shared driveway" agreements, in which both property owners are permitted to use the same passageway to reach their respective lots or homes. In cases such as these, you cannot construct a fence that blocks the other party's right to pass through.

Any existing easements on your property should be noted in the title report of the title insurance company that covers your property. Usually, you will have been made aware of any easements on your property at the time you bought it, but regardless of whether or not that is true in your case, as the property owner it is your obligation to know about any easements attached to your property. If you build a fence that violates the easement, even if you did not know about the easement, you may have to dismantle the fence at your own expense. So if you have any concerns about easements, it might be wise to discuss the matter with a real estate attorney.

Boundary Disputes

In strict legal terms, a boundary (or division) fence runs directly along the property line, and is therefore assumed to be commonly owned by the neighboring property owners. You cannot build such a boundary fence without your neighbor's permission, nor can you remove one.

Boundary fences make a lot of sense and are quite common. After all, if both you and your neighbor want a fence, you could each build one as long as it was set back a required distance from the property line. But it makes a lot more sense to build a single

Bad Fences Make Bad Neighbors

One argument that may not hold much water if you are requesting a variance is that you just do not like your neighbors or feel like punishing them for some wrongdoing by erecting a fence. There is a name for such a structure; it is called a "spite fence," and in many states it can be declared illegal.

Laws regarding spite fences have a lengthy history in the United States. They were often the first legal efforts of communities to exercise some control over local fence construction. As more thorough fence laws have been put on the books, the spite-fence statutes have receded. But they have not disappeared. If your fence is successfully challenged by a neighbor under a spite-fence or "private nuisance" provision, you may be required to remove it and even to pay damages to the neighbor.

In my home state of New York, the state statute says that if my neighbor builds a fence that is over 10 feet tall and has no function aside from annoying me, it constitutes a spite fence and I can seek to have it removed, and maybe even sue for damages. In other states the height limit is 5 or 6 feet. Even in areas where there are no spite-fence laws, a fence can be challenged under the common law. A fence that meets all of the technical rules and regulations, but that nevertheless can be shown to have been constructed out of malicious intent to upset a neighbor or otherwise interfere with the neighbor's enjoyment of their property, may be vulnerable to legal action. This is one more good reason to discuss your plans with your neighbors before you start building. ■

fence along the common line, and then share the maintenance duties on that fence.

So what happens when you want to build a boundary fence, but neither of you knows exactly where the boundary is? Boundary lines are often rather vague; your respective deeds may even be a bit contradictory in their descriptions of boundary markers. You can hire a licensed surveyor to prepare a careful map of the properties, but it can be less expensive simply to establish a legal boundary with your neighbor. Decide between the two of you where you want to put the fence, then put your agreement into writing. If you want to make sure you meet all legal requirements with such an agreement, talk with an attorney.

On the other hand, what do you do if your neighbor puts up an ugly yet legal fence near the boundary line? If your neighbor put up the fence without any effort to talk with you, this might be a good time to try and open up a dialog with the neighbor. But if that fails, you can always exercise the option of building a much nicer legal fence along your side of the boundary line. Then your neighbor will be faced with having to answer the question: Why did you build such an ugly fence next to such a nice one?

A single boundary fence nearly always makes more sense than separate fences built at the setback lines. Before you can build one, though, both neighbors must agree to do so.

In some states, as well as in some Canadian provinces, disputes that arise over boundary fences are handled by elected or appointed officials known as fence viewers.

Rural Fences

In many rural areas, zoning laws and building codes have little, if any, presence. Yet fence laws may still be on the books that, originally at least, were intended to control the movements of livestock. As discussed on pages 4–5, early American colonies and states tended to adopt one of two types of fence laws: those favoring open ranges and those stipulating closed ranges. In states with open-range laws, animals are allowed to roam freely, while property owners are expected to erect fences to keep the animals out of their fields, gardens, and backyards. In closed-range areas, it is the animals that must be fenced in.

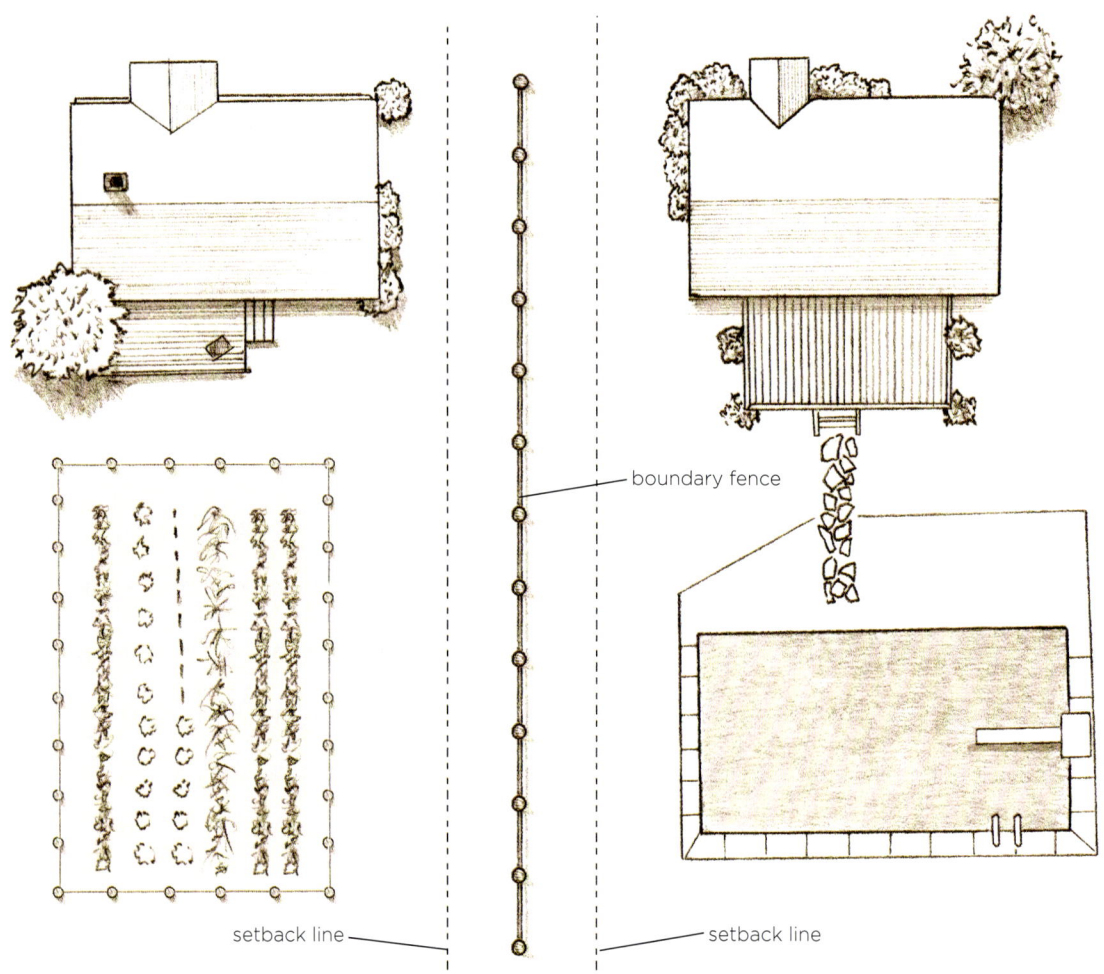

boundary fence

setback line

setback line

In this typical rural scene, separate fences perform separate functions. Without fences, such a neat division of space would not be possible.

Local custom often trumps the formal law in rural communities, but the changing character of rural life in much of the country tends to bring these laws into the spotlight from time to time. The state of Iowa, where I spent most of my youth, has a hybrid form of fence law, known as "fence-in and fence-out." Rather than favoring either the open range or the closed range, this approach recognizes that all parties share in the joint effort to keep animals both on their owner's property and off the neighbors' property. When established farmers lived next to other established farmers, the law was simple both to understand and to implement. But as new, nonfarming residents move into rural areas, confusion and dispute sometimes develop. The basic feature of the fence-in and fence-out law is that when one neighbor decides that he or she wants a fence to contain some newly acquired cattle, the other neighbor is required to build and maintain half of the fence, the assumption being that both parties need and will benefit from the fence.

"Attractive Nuisances"

The legal profession is known for its mind-boggling complexities, but also for its clever terminology. One of my favorite creations from this latter category is the "attractive nuisance." The phrase seems almost perfectly to summarize the yin and yang of life, with its sweet, naughty temptations. Sadly, in legal terms, what it more specifically refers to is one person's backyard swimming pool, which looks so inviting to the youth of the neighborhood that they can almost be expected to try and jump in at some point.

Although the laws vary from place to place, as a general rule property owners are considered to be responsible for keeping curious children away from any inviting, but potentially dangerous objects on their property. If you have a pool or pond on your property, or a trampoline, old car, or pile of dirt, kids are going to be drawn to it, and it is your responsibility to see that they do not get close enough to hurt themselves. Some states no longer use the expression "attractive nuisance" in their statutes, but most still have related rules. Some types of attractive nuisance can simply be carted away, but others require the erection of a fence or other barrier to keep children away. This situation most commonly arises regarding residential swimming pools, and many localities have very strict rules about the type of fence needed to restrict access to the pool area. Even where such rules do not exist, it is hard for me to imagine why anyone would want to maintain a swimming pool that a young child could wander into on their own. General guidelines on pool fences are given on pages 24–25.

Fencing with a Purpose

"WHAT'S THE POINT?" That's the rhetorical question my college professor scribbled in the margins of my written assignments when she detected (accurately, I can now acknowledge) that my words were not supporting my thesis. I think it is a good question to ask yourself (and, of course, to answer) when you start contemplating a home improvement project. The most satisfying projects begin with a clear purpose and conclude with a successful answer.

From selecting a style to choosing the material to deciding on the size, every major decision about fence design is best made with a specific goal or goals in mind. If you have been thinking about building a fence for a number of years, there is a good chance that you have forgotten why the idea first appealed to you. Take some time to journey back through your memory to reestablish a connection with the origins of the idea. This might seem like silly advice, but I often find myself needing a dose of it. Sometimes the anticipation of getting down to work can overwhelm the rationale for doing the work in the first place.

Traditionally, fences have had a job to do, and nothing more. Keeping animals in or out is probably the all-time number-one fence function. Increasingly, however, fences have come to serve less critical needs, such as providing privacy and security, marking boundaries, hiding an unpleasant view, or creating a new visual focus for your property. Fences have become a bit more about providing comfort and less about ensuring subsistence. Often, these needs and functions overlap, and when that happens it becomes necessary to prioritize, and to prepare for compromises. For example, if security is your one and only objective, you might choose one type of fence. If you want security but also want to create an attractive addition to your property, you may be faced with sacrificing a bit of security so that you are happy with how the fence appears.

Privacy and Security

These two goals often go together in fence design, but not always. Privacy fences can substantially expand the usable living space of your house, turning much

In a property with an open expanse of yard [LEFT], privacy is confined to the house interior. Fences effectively expand the living space by moving the privacy realm away from the house [RIGHT].

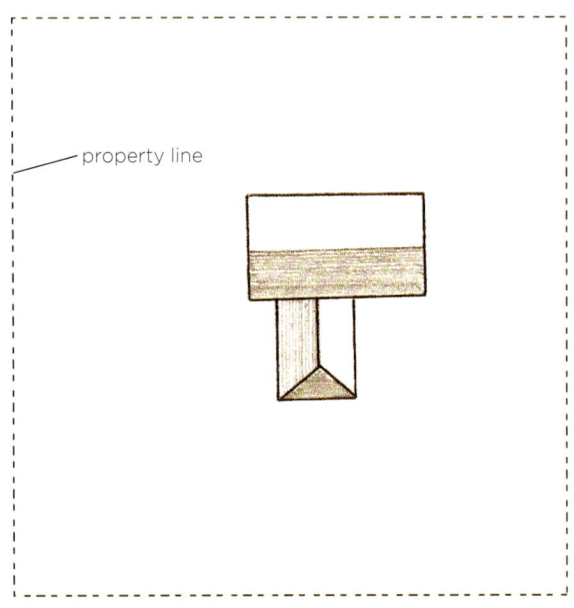

SMALL PRIVACY REALM

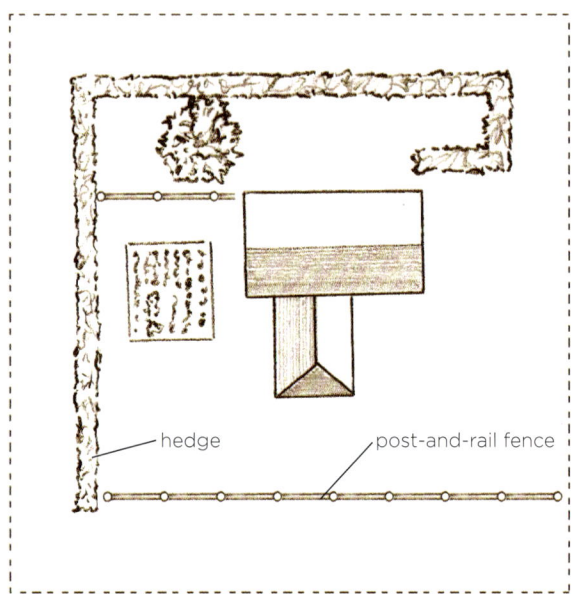

LARGE PRIVACY REALM

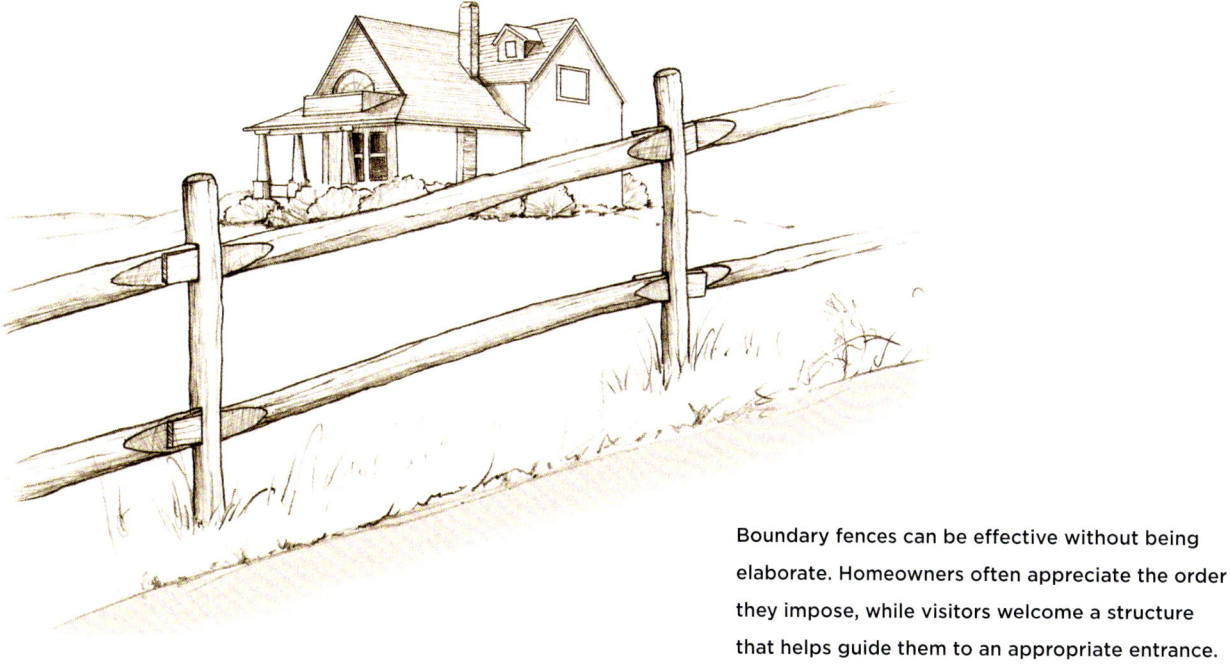

Boundary fences can be effective without being elaborate. Homeowners often appreciate the order they impose, while visitors welcome a structure that helps guide them to an appropriate entrance.

of your yard into functional outdoor rooms where you can sit, read, relax, and converse without feeling on display. A good fence designer will think about privacy fences the way an interior decorator might think about walls, and try to create different colors and textures to help define different "outdoor rooms."

Privacy fences, like decks, are a direct response to ever-growing popularity of the backyard as a site of social activity. Privacy is usually achieved by building a high fence and providing it with a solid or near-solid infill. Board fences are the most common choice, but a thick row of hedges could also fill the need.

A security fence should aim to discourage potential intruders from going over, under, or through the structure. Height and sharp edges of one form or another address the first concern, while strength and ground-hugging construction handle the second. Solid concrete or mortared stone or concrete block are tough to break through, although challenging to build high. Chain-link fences are easy to install and relatively inexpensive (if unattractive) options for security, although a properly equipped intruder could cut his or her way through quickly enough. A wood fence can certainly be built that is strong enough to resist someone trying to kick their way through, but, being wood, it is vulnerable to anyone with a cordless saw in hand. Ornamental metal fences can be tough to get through, although pricey wrought iron is much

stronger than the more affordable (and more DIY-friendly) tubular aluminum or steel. For combining good looks and solid security in a fence without blocking views, ornamental metal is tough to beat.

Gate construction is frequently the weak link in a security fence. The hinges must be strong and fastened securely to both the gate and the post, and the latch should be lockable and jimmy-proof.

If security is your overriding concern, try to think like a burglar as you design your fence. Your instincts might lead you to build a solid fence, which would keep unwelcome eyes from surveying your house and property. But if you were an intruder who managed to get through the fence, you would certainly feel much more comfortable being able to conduct the rest of your business inside of a solid fence, out of view of neighbors and passersby. Right?

Marking Boundaries

Boundary fences offer simple comfort. They block no views and discourage no conversations. They define limits without imposing restrictions and surround without excluding. Typically low and simple, a boundary fence subtly reminds the neighborhood athletes that your front yard is not part of their soccer field, just as it encourages visitors to approach the house via the sidewalk. Picket fences and minimalist post-and-rail fences are classic choices for boundaries marked by wood, while stone fences can achieve the same effect with an even more informal and unimposing presence.

Hiding Unpleasant Views

A fence can provide the outdoor equivalent of the popular problem-solving strategy known as "sweeping it under the rug." Even homeowners who are highly committed to maintaining an attractive yard are often stuck with fixed, ugly realities, such as garbage cans or propane tanks. Some people construct elaborate wood structures to house—and hide—their garbage and recycling containers, but you can achieve the same out-of-sight, out-of-mind effect by building a short fence. Just a single section, consisting of a couple of posts with a piece of wood lattice, can provide a welcome facade for those tanks or containers. Before trying to conceal the propane tanks, however, I would suggest that you discuss your intentions with the dealer to see what kinds of needs or requirements they have on maintaining access to the tank or tanks.

A gateway to . . . well, where, exactly? Short fence sections and fenceless gates can create wonderful landscaping additions in the most unlikely locations.

A Fence to Look At

My grandfather would not have concurred, but it's possible for a fence to serve no function whatsoever beyond looking nice or, more precisely, making your yard look nicer. Just as you hang pictures and paintings on the walls of your house and set plants and pottery around the interior and exterior for the sheer sake of adding beauty, so a fence can be set on your property to add an attractive human-made touch to the natural surroundings.

Many boundary fences are really nothing more than attractive additions to the property, built to create an image or visual impact more than to secure and protect. A simple post-and-rail fence along the street or around the perimeter of the yard is a case in point. But you can create a similar effect just about anywhere. For example, you can break the monotony of a large stretch of flat grass with a 10- or 20-foot stretch of fence, which all by itself automatically creates a new visual focus. An even better strategy is to use the new vertical structure as a background to a small flower or rock garden or as support for some climbing vines or flowers.

Animal Control

Keeping your pets enclosed and those of the neighbors excluded is a major fencing purpose. But there is no "one size fits all" style of fence for pets and other animals. Here are some suggestions on tailoring your fence to the specific animal concerns you face.

Deer

Deer just love my 5-acre homestead. They stroll through the orchard as though it were their own market, wander over to the chestnut trees when they tire of apples, then retire to the stand of pines for some rest. I would rather they went elsewhere, and I suppose if I tried really hard I could encourage them to do so, but the effort does not strike me as worthwhile.

Deer can be tremendously destructive pests, as most people who have to put up with them can attest. A small gang can wipe out a garden or row of newly planted trees almost faster than you can think about it. The only effective way to keep deer away is with a fence. But just any old fence will not do. Deer are great leapers, vertically and horizontally. A regular fence should be 10 or even 12 feet high to keep them from jumping over. And deer can also crawl, so the fence needs to be tight to the ground. Electrifying the fence makes it even more effective.

Unfortunately, fences this high can be expensive, unsightly, and in violation of local ordinances. The best way to get around these limitations without reducing effectiveness is to substitute some depth for the reduced height. While deer can jump both up and out, they do not like to do both at the same time. And they do not like to jump into blind spots. So you can keep the deer out with a 6- or 8-foot-high fence as long as you plant some shrubs or trees in what would be their landing zone.

Another option is to build a 10- or 12-foot-high fence, but then install it at a 45-degree angle, with the high side facing the deer. Keep the bottom of the fence snug and close to the ground, and add electricity for additional security.

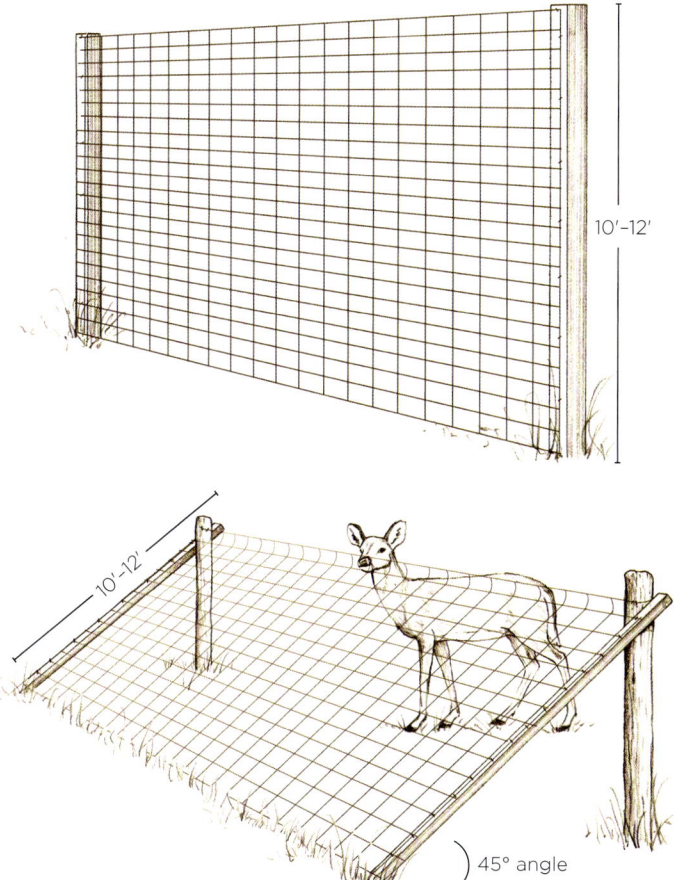

THREE WAYS TO KEEP DEER AWAY.
A fence with great height [TOP LEFT], a fence with modest height and modest depth [TOP RIGHT], and a fence with great depth [BOTTOM LEFT].

Raccoons

Raccoons can climb and dig, so an effective fence needs special attention at the top and at the bottom. This is best accomplished using wire mesh fencing. To prevent burrowing, bend the bottom of the fencing so that about 2 feet extends horizontally, and bury it 3 to 4 inches below ground. To discourage climbing, add an electric wire along the top of a 3- to 4-foot-high fence, but leave the top 18 inches of fencing loosely attached, or completely unattached, so that it falls backward when the critter tries to scale it. Use the same approach to control opossums and woodchucks.

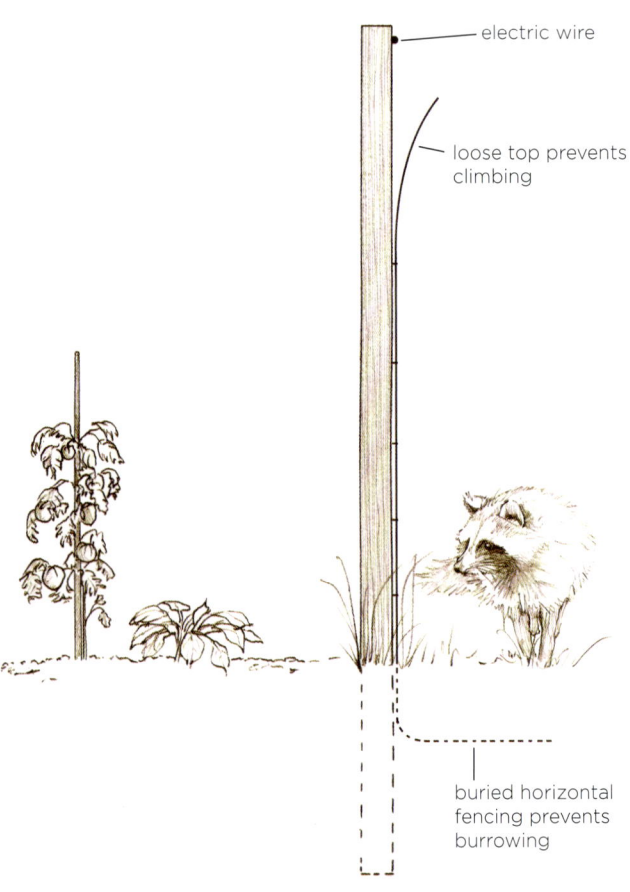

electric wire

loose top prevents climbing

buried horizontal fencing prevents burrowing

Raccoons can be relentless. The best protection prevents them from going under, through, or over the fence.

Rabbits

Rabbits can be kept out with a 2-foot-high fence of chicken wire that has holes not bigger than 1 inch in width or height. To prevent them from digging their way in, bend and bury the fencing as described for raccoon fencing.

Dogs

If you want to build a fence to keep in your dog, it needs to be only as high as your dog can jump. Typically, large dogs require a 6-foot-high fence and small dogs a 4-footer. If the purpose of the fence is to keep other dogs out of your yard, then you should plan on a 6-footer. And remember that dogs can be aggressive diggers, so create a below-grade extension of the fence as described for raccoons above.

Some people swear by the effectiveness of hidden electronic ("invisible") fence systems for keeping their dogs contained. These systems rely on wires buried around the perimeter of the property, a battery-powered collar that goes on the dog, and a controller that sends an electrical shock to the dog when the dog approaches the property line. The "fence" is not seen, but theoretically it serves the same function as a full-sized dog fence. This type of fencing is usually installed by specialized contractors (search online for "invisible fence"). I do not have any personal experience with either installing or using an electronic pet fence, but I do have experience keeping dogs as pets. And based upon that experience, I must confess that I cannot imagine wanting to submit my dog to electric shocks when a well-secured chain or securely built fence would be just as effective (as well as harmless).

If you decide that you do want to install a hidden electronic fence, do as much research as possible. There are several national manufacturers of these systems. Talk to as many of them as you can, ask for the names of people in your area who have installed the systems, and talk with the customers directly to see how they and their dogs have fared. Spend some time on the computer, where you should be able to access a trove of information from websites and discussion groups devoted to dog and other pet matters. The hidden fence systems I have looked at seem easy enough to install, but you will want to follow the manufacturer's instructions carefully.

Horses

A good horse fence must be strong, visible to the horse so that it does not run into the fence when spooked, and at least 4½ feet high. Some locales require that fences for horses be 6 feet high; this is also the recommended fence height for stallions.

Horse fences traditionally have been made with wood posts and rails. This can be an expensive option for a large pasture, however, and there are several less expensive options. The most economical fencing choices are the various electrified products. Electric tape, electric rope, and electric braid are all easy to see and to install. They have wires running through them that, when attached to a charger, shock the horse (and anything or anyone else) when it comes into contact with them. The fencing is strung between wood or metal posts in several rows. Electric tape is the most likely to start loosening and sagging in windy areas, but this problem can be offset by spacing posts closer together. Check with the manufacturer of your specific product for installation instructions. As a general rule, the bottom row is usually installed 18 inches aboveground for foals or 24 inches if only adults will be enclosed. Upper rows should be spaced about 12 inches apart. Posts can be spaced up to 50 feet apart, although 25 to 30 feet is more appropriate on rough terrain. The charger and wire tension need to be checked regularly.

High-tensile wire is another low-cost option for horses. It can be electrified, but it is also effective by itself. If you plan to attach the wire to a charger at some point, however, be sure to mount the wire on insulators. Coated wire is more visible than plain galvanized coatings. Adding springs to each row of wire fencing will add sufficient flexibility to minimize the chance of the fence breaking when a tree falls on it. Keep the bottom row 18 inches aboveground, and space additional rows about 7 inches apart.

Wood horse fences are usually built with 4×4 posts, spaced about 8 feet on center, with 16-foot-long 1×6 or ¾×6 boards fastened to the horse side (that is, the inside) of the posts. (Attaching the boards to the outsides of the posts may be more attractive, but it increases the possibility of a horse pushing the boards and their fasteners out of the posts.) The bottom rail is spaced 18 inches from the ground, with upper rails having about a 12-inch gap between them. White vinyl or vinyl-coated wood products can mimic the look of a wood fence, at least from a distance, and reduce some of the maintenance. (See Chapter 3.)

You can keep horses from chewing on the fence or leaning against it by attaching an electric wire to the top rail. This modest increase in cost will more than pay for itself in fence life.

Woven-wire fencing is very effective for horses. "No-climb" (with 2-inch × 4-inch rectangles) and diamond-shaped mesh are both available. This type of fencing keeps small animals out, is strong and long-lasting, and is very safe for rambunctious foals. It is available in various heights and is generally attached to wood posts. A 1×6 board can be attached along the top to increase visibility. Let the wire run an inch

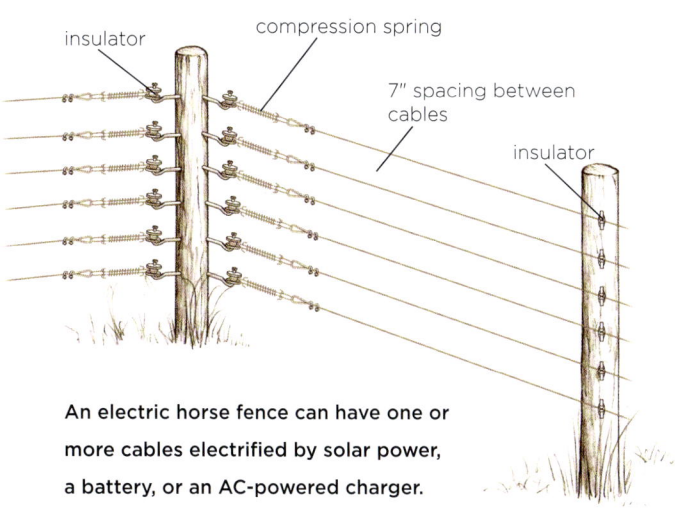

An electric horse fence can have one or more cables electrified by solar power, a battery, or an AC-powered charger.

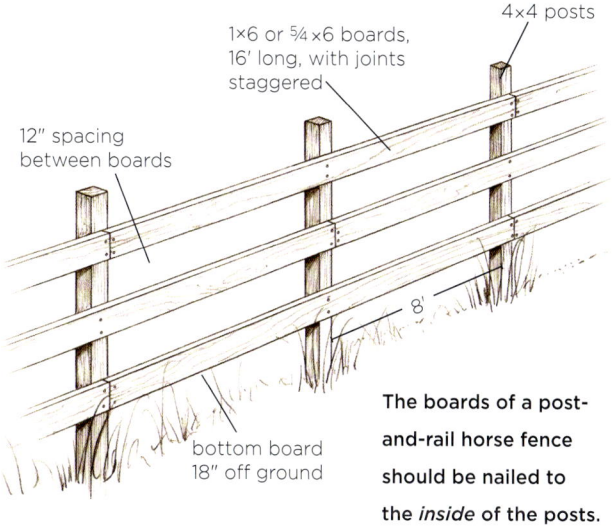

The boards of a post-and-rail horse fence should be nailed to the *inside* of the posts.

or so above the top rail to discourage horses from chewing on it. Woven wire is great on flat surfaces, but it is more trouble than it is worth on uneven terrain.

Some horse owners like to build a traditional wood fence, then attach woven wire to the inside. This approach combines maximum functionality with attractive appearance. It also discourages horse theft. Thousands of horses are reported stolen every year in the United States, and one of the quickest ways to grab a horse is to cut through wire fencing. Using boards, as well as locked gates, adds to the work involved in gaining access to the horse, and may make potential thieves think twice.

A horse fence must be strong and easily visible to the horse, but it does not have to have particularly solid infill. Widely spaced, well-fastened boards create a traditional look and a fully functional enclosure.

Sheep

Sheep fencing must keep the sheep contained, but also keep coyotes and dogs out. Woven-wire fence, sometimes sold as "sheep fencing," is usually the best choice for keeping the sheep in place. To discourage predators from attempting to penetrate the fence, place an electric wire just outside of the fence, about 8 inches off the ground, and another electric wire along the top edge of the fence.

Goats

Goats can be tough to fence in. They will try to crawl under, jump over, or push through any kind of fence you erect. They will also try their best to open the gate. To frustrate their escape efforts, construct at least a 4-foot-high woven-wire fence with mesh openings no larger than 6 inches, much as shown in the illustration at right. For Nubians and other active breeds, set the fence 5 feet high. A strong connection with buried wood posts is vital (see pages 74–89). Set the posts every 8 feet.

For added security, place an electric wire on the inside of the fence about nose high and another about 12 inches off the ground.

Chickens

Chicken wire ("poultry netting") is very effective and inexpensive, although it is not particularly durable. For longer service, thicker woven-wire fencing with 1-inch mesh along the bottom is recommended. The fence should be at least 4 feet high, or higher if you have a breed that can fly that high.

Geese and Ducks

As with chickens, a 4-foot-high woven-wire fence will usually suffice for nonflyers, while 5 or 6 feet will be better for flying geese. To confine flying ducks, you will need to clip their wings or drape netting over their yard. A strip of small mesh netting along the bottom of the fence will contain hatchlings. Control predators with an electric wire along the top edge of the fence.

Cattle

High-tensile wire fences are a good low-cost option with cattle. Wire mesh is even better, as it will keep the cattle from sticking their noses through and trying

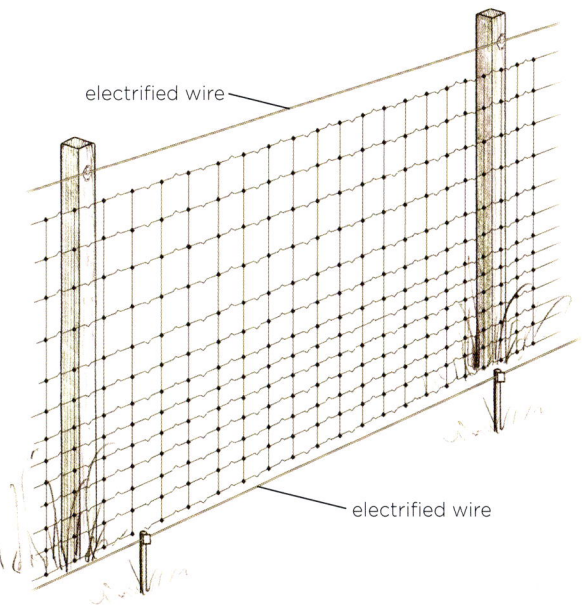

electrified wire

electrified wire

Electrified wires at the top of a fence and outside the fence perimeter discourage predators from approaching the fence (and the animals contained therein).

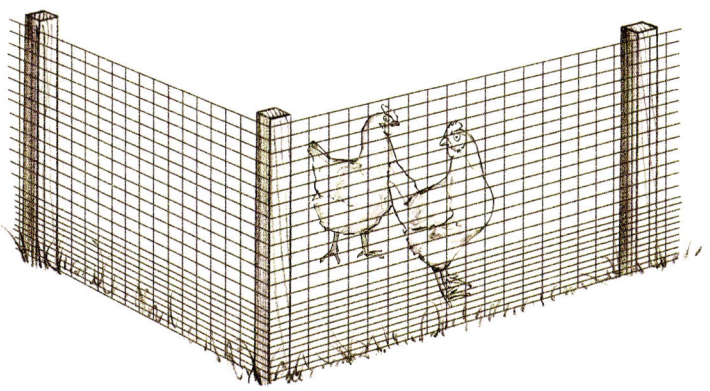

Woven-wire chicken fencing (shown here) is more durable than standard chicken netting.

to graze on the other side, which in turn will keep them from leaning against the fence regularly. An electric fence will also discourage the cattle from touching the fence. Barbed wire is still used with cattle, although it is often not allowed in residential areas and should be avoided if uninitiated children are likely to wander by.

Controlling the Elements

Fences are frequently built to control those environmental factors that do not move around on legs. Wind, snow, and noise can all be subdued with the right type of fence, properly located. When used as part of a broader landscaping scheme, fences can combine with trees, hedges, and walls to reduce a home's energy consumption.

Fences for Shade

The use of shade trees to block summer sun, and thus cut cooling costs, is well enough understood (and beyond the scope of this book). Fences can produce similar results, albeit on a smaller scale. A built or planted fence can shade a driveway, sidewalk, or patio, thus reducing the buildup of heat in the solid mass. They can also shade the side of the house.

Windbreaks

In cold climates, there is a direct relation between heat loss and wind velocity—the higher the latter, the quicker the former. A fence, built or planted, can reduce the velocity of the wind striking your house and thus reduce heating bills.

Contrary to what you might think, a solid fence is not necessarily the best type to use as a windbreak. It is better to let some wind pass through. The ideal design for controlling wind uses louvered boards, attached horizontally and angled down toward the house. But even a standard post-and-board fence can be effective. A hedgerow planted a foot or two away from the wall of a house can redirect much of a cold breeze up and over a low house, creating a mildly insulating dead-air space in between. Even if you are not concerned with energy loss, at least as far as fencing goes, a modest windbreak fence can slow the breeze enough to make it much more comfortable to sit on your patio or deck.

Snow Fences

A good windbreak fence will usually help to control snowdrifts as well, but style and location are even more critical. Solid fences and porous fences have different effects on drifts. A solid fence acts like a more powerful magnet, pulling larger amounts of snow near to it, while an open style of fence creates more gradual drifts, with less snow accumulating on the downwind side. Since both types create concentrated drifts, however, do not place either too close to a driveway or sidewalk. The general rule is to locate a wind or snow fence at least as far from the passageway as the fence is high.

In my neck of the woods, some homeowners erect temporary wood-slat fences every winter in an attempt to reduce their shoveling load. These seasonal structures can be effective, as long as they are attached securely to posts buried solidly in the ground and are oriented perpendicular to the prevailing winds. Too often, however, I see hastily erected fences of this nature that cannot even stand up to the first major snowstorm.

Researchers have actually studied the performance of various types of fences in controlling snowdrifts. They have discovered that a particularly effective style is a carefully constructed post-and-board fence that

HEDGEROW WINDBREAK

LOUVERED-FENCE WINDBREAK

maintains a porosity of approximately 50 percent. This can be accomplished by attaching 1×6 boards to 4×4 or 6×6 posts, set every 8 feet. The posts should reach at least 5 feet aboveground. The bottom rail is then attached 10 inches aboveground, with remaining rows allowing a 6-inch gap.

Most of the research into snow fences is driven by efforts to keep snow from drifting onto roadways, and thereby reducing the costs of snow removal. The state of Minnesota has implemented a program that compensates farmers and other landowners for constructing living snow fences that capture snow before it starts piling up on the road.

Remember that a fence meant to control wind or snow will be effective only if it is placed perpendicular, or as close to perpendicular as possible, to the pre-

vailing breeze. That typically means on the north and northwest sides of the house. Obviously, this often means that the most effective structure would have to be positioned awkwardly in your yard, jutting out at a weird angle from the driveway or side of the house. Although you may want to experiment with a landscaping scheme that could comfortably accommodate such an irregular feature, most folks will probably opt for the "as close as possible" approach.

If you live in snow country and rely on a plow to clear your driveway, do not build a fence that will restrict the plow's work. The removed snow needs someplace to rest, and if there is a fence or other objects close by on both sides of the driveway, it will be difficult to clear.

Fences to Block Noise

Fences can reduce some of the noise that reaches your house from the street and from next door. However, building a fence for the sole purpose of blocking noise may well prove to be disappointing. The best strategy for keeping the exterior roar from disrupting life inside your house is to soundproof the house. Noise is nothing more than airborne vibration. We experience noise when those vibrations reach our ears, so blocking or deflecting those vibrations will aid the quest for quiet.

An entire wing of the construction industry is devoted to soundproofing. Contractors, engineers, and specialized products are all available for that purpose. But the basic approach should be to close gaps and increase mass. Caulking and sealing around doors and windows is the first course of action for closing gaps. Mass can be added to walls with insulation and an extra layer of drywall or acoustic board. Thicker exterior doors and new windows or storm windows can also help. Beyond that, a standard wood fence or hedgerow is likely to have little effect on the sound waves traveling indoors. The psychological "out of sight, out of mind" effect, however, can be beneficial.

To make a real dent in noise with a fence, go for maximum density, such as with a poured concrete or brick wall, and build it as high as possible. A solid wood fence would be the next choice. Of course, the fence would need to be located between your house and the source of the noise. A fence is obviously not going to affect the sound of planes flying overhead.

SOLID SNOW FENCE

deep drifts

OPEN SNOW FENCE

shallower drifts

A solid fence [TOP] encourages deep drifts on both sides. An open fence [BOTTOM] produces longer, shallower drifts.

Pool Fencing

Year in and year out, thousands of young children in the United States drown or suffer related injuries in and around swimming pools. Most of the drowning victims are under three years of age. I believe that these accidents can be avoided if children are allowed access to the pool only under close adult supervision. Fences, and secure gates, are the best way of controlling such access.

If you have a swimming pool (or outdoor spa, hot tub, or water garden), or plan to install one, your locality may require that you surround it with a fence. The United States Consumer Product Safety Commission (www.cpsc.gov) has developed guidelines for safety-barrier fencing that, if followed, will keep curious, unsupervised children away from the water and out of harm's way. These guidelines have been incor-porated into many local building codes, and I strongly recommend that you follow them even if you are not required to do so.

A good pool fence will prevent a child from crawl-ing under, climbing over, or squeezing through. To prevent climbing, the fence should be at least 48 inches high, with no objects near the fence that would facilitate climbing (garbage cans, for example). Fence styles that create a ladderlike effect should be avoided. If you use any horizontal rails on the fence (to hold vertical pickets, for example), keep the rails spaced at least 45 inches apart and place them on the inside (that is, the pool side) of the fence. Keep spac-ing between pickets no more than 1¾ inches. If you use chain-link or lattice-type fencing, make sure that the spaces do not exceed 1¾ inches.

To prevent crawling under or squeezing through, keep the bottom of the fence no more than 4 inches aboveground (some codes require no more than 2 inches). The gate of the fence should be self-closing and self-latching. The smartest design has the gate swinging out, away from the pool. The latch should be out of a child's reach.

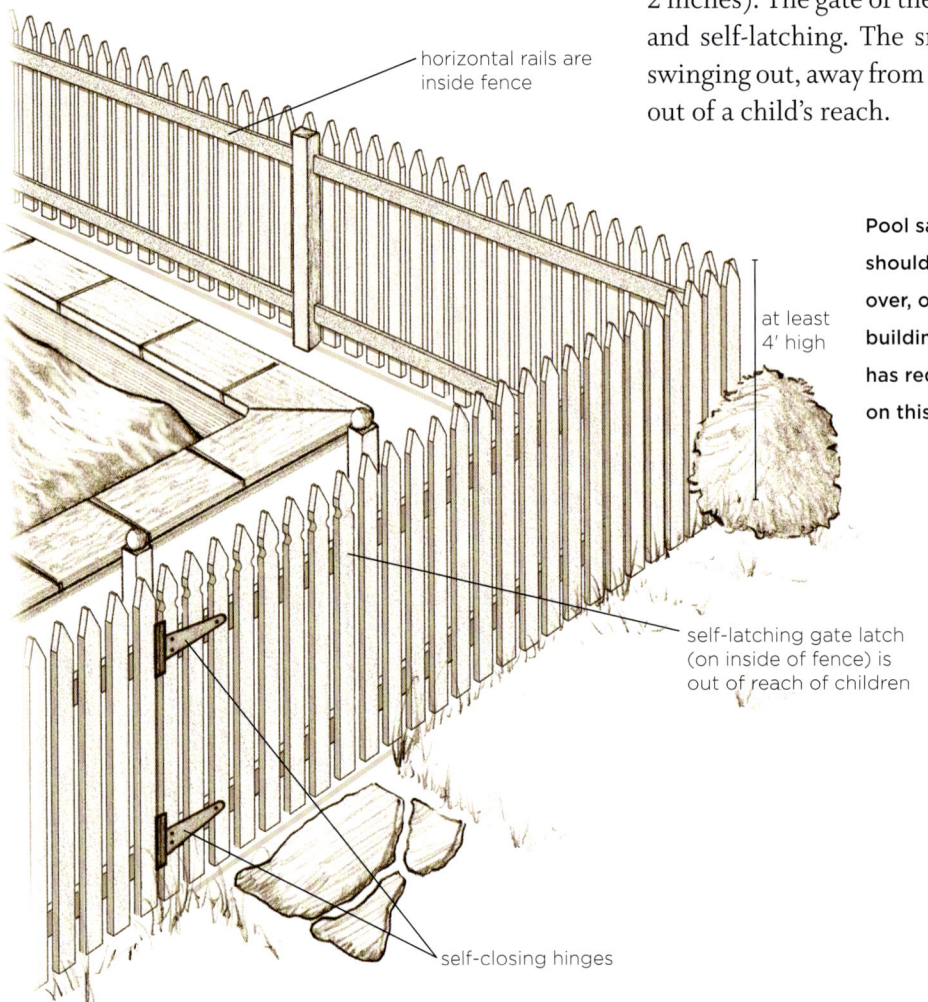

horizontal rails are inside fence

at least 4' high

Pool safety begins at the fence, which should be hard for kids to open, climb over, or crawl under. Check with your building department to see whether it has requirements that differ from those on this page.

self-latching gate latch (on inside of fence) is out of reach of children

self-closing hinges

Out by the Pool

First and foremost, pool fencing is about safety and carefully meeting, if not exceeding, all legal requirements. Beyond that, most people also want a fence that provides plenty of privacy for the pool area. Fortunately, it is not difficult to meet both of these needs.

Materials

FENCE STYLES do not seem to change much over time, but the materials that are used to build those fences have undergone some dramatic changes in recent years. That lovely picket fence you noticed the other day may well be built with extruded polyvinyl chloride (PVC), just like the siding that is going on many houses these days. Posts and rails on fences that you would swear were solid wood may actually be manufactured composites, created in factories out of a mixture of recycled plastic and wood fibers (that is, the same material that is becoming so popular on decks). And tubular metal fences are available that look exactly like classical wrought-iron specimens. Even the stone fence may not be what it appears, as manufactured stone shows up more and more in construction products.

This book is organized into chapters that focus on specific fence materials, with plenty of room allotted to their look-alike imitators. This brief discussion is not intended to duplicate the much more thorough discussion of specific fencing materials later in the book. Rather, here we will take a brief look at some of the attributes, pro and con, that define each material. You may already know what material you want to use on your fence, or may feel suitably swayed by this compressed discussion, but I would urge you to read the later chapters carefully to learn more about the characteristics of different materials, and perhaps more importantly about the construction techniques that each employs. It is not unreasonable to suggest that you may have the taste for one type of fence, but the building talents, tools, and budget for another.

Wood is usually the first material I think about when planning a fence or other outdoor structure. It is always available, affordable, easy to work with, and adaptable to an almost endless number of styles and decorative touches. Fasteners and finishes are also close at hand and easy to work with. Most do-it-yourselfers have some basic tools for working with wood, or are happy to buy them because they know that there will be additional woodworking jobs in the future. In short, you can do just about anything with wood, mistakes are easy to fix, and repairs require no special talents. If you want to learn some basic carpentry skills, building with wood is the only way to go.

Cedar and redwood are wonderful choices for fence construction. Both are naturally decay-resistant, at least to some extent. Much of the reputation that cedar and redwood have earned for durability in outdoor construction is based upon the performance of heartwood cut from old-growth trees. Much of the lumber available today, however, comes from newer, faster-growing trees, and contains significant quantities of sapwood. These latter lumber products may well prove to be less effective in outdoor construction. Regular supplies of cedar and redwood are limited to certain regions, and the cost can be quite high.

Affordable and readily available, wood can be used to build nearly any style of fence. With regular maintenance, a wood fence will last for decades.

Pressure-treated pine is widely available and is a good, affordable choice for wood fences, especially for posts that will be buried in the ground and for any components that will be in close contact with the ground. Although pressure-treated lumber still must be finished and maintained, it can last as long as or longer than naturally decay-resistant wood in outdoor building projects.

Natural stone is another material that has been used in fence construction for hundreds of years. In fact, a decently built stone fence will last for centuries, and then some. Stone is heavy, and thus hard to transport, and it is by nature highly irregular in shape, making it tricky to work with. But it is also durable, attractive, and virtually maintenance-free. Buying stone is not easy on the wallet, however, and it is usually best utilized on relatively low boundary fences.

Manufactured stone, concrete block, and brick can also be used to build solid, natural-looking fences. Concrete block is inexpensive and easy to assemble.

If your back can handle the labor, a well-built stone fence will require little if any maintenance and should long outlive everyone you know.

Although it is several degrees short of attractive on its own, the block can be covered with a veneer of manufactured stone to closely resemble a natural stone fence (at less cost), or it can be coated with a stucco-like surface to give it some character and color. Brick can be used to build a full wall or to create unique, decorative features in a fence or other type of wall. Brick posts look great with fences that use wood infill, in my opinion.

Metal fences can be roughly divided into two categories: ornamental and utilitarian. Chain link is a principal example of the latter, offering outstanding security at a reasonable price. Chain-link fences are not pretty to look at, however, although colored inserts and vinyl coatings are available to dress them up a bit. It's easy enough to use a chain-link fence as a trellis of sorts, though, and when a chain-link fence becomes hidden by or covered with plantings, it really does start to disappear into the landscape. A wide variety of wire fence materials is available for specific types of fencing needs.

Wrought iron is the classical example of an ornamental fence material. Today, tubular aluminum and steel products are widely available, offering attractive styles and color choices along with easy construction and low maintenance.

Vinyl seems to be showing up in increasingly diverse fence products. Vinyl fence kits can be assembled with few tools. Good-quality vinyl stands up well to the elements, requiring very little maintenance over the years. Most vinyl fences resemble classical wood fence styles, and you can now even buy vinyl trellis for use as infill or a topper.

There are two types of costs involved with any type of construction material—up-front costs and long-term costs. Often, spending more initially will actually save you money in the long run if you choose a durable, low-maintenance option.

Hedges are not often thought of as fences, but they can be the best choice in many situations as long as you can afford to wait for them to grow to the necessary height and depth.

When put to creative use, brick can be an ideal material for building a solid, low-cost fence [TOP]. Traditional wrought iron [BOTTOM] is beautiful but can be expensive for residential purposes; much more affordable tubular metal products are available for building strong fences that need little maintenance.

Before You Start Digging

MOST FENCES REQUIRE that you dig some holes for the posts. Before you pick up your shovel, however, make sure that you know what you are likely to encounter underground. Drainpipes; water and gas supply lines; septic tanks and drainage fields; and cables for electrical supply, telephone, and cable or satellite TV are all frequently buried. Knowing if you have any such underground obstacles on your property—and, if so, exactly where they are—is information that every homeowner should have, regardless of whether they are building a fence. While there is not necessarily any problem with building a fence that passes over a buried cable or pipe, you certainly do not want to encounter such objects when you dig your postholes.

If you are not sure about the presence of such obstacles in your yard, contact each of your utility companies as soon as possible. If you find that you do have buried lines to contend with, mark their locations on your site plan drawing (see pages 53–55), and adjust your post layout accordingly. Any damage caused by digging can be both expensive and dangerous.

Assessing Your Limits

As someone who writes about and practices home improvement, I speak to a lot of carpenters, electricians, plumbers, masons, and other skilled members of the building trades. When I tell them that I write how-to books, I am often greeted with gratitude. Now, you might think that someone who produces books for people who want to do-it-themselves might be looked upon with some suspicion by people who earn their living doing the same type of work for others, for pay. The sad truth is, however, that many do-it-yourselfers begin projects that they cannot complete, creating a job opening for a professional, called in desperation to finish a job that might never have existed if the ambitious homeowner had not begun the project in the first place.

I cannot emphasize enough how important it is to honestly assess your own skills, abilities, endurance, and patience before launching a home improvement project. Rest assured that it will almost certainly take more time, cost more money, and produce more backaches than you expect. Even seasoned pros and experienced do-it-yourselfers fall short from time to time in estimating how long a job will take, how much it will cost, and how many unexpected problems will pop up along the way to frustrate the game plan.

That said, I also want to add that there is simply no greater satisfaction than being able to point at a completed building project and say, "I did it myself." If you are a relative novice at building things, an outdoor project is the best way to start learning a few things about construction. Building a fence (or a deck or shed) can be done without disrupting life inside the house. Everyone gets to keep sleeping in their regular

bedrooms, the kitchen remains fully accessible, and the electricity and water do not have to be shut off for periods of time. The noise, the dirt, the materials, and the tools can all stay outdoors, and the skills you pick up along the way will translate into greater efficiency if and when you decide to tackle an indoor project. The key, to be redundant, is to be honest with yourself and to set reasonable objectives and schedules.

There are professional fence builders who create spectacular structures, utilizing talents and instincts and experience that you and I can only dream of. But with many projects, including fence construction, the thing that most separates the pros from the do-it-yourselfers is time. Both can create identical fences, but a pro is going to do it quicker because that is the only way to make a living. Amateurs, on the other hand, make their living at other things. Home improvement projects are what they do in their spare time, because they want to or because they have to. If you do not mind taking the time to learn some new skills, and to work patiently and safely, you can probably surprise yourself with the fence you complete.

Doing it yourself does not have to mean doing every little bit of the job yourself. Even experienced do-it-yourselfers know that there are times when it is smart to bring in a professional. If you feel confident about your building skills but weak with design matters, consider hiring a design professional to draw up a plan or offer some guidance. Most fences require digging holes for posts. If you have a lot of holes to dig, and you are digging deep, you may well discover that hiring someone else to dig the holes is the single most important decision you make on the project. And, once the building is all done, if you are feeling like you have run out of steam or time to handle the final stages, do not hesitate to hire someone else to apply the paint or stain.

As my kids are accustomed to hearing, you can play a perfectly fine round of tennis wearing old clothes and using a less-than-state-of-the-art racket. Likewise, you can build a fence without having to go shopping for a pile of new tools and accessories. A mortarless stone wall can be built with virtually no tools at all,

although a strong back or two will be required. Metal and vinyl fence kits can be assembled with minimal hand tools and few, if any, power tools. A simple wood fence can also be built with hand tools, although a circular saw is almost mandatory for the job. On the other hand, if you want to cut your own pickets and add some decorative touches, then a table saw, power-miter saw ("chop saw"), and jigsaw will certainly be welcome additions.

For many of us tool junkies, new projects are sometimes excuses for picking up a new tool or two. Occasionally, being junkies, we buy tools we really do not need and rarely use. It is usually a bad idea to buy good tools that are going to be used on only one project. Most types of construction tools can be rented, and you can usually borrow tools from friends and neighbors for short-term use. As you plan your fence, and read through the chapters of this book that deal with the type of fence you want to build, think about the tools being used and whether you want to make that additional investment at this time.

And if you do decide to spring for some new power tools, do also take the time to learn how to use them. A friend once told me that she had to take her husband to the emergency room the day after Christmas because he had managed to slice off the tip of his finger the first time he tried to use his new table saw from Santa. While sitting in the ER, she discovered two other waiting patients with major injuries suffered from using new power tools. In the apprenticeship program I went through to become a machinist, we were taught to respect what your tools can do *for* you, but also what they can do *to* you. Amen.

Design and Layout

I CANNOT SAY WHERE OR WHEN it first happened, but I feel pretty comfortable in speculating that somewhere, at some time, someone was the first to look at a fence and say, in his or her native tongue, "I can do better than that." Thus was born the idea of fence design. Once freed of the need to use only local materials and primitive techniques, fence builders began to invest thought and creativity into their craft.

I tend to think of the design phase as playtime. This is not to imply that it is not serious work (good play, after all, is important business), but that it can be fun, sometimes frivolous, and easily adjustable to the elements at hand. Design is largely about the paperwork, the preparation of plan drawings, the creation of a shopping list, and the check writing. Make a mistake at this stage of the game, and it can be corrected with an eraser. Which leads me to reiterate a point made in the previous chapter, that one of the big advantages that a do-it-yourselfer has over a professional is time. Hire someone else to do the planning and the building, and every minute costs money. When you do it yourself, time is not so precious, and you can take as long as you need to work out the details before lifting a shovel.

Fences can be cold barriers that say "go away," or they can serve as welcoming gateways. They can inject life and beauty into a landscape, or they can nearly destroy one. Fences are relatively simple structures; yet because of their vertical nature, they are significant, highly visible components of the built environment, sometimes even surpassing the house itself as the defining characteristic of a property. Design is really about comprehending the scope of the environmental alteration you will be undertaking, and then making decisions that support that vision.

Design is almost always an exercise in compromise, pitting dreams and expectations against budgets, schedules, and abilities. A successful design is not necessarily one that causes jaws to drop. Rather, it is one that fulfills or exceeds your basic needs without breaking the bank account, or your back. If you do not have a lot of experience with home improvement projects, you may not have a good sense of just how much money you will need to construct a basic fence. And there is no reasonable way for me to tell you how much your fence will cost. Fence costs can run the

A simply designed fence can be dressed up substantially by a more elaborately designed gate. Here, the arched entry to the house was duplicated in the gate, uniting the two in a common theme.

gamut, from almost nothing for a simple structure made largely from materials gathered on site to astronomically expensive creations of intricate detail and high-priced materials.

Broadly speaking, design is the "theory," and construction is the "practice." In the big world, these functions are often quite separate. Construction crews build what architectural firms design, with different groups of people knowing little about what the other does. True, separate skills and knowledge are required in these different occupations, but everyone benefits when there is some unity in vision, some agreement about the goals. Good design is hard to grasp in the abstract; as long as it exists on paper, it is still incipient. Once executed, we are able to judge how it appeals

Funky, rustic, and solid, this fence combines simple, common materials and a complex pattern to create a unique structure.

to our senses and expectations. Good designs can be ruined by poor building, and bad designs can be rescued by competent builders. But the best projects spring from a happy marriage between the two.

Some people think that design is a lot of mumbo jumbo, that it is oppressive and undermines spontaneity and creativity. Improvisation and free association, they maintain, are better paths to true beauty and innovation. Well, I love jazz about as much as anyone reading this book, and one of the things I most like about it is improvisation. When, for example, Sonny Rollins steps on stage with his tenor saxophone

"First, Do No Harm"

Many people mistakenly believe that the above dictum is a part of the Hippocratic Oath, which is taken by many doctors when they start practicing medicine. The phrase is not in the Oath, although the sentiment is. But my first encounter with this line was not related to the medical profession. In my early days as a book editor, one of my mentors insisted that "first, do no harm" was the most important lesson that one could take in editing the writing of someone else: Do not make changes that subtract from the intention, vision, and voice of the author. I also think it is a good way to approach landscape design. While you may own the chunk of property upon which you plan to build a fence, you do not own the environment, the neighborhood, or the community. True, you can think of your property as your property, and nothing but your property, so help you God, and many people do. But you don't have to. You can also adopt the attitude that your property is just one small part of a larger whole. You can assume some share of responsibility for that whole with your additions or subtractions to the landscape. You can think about a fence design as a feature of the neighborhood, and not just of your yard. You can look closely at the materials you use in constructing the fence to understand where they came from and what harm may have been incurred in their processing and transfer to you. You can look at the changes in the landscape that you may be contemplating in order to create your fence (cutting down a tree, digging up a garden, blocking a view) and wonder whether the benefit is worthwhile. True, it's only a fence, but you have to start with some principles, and to my mind "first, do no harm" is as useful as any other one. ■

and his band, without a script, and starts composing a tune on the spot, exchanging phrases with his percussionist, it can be breathtaking, creative, memorable, and certainly spontaneous. I've known a few carpenters of Sonny Rollins's stature, who could build on the fly, figuring out as they go along what to include, and where. But they are able to do this because they have been in similar situations hundreds of times before. They know what works, and they know what doesn't work. For those with lesser talents and experience, though, improvisation is an unmastered foreign language, and often just an excuse to avoid doing some homework. Design is that homework, and I doubt very much that you will regret the time you put into completing the assignment.

What is good design? I can't really say. At an elementary level, if you build a fence for a specific purpose, and it serves that purpose, it has met at least one objective of good design. But if, in the process of satisfying that purpose, it is nevertheless seen as an eyesore by neighbors, it has presumably failed another objective of good design. And what if it has no practical function? I have seen fences that I thought were wonderful, spread out along a landscape and painted in a bright color, serving no purpose whatsoever beyond adding some unexpected surprise and beauty to its surroundings. The landscape artists Christo and Jeanne-Claude may well be an inspiration for these structures. Best known for their "wrapped" objects, such as their wrapping the German Reichstag in fabric in 1995, Christo and Jeanne-Claude erected an 18-foot-high and 24-mile-long "Running Fence" of nylon ribbon through California's Marin and Sonoma counties in 1976. Say what you wish about these temporary outdoor art projects, but I think it would be a mistake to dismiss the natural environment as a canvas upon which you can add a thoughtful and personal statement. Good design fulfills a function, but that function can be aesthetic rather than practical.

The best answer to the "what is good design?" question that I can offer is the answer once provided by Supreme Court Justice Potter Stewart when trying to define pornography: "I know it when I see it." In other words, it is difficult to quantify good design before it has been experienced as a sequence of design, then practice, then assessment of the result. Likewise, it is

Christo and Jeanne-Claude's "Running Fence," which stretched for 24 miles through Marin and Sonoma counties, California, in 1972–76, turned fencing into a temporary art form.

hard to provide a list of hard-and-fast rules that, if followed correctly, will guarantee a good design. So I will not attempt to tell you what you should or should not include in your fence plans. Instead, I will summarize some of the key ingredients that designers include in their planning arsenal. Give 10 cooks the same ingredients, and you are still likely to wind up with 10 different dishes. The same can be said with fence design: Ultimately, the final design will rise from your own interpretation of the given ingredients.

The much admired and oft-reprinted 1892 book *Fences, Gates, and Bridges* is certainly true to its subtitle: *A Practical Manual*. The book was intended almost exclusively to appeal to and instruct readers who made their living off the land. Aesthetics and ornamentation consumed only a couple of pages. Today, that ratio of priorities has flipped. Few Americans now make their living in agriculture, and most people looking to build their own fence are at least as interested in creating something attractive as they are in making something useful. Fences have increasingly become ends in themselves, intended to serve no function other than to add a design feature to the landscape. Proof that this phenomenon is real can be found in the fact that the fencing industry has coined a name for it: "fencescaping."

Fences have not lost their functional heritage, but their persistent presence in nonfunctional settings reminds us that they fulfill some deep emotional needs for humans, rooted in animal instincts, by offering a sense of safety and privacy without compromising our access to the outer, public world. Many of the houses and landscapes designed by Frank Lloyd Wright strike me this way. The deepest parts of the interior feel like a safe, womblike sanctuary that slowly, gradually unfolds into the outer world. Looking at Wright's houses from all directions, you often see patios, balconies, terraces, vegetation, and walls in the transitional spaces between inside and out. These devices manage to limit how much the outer world can see in (while nevertheless inviting those on the outside to look at and appreciate the structure). Residents are offered strategic spots from within to gaze out, or to remove themselves from view, as their mood dictates. In Illinois and California, Wright built houses that sit on busy city streets, and that almost demand attention from passersby, yet manage to offer privacy and a retreat to occupants.

A well-designed fence can create that kind of comfort and challenge in your own environment. It can expand the control you exercise over what you see and what is seen of you. Skip or scrimp on the design phase, however, and you effectively give up some of that control.

Exterior Design

SOMETIMES THE PRIMARY GOAL is the fence itself, as an architectural element in its own right. Here, the idea is to draw attention to the structure, or to the immediate area around it. More often, however, the fence is a means to an end, an enclosure or a barrier, and is most appreciated for the space it surrounds. The 1977 classic work on architecture and planning, *A Pattern Language* (by C. Alexander, S. Ishikawa, and M. Silverstein), offers this assessment of the outdoors:

"There are two fundamentally different kinds of outdoor space: negative space and positive space. Outdoor space is negative when it is shapeless, the residue left behind when buildings—which are generally viewed as positive—are placed on the land. An outdoor space is positive when it has a distinct and definite shape, as definite as the shape of a room, and when its shape is as important as the shapes of the buildings which surround it" (p. 518).

A positive outdoor space is one that attracts people, and allows them to feel comfortable. It is enclosed, or at least feels as though it is enclosed. It clarifies boundaries, and thereby dictates the kinds of activities that can take place within the space. The outdoor space becomes positive, therefore, when it becomes more like an indoor space, or when it starts to feel more as an extension of the indoor, private realm, and less of the public space that lies beyond.

There can be a strong emotional component to this idea of enclosure. Some years ago I spent a week with a dear friend at a remote coastal village in southern Mexico. The beach was extraordinary, with beautiful sand stretching as far as the eye could see. Even better, we were all alone. We swam for several hours without seeing another living creature, aside from some birds. But when we stretched out our towels, opened the beverages, and curled up with our books, we quickly found ourselves feeling uncomfortable. Even though there were not prying eyes nearby, much less traffic or noise or any of the things we try to hide from, we felt exposed. In the middle of a vast expanse of deserted beach, we were in "negative space." So we created some sand piles to serve as an enclosure, and suddenly found comfort.

As much as we appreciated the setting and wished to remain in it, we were not comfortable merely being swallowed by it. Our low walls created rooms without eliminating the sun, sand, water, and fresh air. And this is why people love outdoor rooms even when they live in the country, away from neighbors. Enclosure can create comfort.

A wide-open yard in a residential setting [ABOVE] is an uncomfortable place to relax. A fence [RIGHT] can turn this negative (or little used) space into positive, well-defined space that invites sitting and relaxing.

Field Research

GOOD DESIGNERS ARE GOOD OBSERVERS. They pay close attention to what already exists, and then use that knowledge as the basis upon which to develop their own variation or innovation. Like music, fashion, and cooking, architectural design involves a bit of theft, which is just another way of saying that each new design expands upon our existing environment and heritage. Being a careful observer will make you a better designer, and there is really almost no job of observation easier to conduct than looking at fences. A walk or casual bike ride through a few residential neighborhoods, or a drive along some rural areas with the single-minded purpose of observing fences, can pay big dividends. Take along a camera or a sketch pad. Develop a self-quiz, perhaps by rating each fence on a scale of 1 to 10, and then adding some comments about your rating. Does the fence look good or bad in its setting? Why? Would it look better if it were more open, or more closed? If the posts were more prominent? If the infill were more horizontally oriented? If it were a different color? If it were higher or lower? If it didn't hide so much of the view of the house? Once you analyze the results of your research, you may discover a few things about your instinctive reactions to fences that had not previously occurred to you. Perhaps you found yourself looking closely at post details, gate designs, unique decorative elements, or the landscaping around the fence more than the fence itself.

Looking through magazines can also be helpful, but maybe not as much as you might hope or think. When I thumb through most of what are known in the publishing trade as "shelter" magazines (that is, magazines devoted to houses and the design of their interiors and surrounding exteriors), I almost never see houses that look like mine or projects that appeal to me or my budget. The gorgeous houses, exquisite yards, fanciful fences, and perky faces, all suitably prepped by a trained photo stylist and captured in seductive light by a first-class photographer, do not usually stimulate my senses. I tend to look at those images rather like I look at the New York Yankees in recent years, concluding that it is always easier to generate a winner if you can throw seemingly endless amounts of money at the effort.

What impresses me more are those designers and builders (not to mention sports teams) who find ways to integrate good quality and winning results with more modest budgets, in real neighborhoods. When I start designing an addition or renovation, I usually begin with an open mind. I try to appreciate what others have done and assess how I react to different environments. But when decision time rolls around,

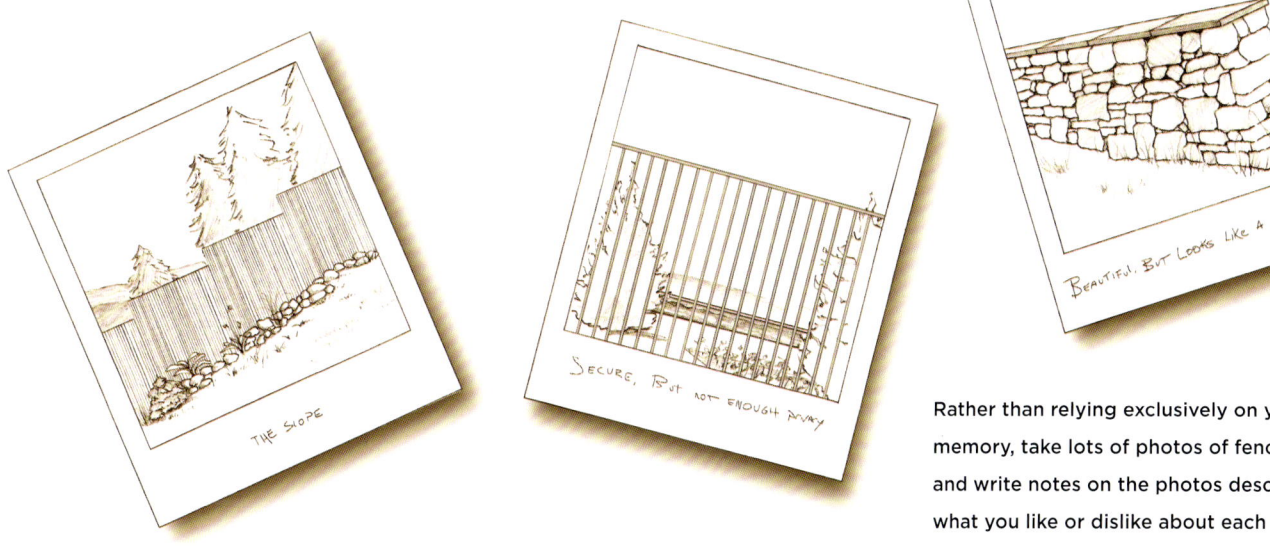

Rather than relying exclusively on your memory, take lots of photos of fences and write notes on the photos describing what you like or dislike about each one.

as often as not I resort to my design default mode, which is to keep it simple and do it well. Nice execution on a simple theme will almost always satisfy most people (though, admittedly, not all). And for do-it-yourselfers, this approach offers the great advantage of generating a project that is more likely to be finished, that does not get bogged down in time-consuming petty tasks and hard-to-execute details.

So, look at photographs and read all of the literature you can find on fences, but then also get out and look at real fences in real yards. Don't be afraid to stop and talk to homeowners about their fences. Do-it-yourselfers, I have found, are almost always more than happy to talk about their creations and to share information on their expenses, frustrations, and lessons. With the owner's permission, inspect the fence up close, examining the hardware and joinery. Test the strength: Do the posts feel solidly grounded? Does the gate swing easily and close securely? How has the finish on the wood stood up over time? You may be amazed at how much useful information you can gather this way.

Choosing a Style

AFTER ESTABLISHING A PURPOSE for the fence and assessing any legal limitations on what you can build, most planners turn their attention to the style. House styles tend to fall into categories such as Rustic, Colonial, Romantic Revival, Victorian, Modern, or just plain Eccentric. And each of these broad groupings incorporates numerous subcategories, with variations on each running every which way. Selecting a style is about deciding what kind of message you want to send regarding your fence, what kind of time period you wish to invoke, and which materials can best accomplish that goal. If you are concerned about the style of your proposed fence, a good place to start is with an understanding of the style of your house. Do you even know what it is? If not, or if you would like to learn more about house styles, I would recommend that you pick up a copy of *A Field Guide to American Houses* by Virginia and Lee McAlester, which was first published in 1984. This book is a classic resource that breaks down American house styles into several dozen principal categories and then uses text, photographs, and illustrations to explain and define features and variations on each. I always keep my copy close at hand; even my kids have found it to be a useful tool in preparing their architectural projects in elementary school.

Perhaps the first piece of advice that most designers would offer regarding fence planning is to view your planned project as an extension of your house. That suggestion most often begins with trying to

The high, solid, and more contemporary-looking fence at top both obscures the house and contrasts with the house style. The lower, more open, and more historically matching fence below is a better fit.

match the style of the fence to that of the house. The McAlesters are not going to tell you much, if anything, about fences, although many of the photographs of houses in the book include fences. But they will give you some important clues about the major characteristics of a house that may be useful in designing a complementary fence. You might, for example, be reminded of the trim around windows and doors that could suggest a decorative touch with the fence. Dominant posts on a porch may inspire the look of posts (especially gateposts) on the fence. Or you might be inspired by the use of stone in the foundation or porch, or maybe the type of siding. In general, long and flat houses lend themselves to horizontally oriented fences, while a house with a prominent gable or complex roof lines might be better served by a fence with components placed at variable heights.

I think that the proximity of the planned fence to the house should influence just how seriously you try to match styles. A fence that is some distance from the house becomes less "of the house" and more "of the landscape." As such, it can easily be built with less focus on a stylistic match, although it nevertheless remains the case that you ought to try and avoid using a clashing style. Fences that are attached to the house often benefit greatly by appearing as though they are an integral part of the overall house plan. Likewise, finding a good stylistic match with the house is more important with front-yard fences, given their more public presence, than those in the backyard, where basic functionality often prevails.

I would argue that the more distinctive your house, the more care you should take in designing a fence for it. Unique, custom-designed houses are generally not well served by low-budget, generic fences. Likewise, the more generic your house, the less need there is to fuss over details of the fence style. In a subdivision of similar houses, a collection of wildly dissimilar fences may look decidedly out of place.

Fence design should incorporate elements other than the fence material itself. Plantings alongside of the fence line or flowering vines growing in the open spaces of a fence can profoundly affect its appearance.

Open or Closed?

ONE OF THE BIGGEST DECISIONS to be made regarding a fence is whether it will be open, closed, or something in between. The choice deals largely with the nature of the infill, which can range anywhere between a high, solid wall to a barely visible structure that serves some modest landscaping purpose with no effort to block any views. The decision is relatively easy to make if you need a fence that is best served by one of these extremes: For example, a privacy fence would generally be very much on the closed end, while a simple boundary marker might be a skimpy skeleton of a fence. The tougher decisions fall on fences that need to be somewhere in between. With board and picket fences, you need to decide how much spacing to leave between pieces. If animal control is your principal objective, the gaps in the fence must be sized to keep the specific type of animal in or out.

Open fences tend to suit front yards, while closed fences are more common along the sides and back of the property. But these preferences are not universal. In densely inhabited urban areas with very small yards, high and solid fences are very common along the noisy sidewalks and streets in front of the house. Closed fences give you privacy, but they can also cut off your view of the neighborhood. When trying to choose an appropriate degree of enclosure, be sure to take into consideration the views from inside the house as well as from the yard.

A solid, closed fence can hide the street and sidewalk traffic from the yard, while a more open style of gate creates a welcoming entrance.

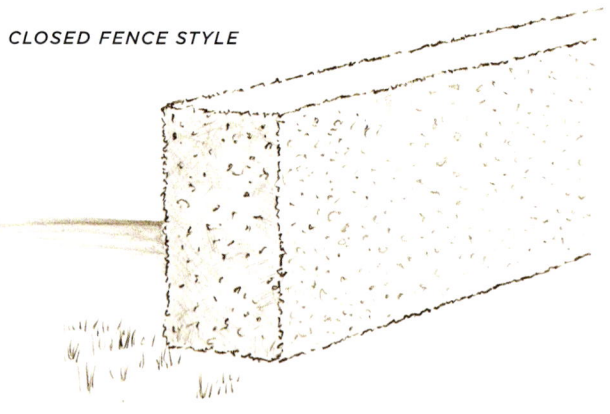

CLOSED FENCE STYLE

OPEN FENCE STYLE

To Gate or Not

Fences intended to offer security and privacy generally require a strong, solid gate with a lock. For other fences, deciding on whether to include a gate is a matter of choice. Gates can stand alone as attractive portals to your property. Often, boundary and purely decorative fences are left gateless. A gate can provide a focal point to your front yard by interrupting a long run of fence with a sudden and dramatic change in style and scale. Gates can also be constructed to blend in almost invisibly with the fence.

Chapter 7 is devoted entirely to designing and building gates, and if you are thinking about including one or more gates in your fence, I strongly suggest that you read that chapter. I raise the subject now, though, because it is important to include the gate or gates in your overall plan. You need to decide on the width of the gate opening, and plan for a hinged gate to have plenty of room to swing open. You need to decide how big the gateposts should be (the bigger and heavier the gate, the bigger the supporting posts need to be). If you are not building a wood fence but do want to include a wood or metal gate, you have to figure out how the gate will be attached to the fence.

Gates are easiest to build and maintain if they are small and lightweight. On the other hand, they need to be wide enough to allow room for people, furniture, appliances, lawn mowers, and whatever other items you may have to move in and out of your house and yard. For most purposes, I suggest that you plan all gate openings to be at least 4 feet wide.

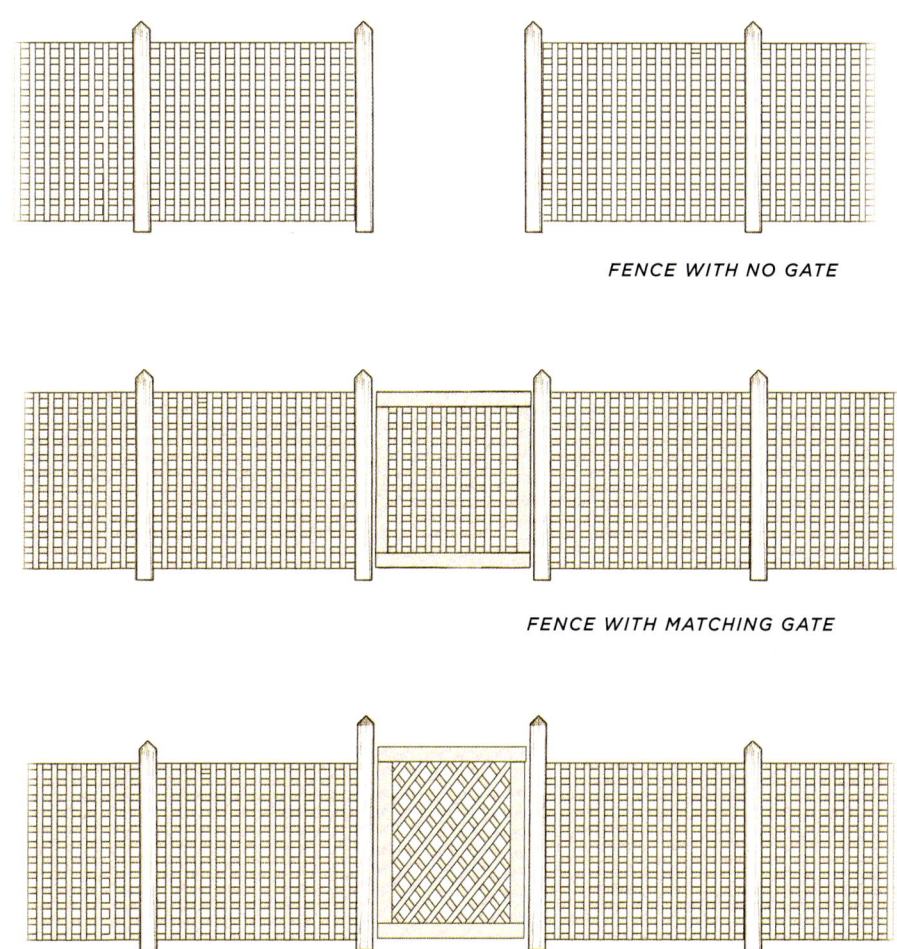

FENCE WITH NO GATE

FENCE WITH MATCHING GATE

FENCE WITH DIFFERENT-STYLE GATE

Designing with Contrasts

DESIGN TALK frequently seems to revolve around promoting harmony and integration, but often what we most appreciate are contrasts, which can introduce surprise and mystery into our environment. In architecture, as in nature, surprises give a place character. When wandering through a natural or designed landscape, or exploring a big old house for the first time, I am always on the lookout for an unexpected scene or experience, whether it be a sunny opening in a wooded forest, a flower bed or water garden in an unexpected setting, or a small cubbyhole of uncertain purpose tucked into some corner of a house.

Contrast can be incorporated into fence design in two primary ways. First, think about the experience of walking through or around the fence. You may want to try and add some differentiation. For example, the opposite sides of a relatively solid fence could be given very different looks. The plain, simple face to the street could be contrasted with a lush, decorative garden running along the fence line interior. More simply, the fence could be a light color on one side and a dark color on the other.

The second way of approaching contrast in fence design is to think about ways that the fence itself could incorporate differing elements. Consider alternating high sections with low sections, open sections with closed, hard surfaces (wood or stone) with soft (vegetation). I think that a fence that mixes different materials (brick or stone pillars amid a wood fence, for example) can be particularly appealing.

Color is an oft-overlooked part of fence design. People tend to think of wood fences as being either white (especially with pickets) or natural-colored. But this sameness of a basic fence style can be made more interesting by using different colors or shades of paint or stain. A fence with pickets stained in the colors of fall leaves or in alternating shades of a single color can be a welcome change of pace.

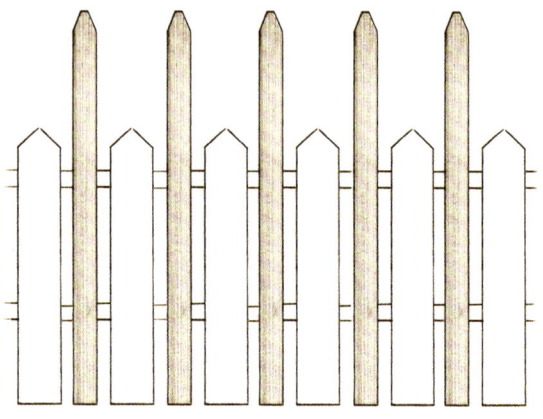

SHAPE AND COLOR CONTRAST

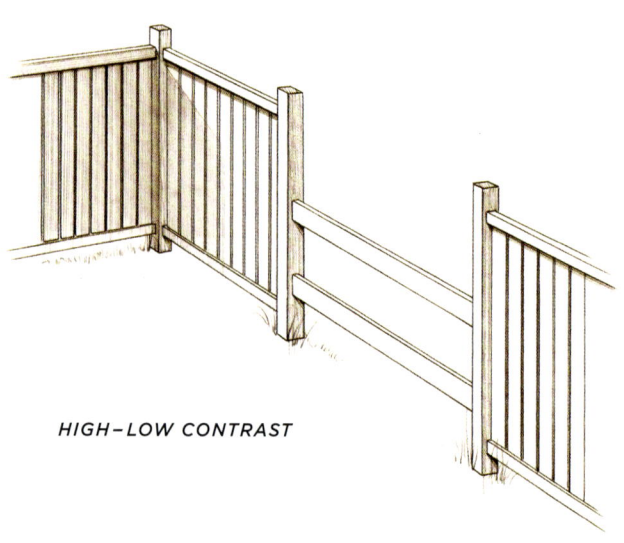

HIGH–LOW CONTRAST

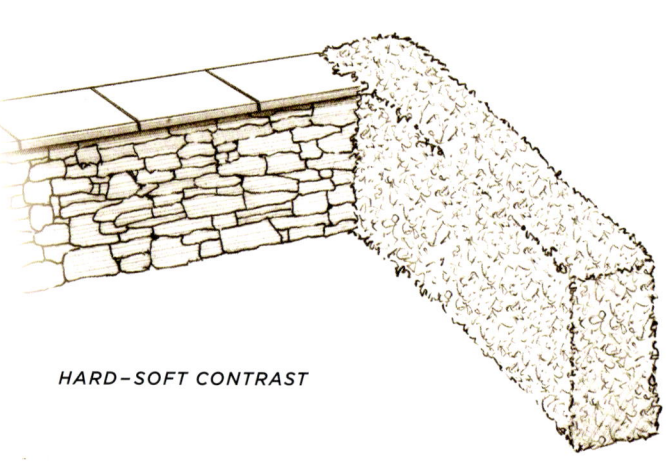

HARD–SOFT CONTRAST

Height and Sight

For many fences and fence builders, height really is not an important factor. Even if that is true in your case, you still must make a decision on the height before you can build the fence. Height is most often related to function, but it can also be affected by other concerns and factors. If you have a particular design in mind, it may be that the components of that design will drive the decision about how high to build. For example, if your property is contoured, you may be able to block an unwelcome view with a relatively low fence located along an upslope or berm. (Chapter 6 has a more detailed discussion on using berms to help achieve fencing objectives.)

Height is often a legal issue. Codes and ordinances frequently restrict how high a fence can be, and if you are building a fence to restrict access to a swimming pool, you will need to pay attention to how high it is required to be. There are also technical matters to consider: The higher (and heavier) the fence, the more securely it needs to be connected to the ground.

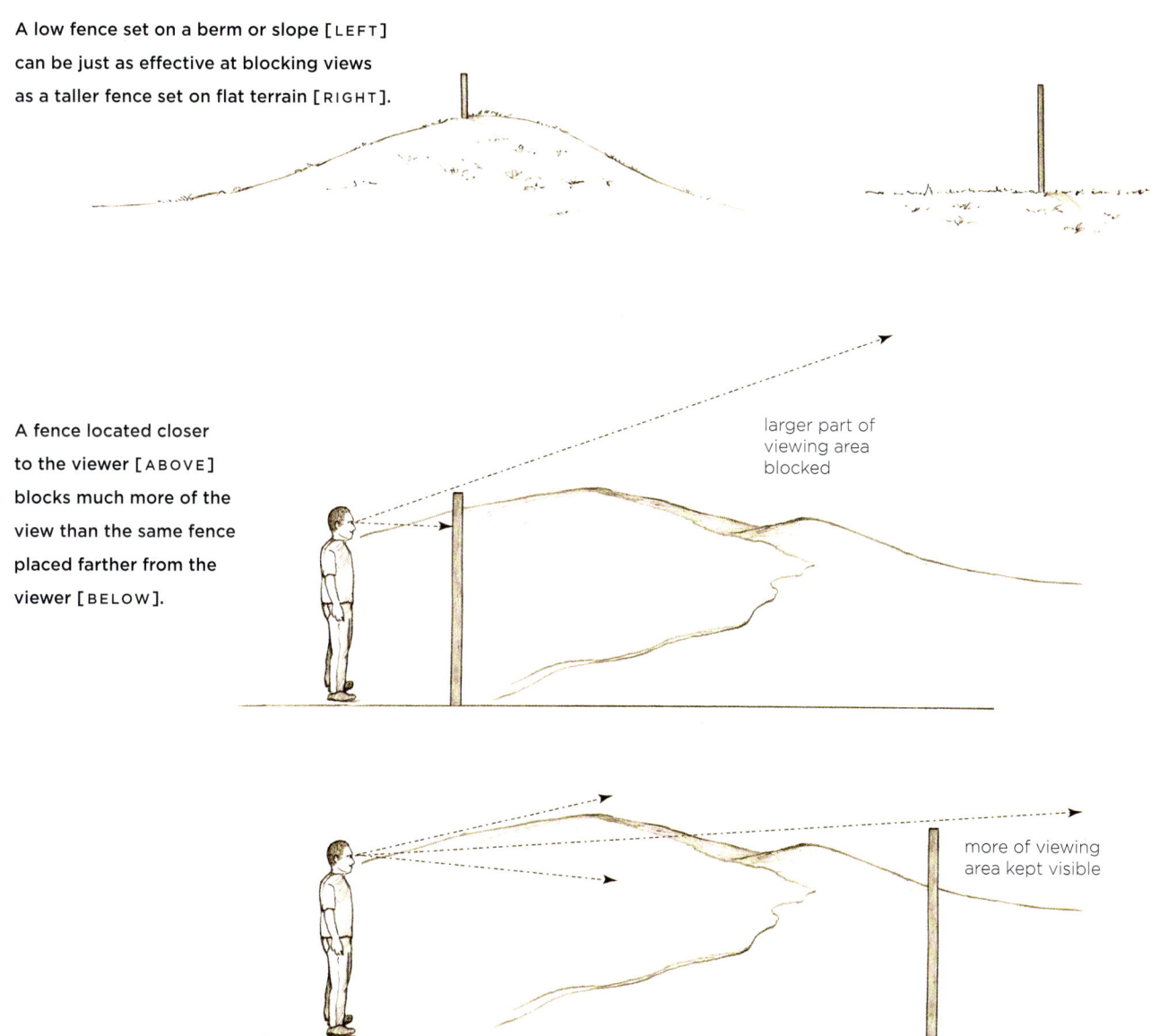

A low fence set on a berm or slope [LEFT] can be just as effective at blocking views as a taller fence set on flat terrain [RIGHT].

A fence located closer to the viewer [ABOVE] blocks much more of the view than the same fence placed farther from the viewer [BELOW].

larger part of viewing area blocked

more of viewing area kept visible

A low fence can help keep foot traffic away from a flower garden alongside the house or sidewalk.

Practically speaking, some general rules of thumb may help you make this decision:

The 2- to 3-foot fence: This type of fence can be used for a basic border around the foundation of your house or garage, to protect flower beds, for example. Keep in mind that low fences can present tripping hazards and should not be placed in the path of foot traffic. Someone carrying a sack of groceries or walking at night may overlook even a colorful and highly visible low fence.

The 4-foot fence: Four feet is a good height for a fence that divides yards without dividing neighbors. It takes a little effort to climb over a 4-foot fence, so it offers some degree of security. And it can offer a fair amount of privacy for someone seated near the fence. Yet it is also a perfect height for adults to lean against while conversing with their neighbors. Four feet is the recommended minimum height for safety barriers around pools and other potential dangers.

The 5-foot fence: Five feet can often be a somewhat awkward height, at least when used as a boundary fence between properties. It's hard to conduct a comfortable conversation over except for tall people, yet it allows partial views of heads that might be bouncing around on the other side. Still, if your priorities place privacy at or near the top, 5 feet would be more functional than a lower height.

The 6-foot fence: A 6-foot fence is often what codes and ordinances mean when they refer to a "privacy" fence. Fences above that height are sometimes termed "spite" fences (implying that their intention is to bother your neighbors) and usually are not permitted.

You will want to make sure that you avoid unintended consequences when determining a fence height. For example, if you want to block the view from your deck or patio of passing street and sidewalk traffic so that you can better appreciate the beautiful hills beyond, don't build the fence so high that it blocks the long-range sightline.

By carefully matching materials and height, a well-designed fence can block unwelcome views while maintaining good ones.

Legal Creativity

If a local code or ordinance prevents you from building a fence as high as you want, there may be ways for you to work around the restriction. The first and foremost option is to pursue whatever means exist for challenging the restriction. As discussed in the previous chapter, you may be able to obtain a variance, which will allow you to build a fence higher than allowed by law. Also, turn your scrutiny to the wording of the code or ordinance that spells out the relevant restriction. For example, you may be restricted from building a wood, metal, or masonry fence or wall over 6 feet tall. If you are trying to block a view that requires a higher structure, you might be well within legal bounds to add some plantings on top of a 6-foot structure. Potted plants could be set on top, or perhaps you would be permitted to use lattice or metal hoops to support flowering vines that reach the height you want. Another option would be to forget about a standard fence and instead plant trees or tall hedges. While the latter approach might require that you wait a few years for the intended result, it would most certainly not face the kind of restrictions that a fence would. ■

If the law says you can't build a fence more than 6 feet tall, you may be able to plant trees or a hedge that grows much higher.

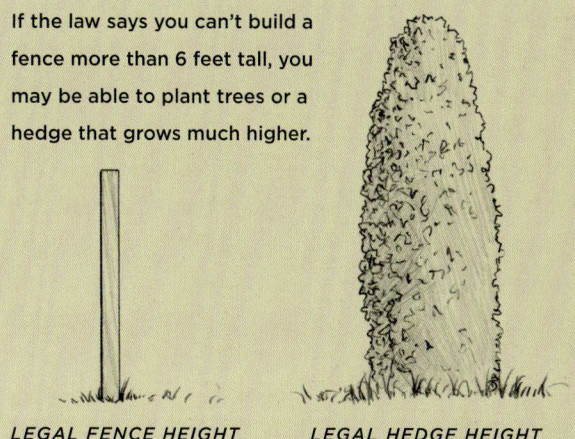

LEGAL FENCE HEIGHT   *LEGAL HEDGE HEIGHT*

Scale and Proportion

IN DESIGN, EVERYTHING IS RELATIVE. Scale and proportion are two concepts that help define and control that relativity. Scale refers to the relationship of the size of the fence to its surroundings (and that usually means specifically the adjacent house). Part of the manner in which we perceive scale has to do with the overall design of the fence, including not just its height but also its mass. A low, solid fence, for example, may feel bigger than a high, open fence in comparable settings. When I find myself looking at a fence and the first thing that comes to my mind is "I wonder why they put that fence there," it's usually because I'm having a knee-jerk reaction to a lack of balance in the scale of fence-to-house. A short picket fence in close proximity to a very large masonry house can produce this effect. So can a fortresslike solid fence surrounding a complex, custom-designed house. Even in cases where a fence is largely ornamental or is intended to be a gentle boundary marker, we will have a more comforting perception of it if the overall scale of the setup feels visually balanced.

Proportion is a similar concept, but with a more mathematical twist. While scale relates to the fit of the fence with its surroundings, proportion deals with the relationship of the parts of the fence to each other. A lack of proper proportion may make a fence appear too top-heavy, or too weak, or too square. The posts may be perceived as too dominant, or the rails as too spindly. While scale can be adjusted by designing the fence with consideration for nearby buildings and landscape, proportion is often addressed by designers with one or more mathematical systems. The best discussion that I have encountered about these matters is contained in a wonderful book by Seth Stem called *Designing Furniture* (Taunton Press, 1989). Stem presents user-friendly summaries of such concepts as geometric, arithmetic, and harmonic progression; the Fibonacci series; the five orders of classical architecture; and others. The more elaborate a design you are contemplating for your fence, the more useful you will find these lessons to be.

For most fences, however, especially those with a sectional pattern (that is, those visually interrupted by prominent posts), the most important proportioning system is the "golden section." This concept derives from the Greeks, as best as I can determine, although the origin of the name itself is unclear. It has also been known in the past as the "divine proportion," a name that suggests how much it occurs in nature, as well as in music, art, and architecture. The golden section is based upon a ratio of 1:1.618, which can be visualized

A grand, formal house is better served by an equally formal hedge [LEFT] than by a small picket fence [RIGHT].

by dividing a line so that the ratio of the long section (B) to the whole line (A) is the same as the ratio of the short section (C) to B. A therefore equals 161.8 percent of B, and B is 161.8 percent of C.

Still with me? Good, because the rest of the description should be easier to grasp. The ratio can also be grasped, without resorting to decimals, as 5 to 8; it's not an exact match, but for our purposes it's close enough. So, how can the golden section be applied to a fence? Well, there are three ways, in my opinion. First, if you are planning a fence with one type of material or design set on top of another, such as a lattice topper over solid boards, the bottom section should be 1.618 times the height of the top section. While this creates a nice vertical balance to a fence, it unfortunately is often impractical, given the height limitations we often face with fence building.

A more useful application of the golden section is in the ratio of height (1) to width (1.618) of individual fence sections, which creates what is known as a golden rectangle. Applied to a 6-foot-high fence, it requires posts set on 10-foot centers (6 × 1.618 = 9.7). Should you want to apply the formula to a fence with posts spaced on 8-foot centers, multiply 8 by 0.618, which results in approximately 5. Any variation on this rectangle will produce proportions that feel balanced to the human eye. But don't get carried away with trying to match the ratio exactly; just do your best to approximate it.

The other potential application of the golden section to fence design concerns the plan view—that is, the ground measurements of a space that will be enclosed by a fence. Again, this is not commonly encountered, but it could be useful for those designing a series of outdoor rooms or a privacy fence in a small backyard. To create a roomlike feel, try to keep the dimensions somewhere between a square and a golden rectangle. The more you increase the ratio between length and width, the more the space feels like a hallway or pathway—a place to pass through rather than to rest and relax. Sitting in a space like this can be disconcerting rather than relaxing. To maximize the intimacy of the space, keep it small.

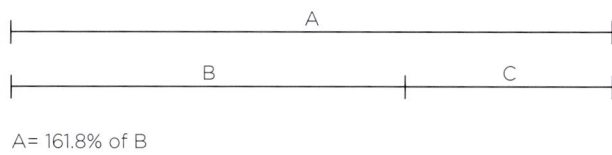

THE GOLDEN SECTION

A= 161.8% of B
B= 161.8% of C

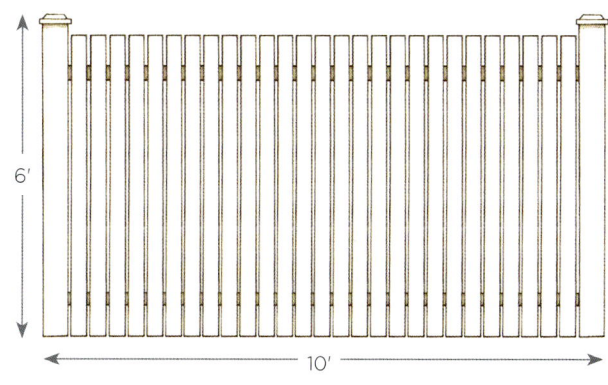

VERTICAL APPLICATION

SECTIONAL APPLICATION

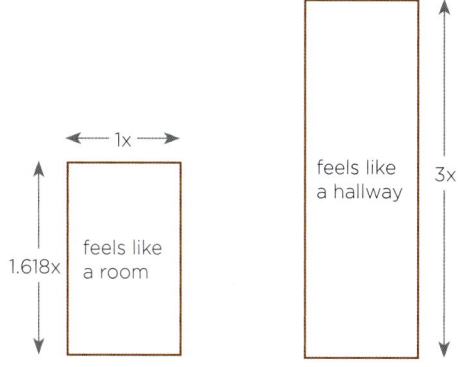

PLAN VIEW APPLICATION

Mixing Materials

THE BULK OF THIS BOOK is divided into sections devoted to specific materials, but great fences can be built using more than one material (which, I suppose, could be interpreted as yet another plea for the reader to digest the whole book before beginning to plan a fence). Mixing materials automatically introduces contrast into a fence design, and it can also add a great deal of character.

In those neighborhoods of grand old historic houses, it is not uncommon to see wrought-iron fences set atop substantial stone or brick foundations. This type of fence presents contrasting auras of solidity and security with openness and elegance. Although it can take a good deal of work, building a fence of this nature is not beyond the reach of an ambitious do-it-yourselfer.

METAL FENCE WITH BRICK POST

WOOD FENCE WITH BRICK POST

STONE POSTS WITH CHAINS

A clever mixture of manufactured and natural materials, this fence uses bricks to create a contrast in color and pattern with the mortared stones.

If you have a good natural stone dealer nearby, you should be able to find granite cut into roughly 6-inch squares that can be used as gateposts, or even as regular posts on a short run of fence. I like the look of a fairly low boundary fence with granite posts and black chain. Granite gate posts also help call attention to the gate opening in a fence otherwise consisting of wood or metal. The tricky part of working with granite (beyond paying for the stuff) is drilling holes needed for hardware and fasteners. If in doubt, ask your dealer for advice on the proper drill and bit. See Chapter 4 for more on working with stone.

Brick also mixes well with wood or metal. Thick brick gateposts are easy to build and look great with a board or picket fence, while alternating or occasional brick posts can break up the monotony of a long run of metal fencing.

Hedges and short trees can be used to great effect with all kinds of fences. By incorporating sections of "living fence" plantings with standard fencing materials, you can add variety in depth, height, and character without undermining the function of the fence. The results can be truly as much about landscaping your property as they are about fencing it in.

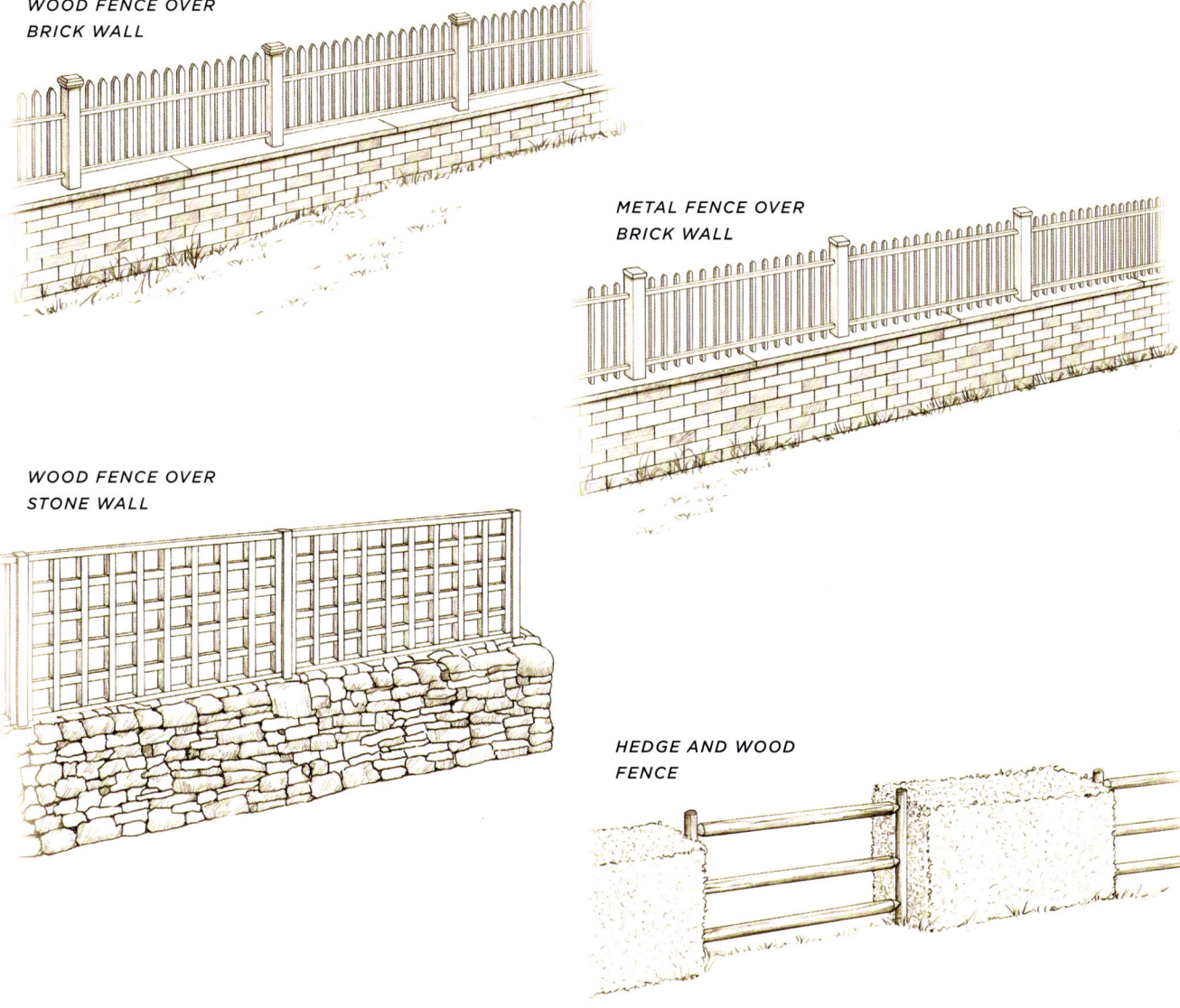

WOOD FENCE OVER BRICK WALL

METAL FENCE OVER BRICK WALL

WOOD FENCE OVER STONE WALL

HEDGE AND WOOD FENCE

Preparing a Plan

AFTER YOU HAVE DONE your research and gathered a sufficient amount of inspiration for your fence, it is time to start narrowing down the options and finding the right design for your property. Earlier, I suggested that you take photographs of existing fences as part of your research. Now, I am going to suggest that you turn the camera on your own property. Shoot a series of photos, capturing the site from different angles and depths (or, better yet, make good use of a zoom lens to adjust the image). Along the intended fence line, drive a few stakes in the ground and stretch a string line between them at the planned height of the fence.

Alternatively, place some furniture or people along the fence line to offer some perspective on how much impact the fence may have on sight lines.

Once the photographs have been developed (a step you can obviously skip if you've gone digital), there are a couple of ways you can put them to use. After selecting the most useful shot or shots, I head for my beloved scanner to create fairly large copies (say 5×7s or 8×10s). Make prints on "draft" or "economy" mode, then use a dark marker to start drawing your fence on the enlarged image. Print as many copies as you need until you feel comfortable with the results. Pay particular attention to the gate or entry sections as well as potential obstructions in the fence line (such as trees or existing gardens).

If you don't have a scanner, place some tracing paper over one or more photos and sketch some ideas. When you find a design you like, tape it in place on the photo.

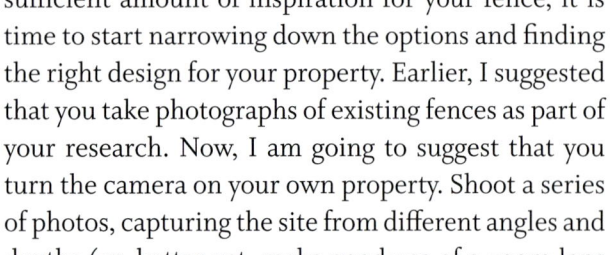

[LEFT] Use tracing paper over a single photo to sketch multiple fence ideas.

[BELOW] With multiple copies of a photo of your house (scanned or photocopied), you can draw different types and sizes of fence to get an idea of how each would look.

Establish the Fence Line

If you haven't already done so, now would be a good time to drive some wood stakes to mark the end lines, corners, and gate locations for your planned fence. If you are planning a wood fence, these stakes would locate the terminal posts (see page 76). If you are constructing a boundary fence and you need to obtain a building permit, you will have to know where your property markers are located. The markers may be metal pins or wood stakes driven in the ground, but they could also be a stack of stones. Their location and characteristics should be noted on your deed map, but if you cannot find them, you may have to hire a surveyor. Be sure to observe any required setback for your fence line; note that this can be a particular problem if your property line contains some zigs and zags. Use a 50- or 100-foot tape measure to measure the length of the fence. Make a rough plan drawing with the dimensions on a sheet of paper.

If you are planning a gate, be sure to note its location and size. With the dimensions of the fence line established, it is time to prepare scaled drawings.

Scaled Drawings

The occasional conflict that arises between designers and builders most often boils down to the simple question, Can this drawn plan actually be built? This is the stage of the process where you need to find an affirmative answer to that question for your own fence. And, although it takes a lot of time that you might think would be better spent out in the yard, I know of no better means of establishing an answer than preparing detailed drawings.

By "detailed drawings" I mean scaled drawings that plot the plan view (the overhead or bird's-eye perspective), the elevation (side view), and details (such as post ornamentation). In some locales, you may need to present scaled drawings to the building department in order to obtain a permit. Even without

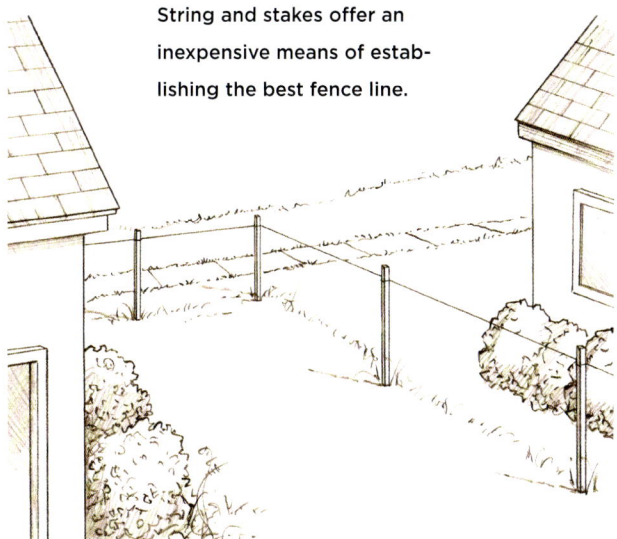

String and stakes offer an inexpensive means of establishing the best fence line.

such an obligation, the drawings will allow you to work out the important dimensions, confront potential obstacles, and calculate the quantity of materials you will need to buy. Scaled drawings are very important with wood fences, and a bit less necessary with other types of fences.

If you are constructing a long fence around a substantial chunk of property, or one with an irregular pattern, you may also want to start with a site plan, which would incorporate the proposed fence line in a sketch of the entire property. A site plan could also be invaluable if your fence was just the first major component of a larger landscaping project, in which case you would want to be able to plan and envision the fence alongside future plantings and other projects. For most fences, however, a site plan is really only necessary if required for a permit. In this case, the plan needs to establish that the fence will not violate any setback requirements.

Graph paper is great for preparing scaled drawings. You can find graph paper with various grid sizes, but the most common type has ¼-inch grids and can be used to make drawings with a scale in which ¼ inch on paper equals 1 foot of actual space. A standard sheet of graph paper would therefore be good for a fence line up to about 40 feet. For longer fences, tape a couple of pieces together. To be safe, use a ruler to draw straight and accurate lines (1 inch on paper will equal 4 feet of actual fence). An alternative (or complement) to graph paper is an architect's rule, which offers you the ability to make drawings using a variety of different scales.

Draw a plan view first. For wood and metal fences, the plan view is useful for establishing post locations and determining how many postholes need to be dug. You will almost certainly run into some problems with post spacing. If you plan to space the posts on 10-foot centers, for example, and your fence line runs 42 feet long, it would be foolish to stick with the 10-foot spacing and add one additional post with a 2-foot spacing. Better choices, as shown on the facing page, would be to make the necessary adjustment equally on both ends, or to recalculate the spacing so that it creates equally sized fence sections, all shorter than 10 feet. The other option is to lengthen or shorten the fence enough to accommodate the 10-foot spacing.

The elevation drawing does not have to include every detail of the entire fence. Especially if the fence sections will be equally sized, or if you are installing a continuous type of fence infill that is uninterrupted by posts, the elevation can be limited to a small section. Elevations contain all of the vertical elements and dimensions. You may want to include the depth of the postholes, and then extend the drawing to the topmost element of the fence. Include dimensions for aboveground post height, locations of rails, and accu-

rately scaled renditions of the pickets or boards you plan to include. Here you can work out the desired spacing between pickets, for example, which will facilitate construction as well as help you determine how much wood to buy and how many pickets to make.

If you plan to add decorative trim to post tops or do some other work that requires precise measurements, you may want to work out the finished dimensions in one or more detailed drawings.

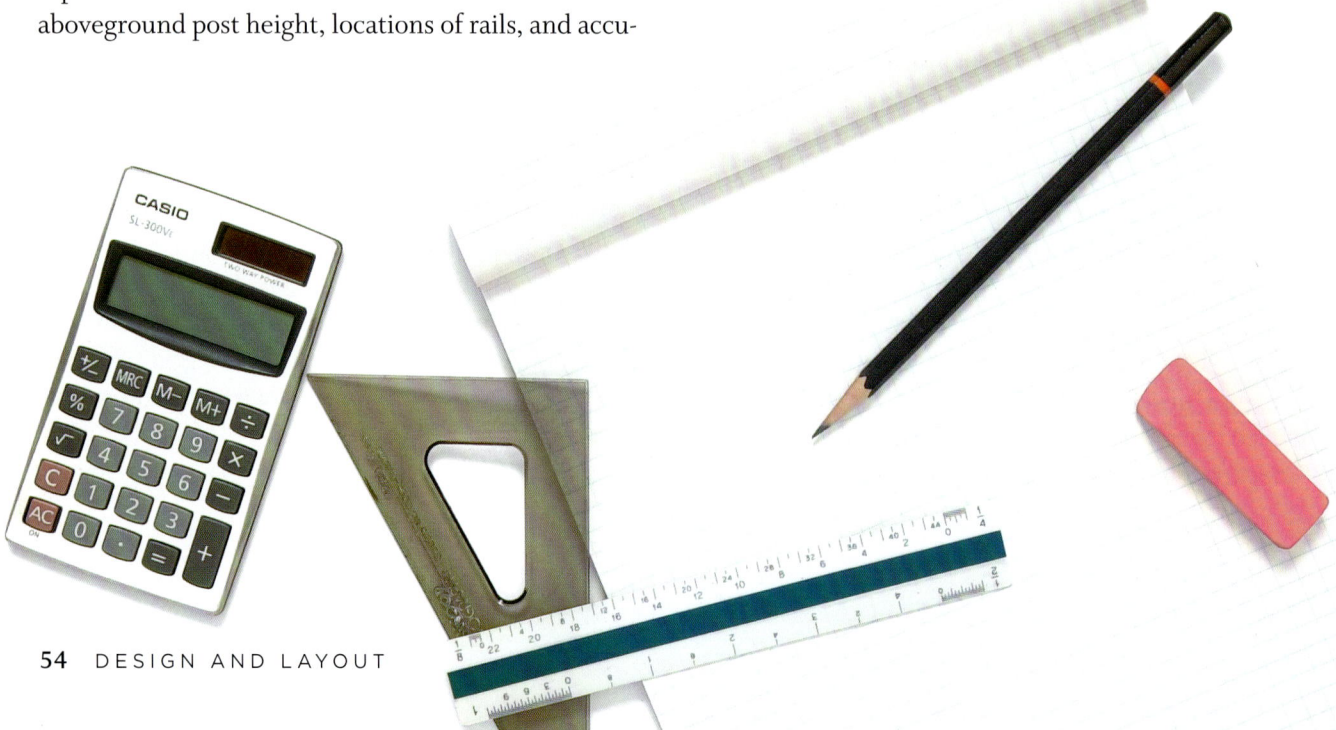

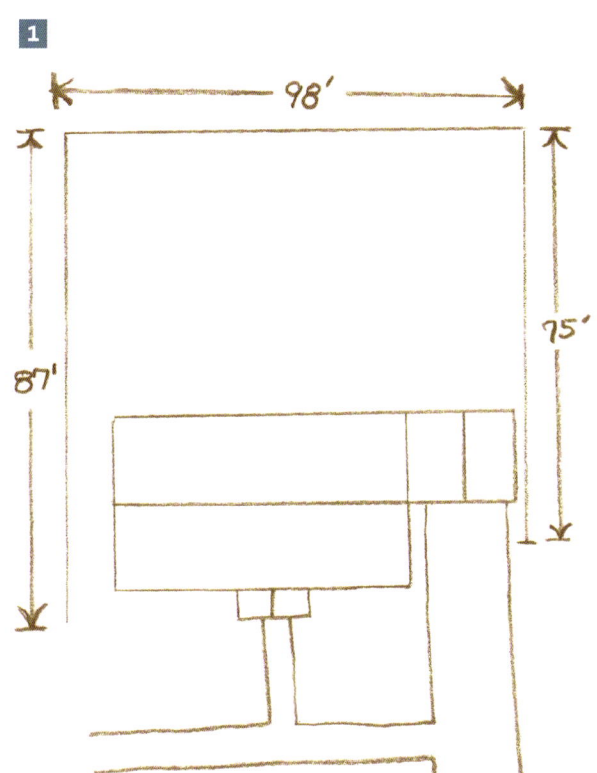

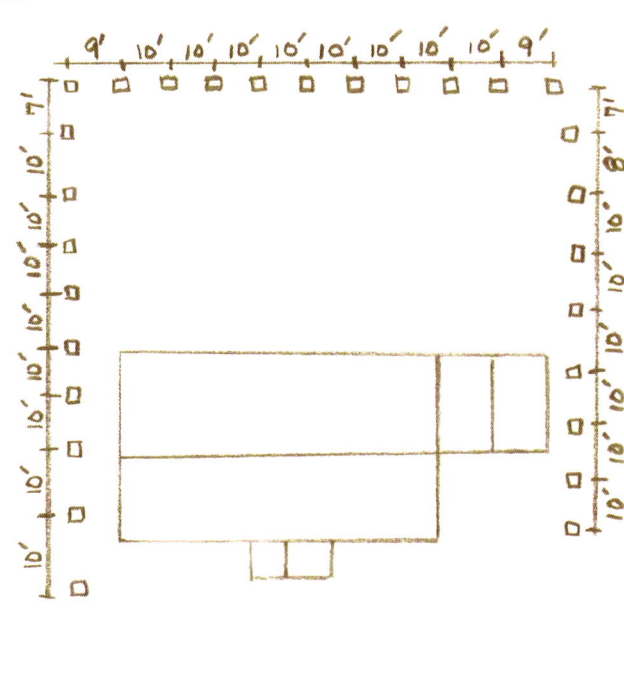

DRAWING PLANS IN THREE STEPS.

1 Begin with a rough outline, establishing the length of the fence. **2** Prepare a plan view, drawn to scale, establishing the post spacing. Note that the standard on-center spacing here is 10 feet, with adjustments made at the back corners. **3** Finally, create an elevation of one or more fence sections, identifying the materials to be used and their length and location.

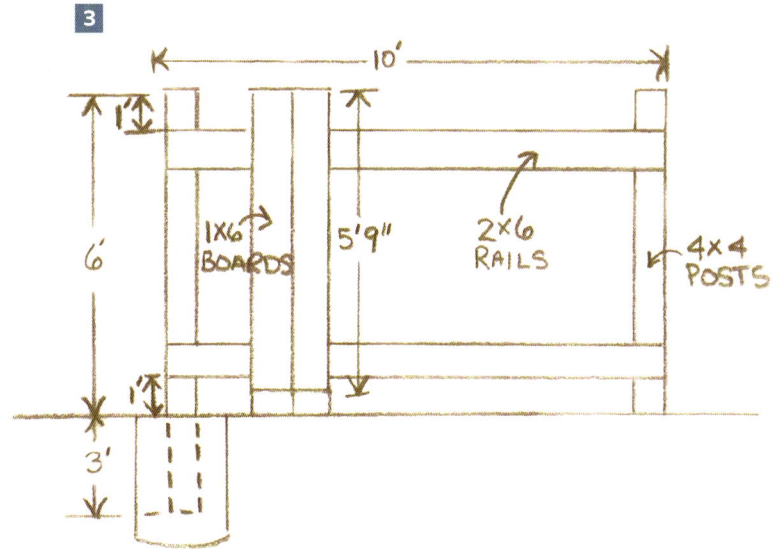

A stepped masonry wall [ABOVE] should follow the same rules of design and proportion as stepped wood fences.

Dealing with Slopes

Elevation drawings also force you to confront potential problems with the site itself. Yards and fence lines are not always flat, and it is often necessary to adjust your plans accordingly. If the slope is pretty steep, or the fence is quite long, it makes sense to measure the slope and include that information on your elevation drawing. For one thing, determining the extent of the slope will help you decide which style of framing to use for the fence. Also, it will help you work out the size of each fence section if you decide to use a stepped framing technique.

For most do-it-yourselfers, the easiest way to gauge a slope is to use a water level, as shown on page 63. Drive a wood stake in the ground at the top and at the bottom of the slope along the fence line. Align the water line in the tube with the top of the stake at the top of the slope, set the other end of the tube alongside the bottom stake, and mark the water line. You can now determine the rise of the slope by measuring from the ground to the reference mark on the bottom stake and then subtracting the same measurement on the top stake. Now measure the run (the distance between both stakes), holding the tape measure taut and aligned with the reference marks.

On a piece of graph paper, draw a level baseline along the bottom, then chart the run and rise on the paper and draw a line to indicate the slope. With a scaled version of the slope now on paper, you can use some tracing paper to divide the fence into sections with equally spaced posts and be able to determine which of the framing methods will work best.

Stepped Framing

With stepped framing, the fence is built in sections, with each section resting level at a different elevation. On gradual slopes, I think it looks best. To determine how much each section should rise, count the number of sections you need based upon your design. Now, divide the rise (in inches) by the number of sections. The result will tell you how many inches each section needs to rise for the fence to step evenly. Yes, this is time-consuming, but the results are worth the effort.

A stepped fence looks best when the distance from the post top to the infill top is uniform and the posts are equally spaced. To achieve this, however, the infill must be cut to different lengths. It can be easiest to cut the infill to length after it has been installed.

Sloped Framing

With sloped framing, the fence is sloped to precisely mirror the slope itself. With steep slopes or rolling terrain, this is often the best approach.

Hybrid Framing

On very steep slopes, the hybrid method is often best. It uses stepped framing, with the infill cut to match the slope.

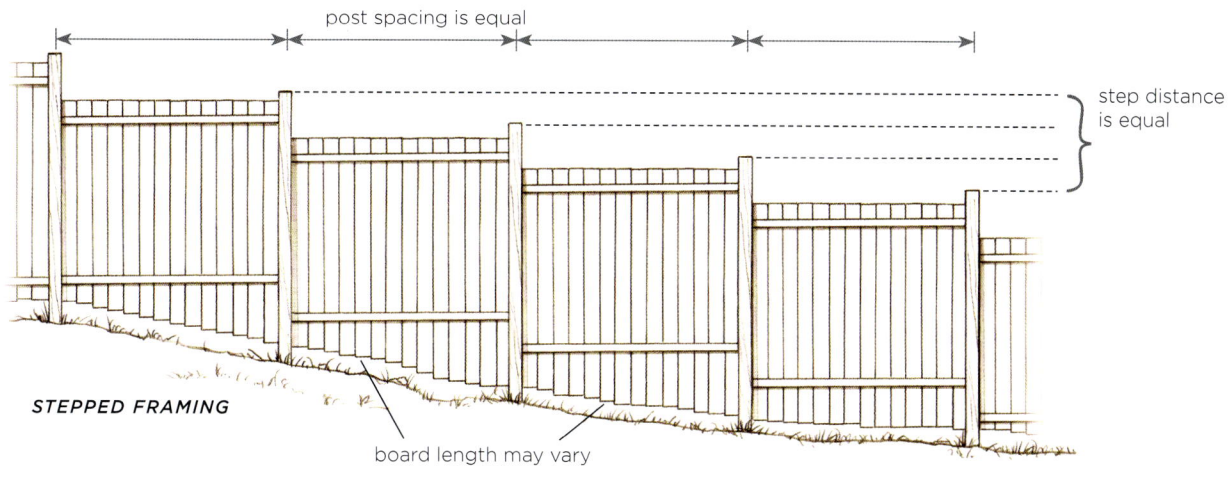

post spacing is equal

step distance is equal

STEPPED FRAMING

board length may vary

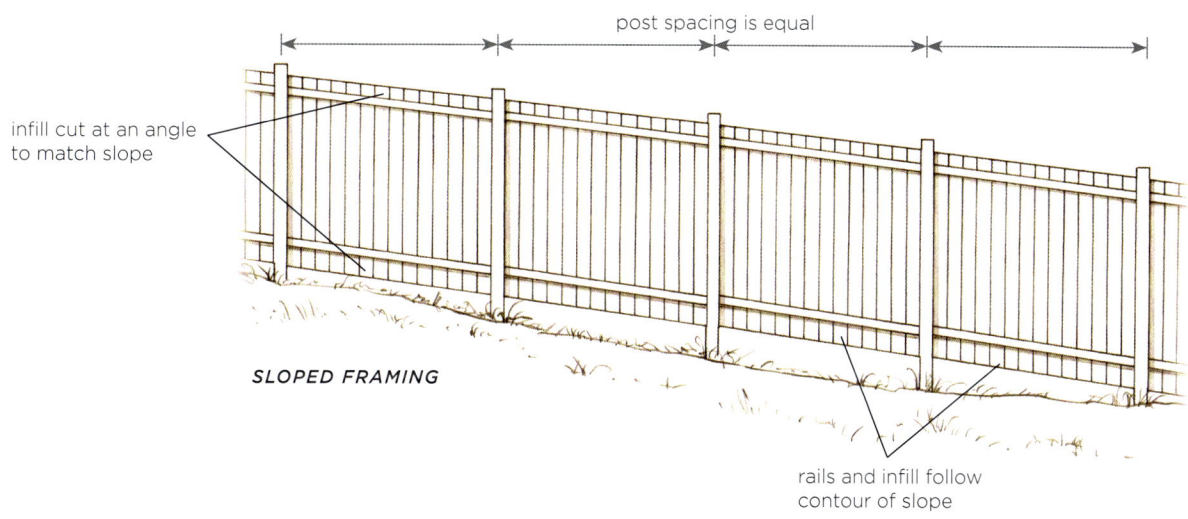

post spacing is equal

infill cut at an angle to match slope

SLOPED FRAMING

rails and infill follow contour of slope

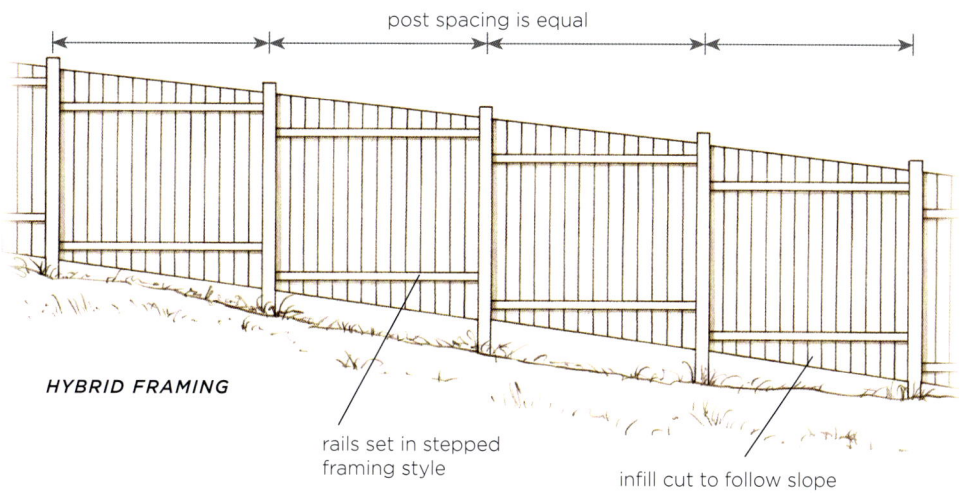

post spacing is equal

HYBRID FRAMING

rails set in stepped framing style

infill cut to follow slope

Trees and Other Obstructions

It is best to try to establish your fence line so that it is not interrupted by trees or other immovable obstacles. When that is not possible, in some cases you can create a little bump-out in the fence line to pass around the obstacle. But if you must include the obstacle in your fence line, and you are building a fence that requires posts, you will find it useful to tinker with the framing plan on a scaled elevation. If a tree is your obstacle, you may be tempted to try and fasten a wooden fence to it, but I would caution against such action. Driving nails or screws into the trunk of a tree can be harmful to the tree itself. And, given that trees grow, tying the fence too closely to the tree can create problems in years to come. It is also not wise to plan on setting fence posts too close to a tree, as digging the holes for the posts will be complicated by roots and could cause serious damage to the tree's root system. The best compromise in this case is to place posts several feet away from the tree trunk, then let the rails and infill overhang the posts enough to fill in the gap. This section of the fence will be somewhat weak, but it may be the best you can do.

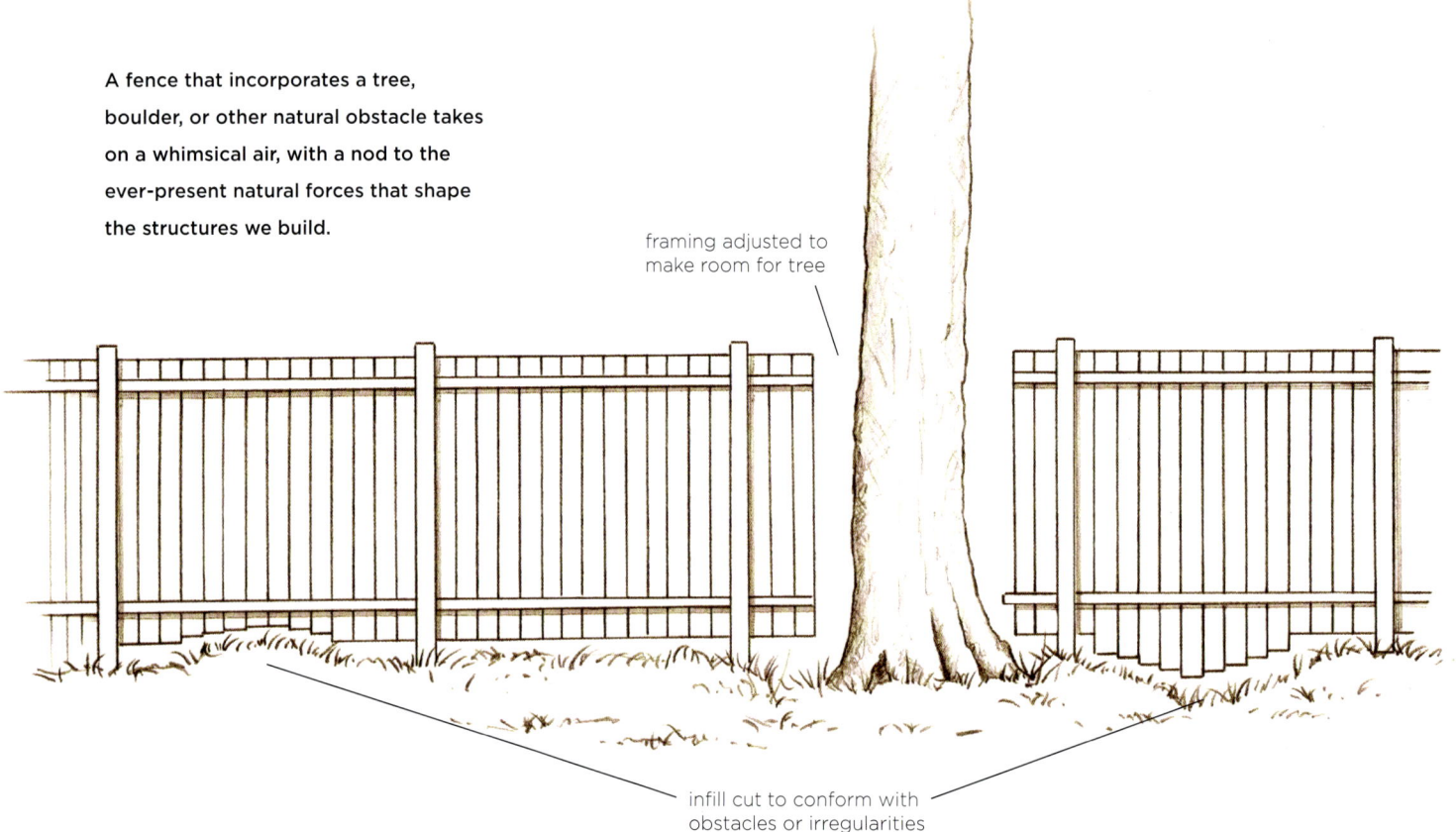

A fence that incorporates a tree, boulder, or other natural obstacle takes on a whimsical air, with a nod to the ever-present natural forces that shape the structures we build.

framing adjusted to make room for tree

infill cut to conform with obstacles or irregularities

Turning Corners

If your fence line needs to turn a corner, in most cases you want the turn to create a perfect right angle. The easiest way to do this is to use the 3-4-5 technique (which, for you math geeks, is carpenter-talk for the Pythagorean theorem: $a^2 + b^2 = c^2$). Measure along one side 3 feet, measure along the perpendicular side 4 feet, and, finally, measure the diagonal between these two spots. If the diagonal equals 5 feet, the corner is square. If it's not, fuss around with the string line until it is. This technique works with any set of numbers that rely on the same ratio, and it will be most accurate on a fence if you use 6-8-10 or even 9-12-15.

A corner will form a right angle when $a^2 + b^2 = c^2$. In this example, $9^2 + 12^2 = 15^2$ ($81 + 144 = 225$), meaning that this layout has a perfect right angle at its corner. If the result is not right, adjust the string line and calculate again.

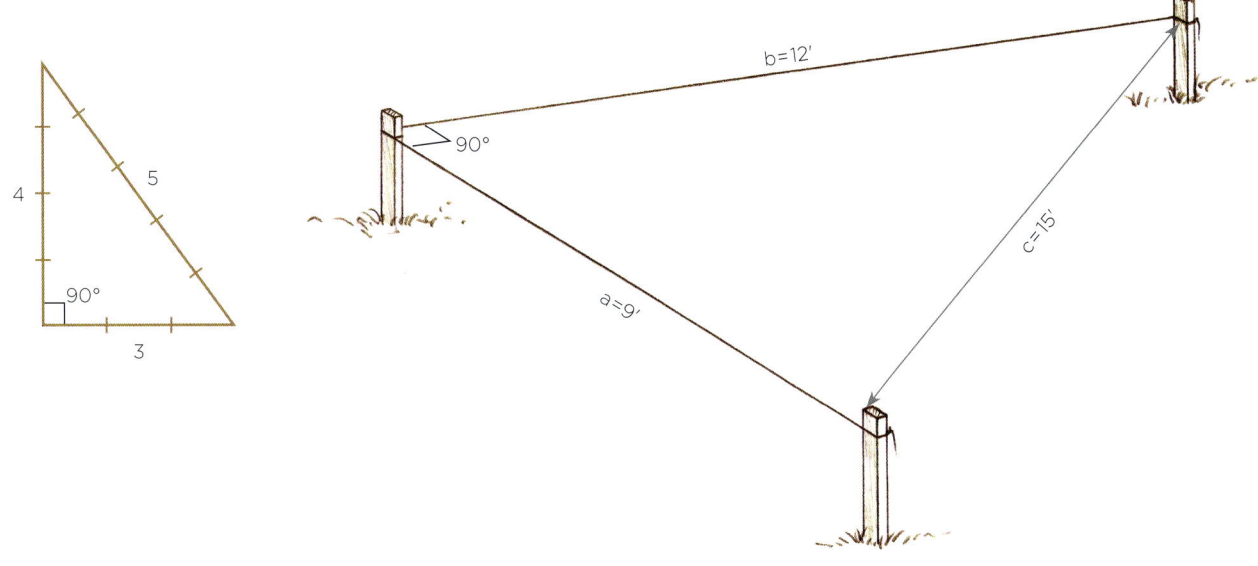

Finalizing the Plan

WITH STAKES AND STRINGS marking the actual site, and scaled drawings that provide the scope and content of your fence, you will have much of the heavy thinking out of the way. You should find that the heavy lifting that follows will proceed more or less according to plans, rather than whims. Take some time to bounce back and forth between the site and the drawings until you are satisfied that they are synchronized and that they represent your ultimate vision.

The discussion to this point has been largely generic, dealing with issues you would likely confront regardless of the type of fence you plan to build. The next steps will depend upon your choice of fencing material and style. You may be able to use the existing string lines to help you mark the locations for posts, which are the principal structural components of wood and metal fences. Or they may help guide you in planting a hedge, or keep you on track in constructing a stone fence. Regardless of the size or type of fence you build, good plans will allow you to estimate accurately how much material to buy, and then ensure that you make the most efficient use of your investment. Although I often grumble about the time spent in planning, I never regret it.

Plumb and Level

You can spend all the money you like on top-of-the-line tools and materials to build a fence, but it will be money wasted if you cannot master the basic carpentry principles of plumb and level.

Plumb refers to something being perfectly vertical. A plumb object is one that maximizes the force of gravity by transferring a load straight down to the earth. Any deviation from plumb weakens that force, and it also looks bad.

Level refers to being true to a horizontal plane (that is, to the horizon).

Square refers to a 90-degree corner, where a plumb object meets a level object.

Checking for plumb and level requires only one tool: the simple carpenter's level. Levels come in all kinds of shapes, sizes, and price ranges. I keep a 9-inch torpedo level in my primary tool bucket, and it gets used quite a bit. I also have 4- and 6-foot levels; the

former gets quite a bit of work, while the latter usually just sits in my garage. The level I use the most is a 2-footer. It's easy to use, and I've learned over time to trust its accuracy.

If you are looking to buy a carpenter's level, I would suggest getting a good-quality 2- or 4-footer. The body of the level needs to be strong enough to withstand a lot of bumps and bruises over time, but the critical components are the vials. All levels should be checked for accuracy from time to time (and you should be able to do this in the store before you buy a new one). Set the level on a flat, level surface and notice where the bubble in the middle vial sits. Now flip the level over end for end so that it comes to rest in exactly the same place (you might want to mark the location with a pencil to be sure). If the bubble is in exactly the same place as before, it is giving you an accurate read. Perform the same test with the two vials used for checking plumb on a wall or other vertical surface. I have also developed the habit of running my hand along the edge of a level before I set it on a surface. A little dab of dried glue or dirt will definitely skew the reading you get and should be cleaned off before you can trust the results.

An important thing to keep in mind when using a level is that the tool is only giving you an accurate reading on the surface it is in contact with. Thus, a 9-inch level held against a post is checking for plumb on only a 9-inch piece of the post. Since lumber is often twisted and warped, this reading from a small section may be misleading. Slide that small level up or down the post, and you may get a different impression about how plumb the post is. Longer levels help offset this potential discrepancy. This is why I am a little leery about small levels of all varieties.

CARPENTER'S LEVEL

[LEFT] The level is plumb when the bubble is centered in the two end vials.

[BELOW] The level is level when the bubble is centered in the middle vial.

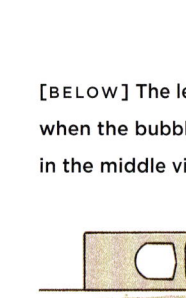

LINE LEVEL

Line levels, for example, are almost never accurate enough to warrant their use except in cases where accuracy is not all that important. A line level costs only a dollar or two, and its only real function is helping to transfer a level line from one object to another, most often from one fence or deck post to the next. To use it, you first establish the reference line on one post. Then you hook the level over a string line held as taut as possible, with one end resting on the reference line. Slide the string line up and down the next post until the bubble rests in the middle of the level, then mark that post and move on. The problems with this approach are that any sag in the string line or breeze in the air will produce an inaccurate reading. And, since the level should be located midway between the posts, it can be hard to read if you and a partner are each holding one end of the string.

Somewhat more reliable results can be obtained from an inexpensive tool called a post level. A post level theoretically will tell you at one glance if a post is plumb on two adjacent sides. It can be quickly strapped to a post and has two level vials and one plumb vial. Once in place, you simply move the post around until the bubbles in all three vials are centered. I have seen one version of this tool that has magnets to hold it to metal posts. I could see using one

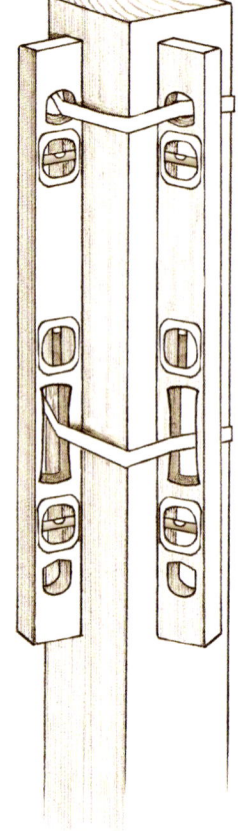

A post must be checked for plumb on two adjacent sides, which is easiest to manage with two levels temporarily attached to the post.

of these post levels if I was installing wood or metal posts for a wire mesh fence or a basic post-and-rail fence. It is certainly simple to use, and very handy if you are working alone. But it still leaves you with the need to figure out how to establish level post tops. I would not rely on a post level to set posts on a nicely designed fence.

I like to set posts using carpenter's levels. One will do, but I find that it's quicker to use two. Using strips of Velcro, I attach levels to two adjacent sides of the post, then keep an eye on all four plumb vials before bracing the post. Velcro makes the job of attaching and removing the levels very easy, although you could also use bungee cords or some tape.

To transfer level lines from one post to the next, I usually strap a level onto the edge of a straight 2×4 (again, Velcro works best). Use a 10-foot-long 2×4 with posts spaced 8 feet apart. Hold one end of the 2×4 at your reference line, move the board up or down until the bubble centers in the middle vial, then mark the spot on the other post.

Sometimes you need to find level over a span too long for the level and 2×4. In these cases, a water level is indispensable. It works on the principle that water in a flexible tube will seek the same level on each end. You can make a water level using any length of ⅜-inch or ⁵⁄₁₆-inch clear vinyl tubing. Fill it with water that contains a little food coloring to make it more visible. Make sure that there are no air bubbles or kinks in the tubing, and keep the ends open. Set one end of the tubing so that the water level aligns with your reference mark. Then place the other end of the tubing in the desired location. Once the water stabilizes, it will be level with the other end. You can also buy water levels that emit a beep when level has been established in the tube, which is a great help when working alone. One warning about water levels: If one part sits in the shade and another part in direct sunlight for long, the water temperature (and, thus, the water density) will differ in the two areas. This will produce misleading results.

For the more ambitious among you, laser levels continue to decline in price and have started showing up in the toolboxes of do-it-yourselfers. Before buying, I would suggest reading reviews in some trustworthy publications, and after buying be prepared to study the directions for a while.

Water in a length of tubing will find its level, regardless of the distance between the ends of the tubing. Align the water line in one end of tubing with your reference mark, and then find the water line in the other end, which will be level with the first one.

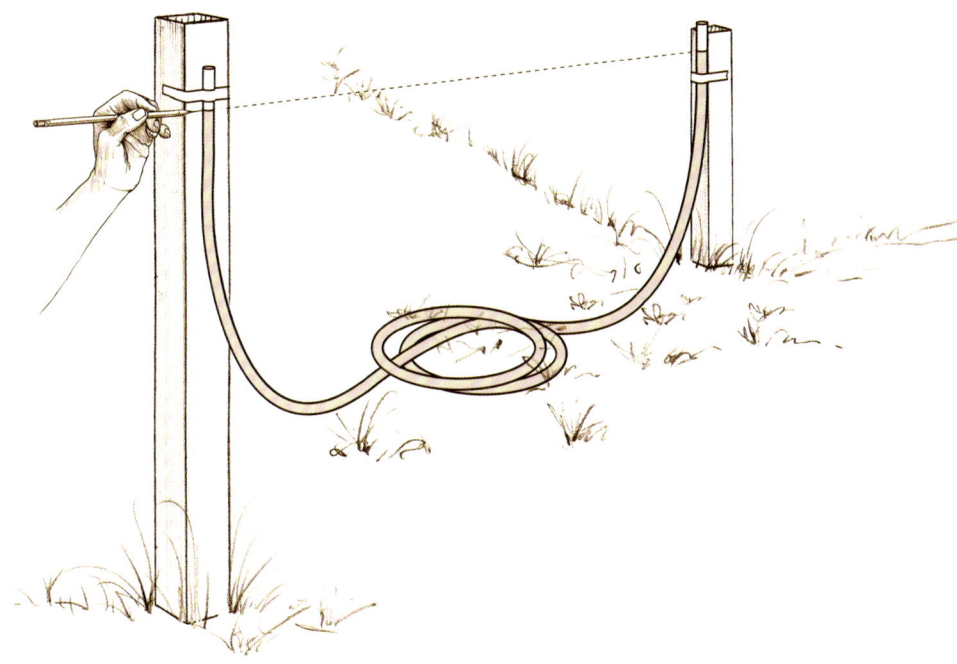

DETERMINING SLOPE WITH A WATER LEVEL.

Slope is a measure of *rise* (vertical distance) over *run* (horizontal distance).

1 Drive wood stakes at the top and bottom of the slope.

2 Align the water line at one end of the level with the top of the stake at the top of the hill.

3 Hold the other end of the level next to the stake at the bottom of the hill, and mark the position of the water line on the stake.

4 Measure the height of the stake at the top of the hill [*a*]. Then measure the distance between the ground and the water-line mark on the stake at the bottom of the hill [*b*]. Subtract *a* from *b*. The result is the total rise of the slope.

5 Find the run between the posts by measuring the level distance between them.

6 Transfer your findings to your layout drawing.

Wood Fences

AWAY FROM RANCHES and farms, the first thing most people think of when the word *fence* is mentioned is wood. And there are good reasons for that identification. Wood is one of those beneficial materials that grows in abundance, with little or no effort, and requires minimal processing before it can be used. There is almost no fencing need that cannot be met with wood, in one shape and style or another.

Humankind has been constructing fences out of wood for thousands of years, and I'm happy to report that this is one of those things that we do seem to be improving on rather than messing up. For most of that history, wood was gathered from trees that were cut or had fallen. Today, we buy our wood as a manufactured product at the lumberyard or home improvement center. Wood is accessible, affordable, and easy to shape and cut and repair. Because wood fences have been built in so many cultures over so many years, they can be easily adapted to just about any style you like. Wood fences can be strong, and with proper planning and care, they can last a long time. Perhaps best of all, building a durable and attractive wood fence does not require an advanced degree in carpentry or architecture.

Some fences, such as this rustic collection of poles, can be assembled from materials found and gathered nearby.

But as much as we love wood, we must acknowledge that we are not alone. The environment is full of insects and fungi that compete for wood's attention and render it a biodegradable substance. Moisture, temperature variation, and sunlight also contribute to the deterioration of wood. In a forest, all of these factors ensure the long-term survival of plant and animal life. In our yards or fields, however, we usually have other objectives in mind. Thus, choosing the right types of wood and tending to regular maintenance are critical factors in fence building.

This chapter will briefly survey the varieties of wood fences that have been—and can be—built, from the most rustic creations to refined and carefully styled contemporary designs. This is the longest chapter in the book because wood fences are an overwhelming homeowners' favorite and because there is so much to say about building them. The choice of wood, selection of fasteners, use of finishes, and adherence to proven construction techniques all involve important decisions that can confuse even an experienced do-it-yourselfer. I can't hold your hand through the process, but I can offer the best advice that I have been able to accumulate to make your decision making easier and more productive.

A low wood fence, cut to mirror the rolling terrain [ABOVE], can almost disappear into the landscape, while a line of buried poles [RIGHT] can create a simple yet effective background for a flower garden.

Perhaps somewhat paradoxically, this chapter will also discuss some fences that are not wood at all but are nonetheless manufactured to look like wood fences. You have probably driven by many fences in recent years that, had you thought to ask yourself, you would have identified incorrectly as being made of wood. Fences made of vinyl, vinyl-coated wood, and a variety of composite materials are becoming increasingly popular, and this chapter will offer some guidance on how you can approach those products. Bamboo isn't a tree; it's a grass. But it is rigid and can be cut with a saw and used in fencing in many of the same ways as wood. So it passes my "close enough" test and will be discussed in this chapter.

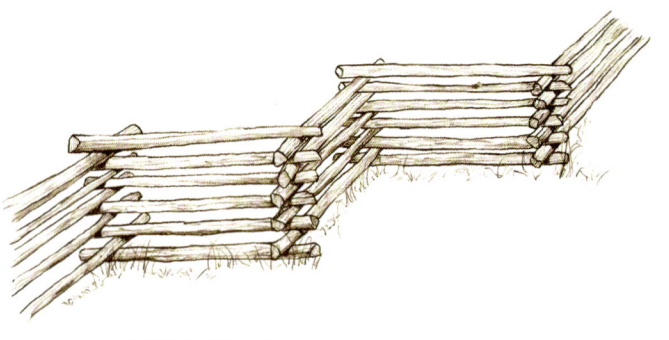

ZIGZAG FENCE

DOUBLE-POST FENCE

From Rails to Pickets

IF A CENSUS of the world's wood fence styles was possible, the results would almost certainly reveal so many styles, sizes, shapes, and construction variations as to defy easy categorization. But categorize we do, despite the inevitable simplifications and generalizations that ensue, because it creates some order out of chaos. With fences, broad categories reflect a combination of influences, including changing needs, expanding production and transportation of lumber products, simplified fasteners and construction techniques, and ever-evolving tastes. The groupings are neither scientific nor binding. With that in mind, it is best to think of the following fence styles as starting points for thinking about your own needs and desires, and not necessarily as polished candidates from which you must choose a favorite.

Rail Fences

First came the rails. Rail fences derive from the practice of constructing fences from trees that were cleared to create fields for farming or building houses. The zigzag (or worm) fence was composed of nothing but rails, stacked horizontally in an interlocking zigzag. This type of fence took up a lot of room and required a lot of wood, but wood was abundant and labor was in short supply in colonial America. Because the zigzag fence could be built without posts, one man could split the rails and assemble the structure relatively quickly.

To address the shortcomings of the zigzag pattern, innovative fence builders began stacking their rails between double posts, with the interlocking pieces tied together with wire or twine. These fences were installed in a straight line, thus conserving space and wood, and the double-post design allowed them to remain standing without holes having to be dug, another labor-saving approach.

Single-Post Fences

The next big step forward came with the use of single posts set in holes in the ground. Rails were trimmed at the ends to slip into matching holes or slots cut into the posts. This "split rail" innovation allowed for constructing strong fences using even less wood than previous styles. The installation sequence was critical: Set one post, then attach rails to that post, then carefully set the next post so that the rails could fit snugly, then move on to the next section.

Sawmills introduced the next wave of fence varieties. When logs could be cut into standardized sizes, construction could also become more standardized, and the post-and-board fence was born. The rails were now relatively thin and were fastened to square or rectangular posts. This milled lumber could be shipped to the site, which remained a sometimes-daunting task. But improvements in log-cutting machinery allowed sawyers to travel from site to site, cutting logs as they went. Mortised-rail fences were a variation on split-rail fences; they used mortise-and-tenon joinery to create tight, strong connections between rails and posts. Notches, or dadoes, could also be cut into the faces of the posts for rails to be inserted and, sometimes, nailed in. As nails became stronger and more weather resistant, fences began to be built with rails simply nailed to solid posts.

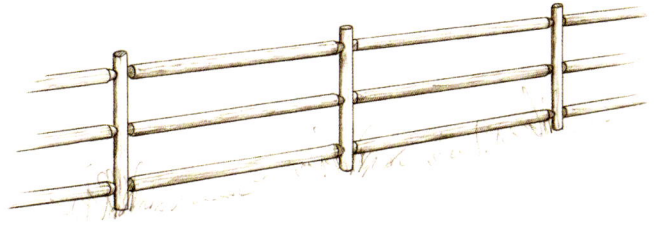

SINGLE-POST (SPLIT RAIL) FENCE

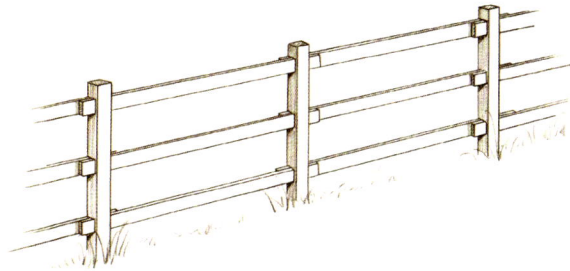

MORTISED-RAIL FENCE

Vertical-Board Fences

All of the above fence types and innovations were driven by the need to control the movement of animals. Building ever-stronger fences with as little wood as possible helped to keep down costs, labor, and maintenance. Vertical-board fences addressed different concerns (privacy and security) for a different group of people (urban dwellers). Using the basic post-and-rail skeleton, these fences were built with an additional layer, composed of closely spaced vertical boards. These fences tend to be higher than the earlier types, and they exist to restrict the view from one side to the other.

Vertical-board fences also opened up a whole new world of design choices. Boards can be installed tightly against each other to create, in essence, a solid wall. The same effect can be achieved with a bit more visual interest by alternating boards of different widths. Boards can be installed on both sides of the fence, which creates a pleasing visual sense of depth and adds the benefits of some ventilation, without necessarily sacrificing any privacy or security.

VERTICAL-BOARD FENCE

PICKET FENCE

Picket Fences

Picket fences are a variation on board fences. Their primary visual orientation is also vertical, but pickets are typically shorter and installed with spaces between them. Thus, picket fences are less called upon to provide security and privacy, though they do offer both to some degree. But what we most think of as far as picket fences go is their simple decorative charm.

They define boundaries and sometimes contain pets and children, but they do so in an understated way. They are the quintessential front-yard fence, functional if necessary but equally suited to existing for the sole purpose of looking nice.

Pickets have long proven to be an irresistible material for creative folks, who seem never to run out of ideas on how to cut and shape them into unique forms. In fact, if you have your heart set on a picket fence, and you want to make it a one-of-a-kind statement, you can do so by simply designing a distinctive pattern or cut to the pickets.

Choosing the Wood

UNDER CONTROLLED CIRCUMSTANCES, wood can last for a long, long time. Many people have wood furnishings that are generations, if not centuries, old. But outdoors, wood decays, and no one wants to invest time and money in a structure that is destined for the compost pile anytime soon.

Early fence builders learned quickly that some types of trees produced lumber that could last much longer than others. White oak, a reasonably durable and abundant species, found its way into many fences. Black locust and osage orange turned out to be exceptionally good woods for fence building. American chestnut, cypress, black walnut, cedar, and redwood were other good options. Some of these woods are available in limited regions, and a few of them were harvested into near extinction at one time or another. Pines, spruces, poplars, maples, and willows, while plentiful, offer little resistance to decay, especially if they are in regular contact with the ground.

The old-growth trees from which the first fences were built produced much better lumber than we get from the same species today. Trees in those virgin forests grew slowly and achieved full maturity with a high percentage of heartwood, which contains the most active decay-resisting properties. Today's commercial woodlots, by contrast, produce fast-growing trees that contain a greater proportion of sapwood, which is far more susceptible to damage from fungi and insects. Wood from trees that grew slowly can be identified by tight annual growth rings, which help produce strength and stability. Fast-growing trees produce fewer rings per inch, which means that their wood is weaker.

The most widely available decay-resistant woods today are cedar and redwood. In California, many people have built redwood fences and decks that have lasted for decades. Having spent most of my adulthood far from California, I almost never see redwood stocked at lumberyards or home improvement centers. But with a little work, you can usually find a source for redwood if you want it. Wood dealers and lumberyards can special order a supply. Cedar grows over a wider geographical area than redwood and can

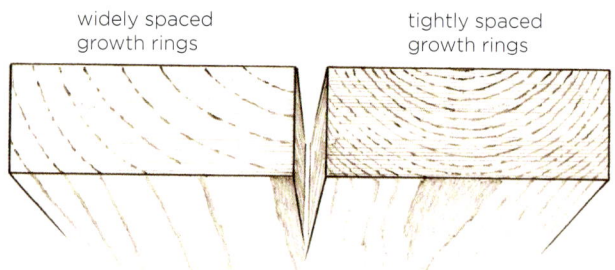

widely spaced growth rings — tightly spaced growth rings

The more widely spaced the growth rings are, the more likely the lumber is to warp.

be found at most lumberyards and home centers.

Though you often pay a premium for cedar or redwood, the fact of the matter is that you may not be getting what you think you are paying for. Even well-meaning lumber dealers tout the rot-resistance of these woods as though it was a given fact, but it isn't. To understand this paradox, it is necessary to understand a few things about the nature of wood.

Heartwood versus Sapwood

Most of us who successfully made our way through elementary school learned about growth rings in trees: Count the rings and you can determine how many years old the tree is. But trees live and grow through the action in the outer, newer rings, which contain living cells that provide passage for sap. As the tree matures, the core of the tree changes from sapwood to heartwood, which means that it no longer conducts sap nor contains living cells. In most types of trees, the heartwood becomes darker than the newer sapwood, and in some trees this heartwood develops extractives that are repellent or even toxic to fungi. Heartwood from these trees can therefore survive harsh outdoor conditions better than heartwood from other trees. Notice the qualification here: It is only the heartwood that provides this natural resistance to decay. The sapwood of cedar and redwood is really no more resistant to fungi and insects than any other wood species. If you want decay resistance, you need to make sure that the lumber you buy contains some heartwood.

So, if you want to build with a wood that is naturally decay resistant, how do you know if you are buying boards and posts that contain heartwood? There are two ways. First, find out if the wood you are buying is graded by a reputable agency. Redwood that comes from mills that are members of the California Redwood Association will be graded according to appearance and durability. The latter category reflects the quantity of heartwood in each piece of wood.

Cedar grading can be a more complicated matter, perhaps because there are so many types of cedar available, and there is no single trade association overseeing the whole industry. Western red cedar is a popular choice for outdoor construction products, and for good reason. It has great dimensional stability, meaning it resists shrinking, twisting, and warping better than other softwoods. It is also lightweight and attractive. As a general rule, cedar should not be used where it will be in contact with the ground, and I don't think cedar is strong enough to serve as structural posts in a fence. But I can usually find cedar boards, with one rough face and one smooth face, priced competitively with pine boards of comparable quality. For its superior dimensional stability alone, this would make cedar a good choice, provided you planned to apply a protective finish.

The second way to determine whether a board contains heartwood is through a visual inspection. With many types of wood, including redwood and Western red cedar, you need only look at the color: The heartwood is darker, in the reddish to purplish to brownish range, while the sapwood is more cream colored. Unfortunately, Mother Nature felt it necessary to toss on some exceptions to that rule. Northern white cedar, for example, has little if any color differentiation between sapwood and heartwood, yet the heartwood still provides decay resistance. In this case, examine the end of the board for some clues. By studying the shape and orientation of the rings, you should be able to guess how closely the board was located to the heartwood.

But natural resistance is only part of the question when assessing the qualities of any particular piece of wood. Perhaps more important are the matters of where the wood will be located and how it will be used. Wood that is kept away from ground contact and

Tropical Hardwoods

A variety of exceptionally attractive and decay-resistant hardwoods are produced in the tropical forests of the Americas and elsewhere. Some of these woods (such as ipe, cambara, and meranti) are also very durable and virtually maintenance-free, making them popular choices for big-budget decks. Mahogany is a tropical hardwood that is easier to work with than many others in this category; it is often used in boat building as well as cabinet making.

Unfortunately, these woods can be extremely expensive, and you may have to look hard for a supplier. Also, the processing of many tropical hardwoods has been a major factor in the deforestation of many areas. Though tropical hardwoods tend to be too expensive to use as standard fencing material, they could be a viable choice for a small fence, utilizing relatively little wood, that was built in a highly visible location (these are fences you would want to show off). Before buying tropical hardwoods for any purpose, however, I strongly urge that you research the subject of sustainable forestry practices and make sure that the wood you buy comes from a reputable and certified source. ■

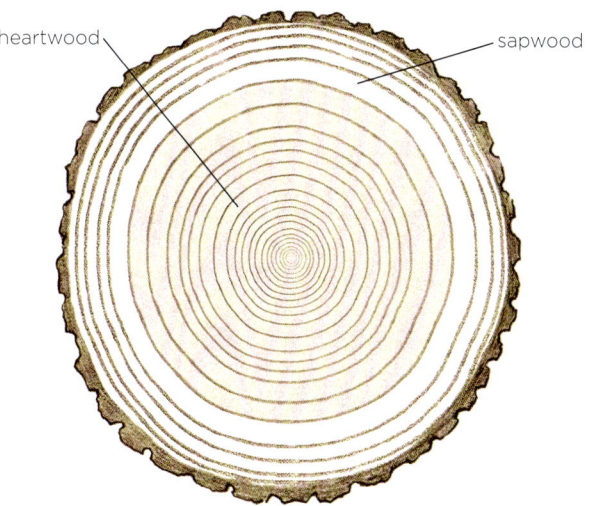

The sapwood provides life to a tree, but the darker heartwood of some species is far less likely to decay.

that is cleaned and coated with a water-repellent sealer regularly will last much longer than wood that is set on the ground and left unfinished. For most fence styles, that translates into this simple lesson: When shopping for posts, place decay resistance at the top of your considerations. For other parts of the fence, balance your finish and maintenance commitment with the level of weather resistance offered by the wood. I have often used plain old pine boards on outdoor structures with no regret.

Another important factor is where you live. Wood deterioration is a far more serious problem along the humid Gulf Coast of the United States than it is in the arid Southwest and Great Plains.

Pressure-Treated Wood

People have long strained to lengthen the lifespan of wood used in outdoor building projects by brushing on any number of coatings. But it was the development of the process of pressure-treating wood that really made it possible for decks, garden structures, and fences to be built with good assurance of longevity. Pressure treating takes place in factories equipped with pressure containers, where a mix of chemical preservatives are pushed deep into the wood cells to effectively immunize the wood from many of its major enemies. A variety of waterborne preserva-

Black Locust Posts

I occasionally see people building fences with black locust posts, which can be bought at farm-supply stores. These posts are fairly cheap, but they are also pretty raw, often sold with the bark left on and the sizes and shapes on the inconsistent side. Locust can be a bear to work with, as I first learned years ago when I struggled to cut some old locust trees with a chain saw. Still, for agricultural purposes or for a rustic-looking fence, these natural fence posts are worth looking into. ■

tives have been used over the years in lumber intended for residential and recreational use, while a different group of nasty substances (creosote and pentachlorophenol) have been used on railroad ties, utility poles, and other industrial materials. With pressure treatment, widely available and easy-to-grow wood species with poor performance in outdoor applications were suddenly transformed into reliably decay-resistant products.

For many years, the most widely available product was pressure-treated Southern pine containing chromated copper arsenate (CCA). In 2002, however, fears about the long-term effects of the arsenic in the treatment resulted in a plan to quickly phase out CCA

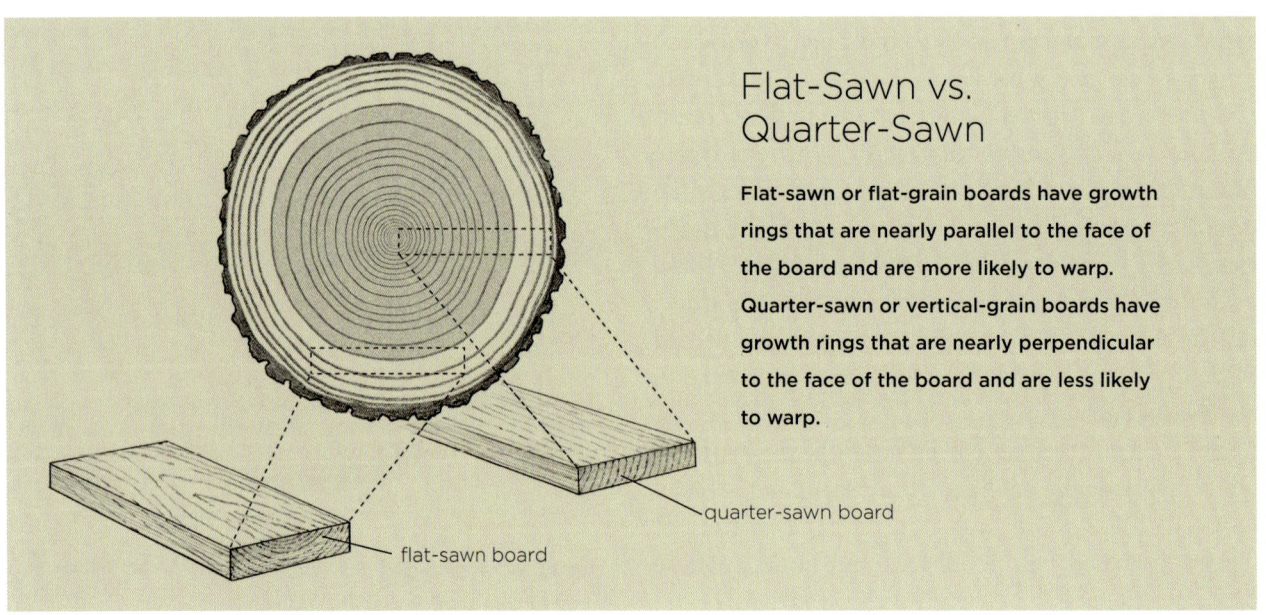

Flat-Sawn vs. Quarter-Sawn

Flat-sawn or flat-grain boards have growth rings that are nearly parallel to the face of the board and are more likely to warp. Quarter-sawn or vertical-grain boards have growth rings that are nearly perpendicular to the face of the board and are less likely to warp.

quarter-sawn board

flat-sawn board

lumber for most consumer and residential purposes. This phase-out quickly produced alternative products that promise similar service with less environmental and health impacts.

CCA-treated wood was considered a toxic material, and working with it safely entailed following a host of precautions to prevent ingesting the chemicals or inhaling dust or smoke from cutting or burning the wood. The newer products that I have seen do not carry such warnings, but since I cannot predict what the product lineup is likely to be in coming years, I would strongly urge you to check with the manufacturer or dealer before you start handling any pressure-treated wood on your fencing project. The long-term performance of these products will remain uncertain for some years, while specific suggestions on handling, cutting, and disposing of them will become clearer as they become more common.

Shopping Decisions

Have you ever tried to decipher one of those complex grading stamps on a 2×4, only to mutter, "Huh?" Well, you are not alone. In fact, I've learned to pretty much ignore the stamps altogether, letting my eye and the size of my checking account make purchasing decisions for me.

Grading stamps can tell you a bunch of things almost no one cares about, such as the identification number of the mill and the names of the relevant inspection agency. Still, there are a few reasons why you might want to take a look at the stamp before buying. On pressure-treated lumber, the stamp will tell you the "retention level" of the preservative used, which relates directly to where that piece of wood should be used. Lower retention levels are intended

for aboveground use only, while higher levels provide additional protection for ground contact, and an even higher level is used for wood that will be placed in or under the ground. Though it may be harder to find, this latter category is the best choice for buried fence posts.

When shopping for fence boards (typically 1×4s, 1×6s, or 1×8s, all of which do not physically carry grading stamps), all you really need to know is that looks matter. Straight, clean boards with a few small, tight knots will cost more than boards with large, loose knots, holes, and a bit of crookedness. Buy the best boards your budget will allow, unless of course you are not particularly concerned with looks. If parts of your fence will be more visible than others, consider buying a better grade of boards for that section and lesser grades for the less-noticed sections.

Lumberyards in my area also sell inexpensive "barn boards," which are usually rough-sawn hemlock that, as the name implies, are used as siding on barns and frequently are coated with red stain. You sometimes see these boards used as board-and-batten siding on houses and garages (usually of the "ownerbuilder" variety). The boards also can be used on fences, either as is or planed for a smoother appearance.

Estimating Lumber Quantities

The bigger the fence, the harder it can be to estimate how much lumber to buy. I always find this task is simplified when I prepare detailed scaled drawings, which makes the task of counting posts, rails, boards or pickets, and other major components straightforward. Unless you live next door to the lumberyard, I recommend buying 5 to 10 percent more lumber than you think you will need. Invariably, you will find some wood too damaged to use, or you will cut a few pieces too short, and having the extra lumber on hand will save a repeat trip to the lumberyard. My local lumberyard will take undamaged returns with a full refund, which makes it easier to buy a little extra.

Outdoor Storage

Some lumberyards keep their supply of treated lumber stacked outdoors, uncovered. I stay away from that lumber, unless it is "stickered," that is, stacked with shims or small strips of wood between the boards that allow air to circulate. ▪

Fasteners for All Seasons

WHY ANYONE WOULD try to cut corners when it comes to the fasteners they use on a fence or other outdoor project is beyond me. Even using the best-quality fastener available is going to account for only a small percentage of the overall budget. But cut corners they do, and the long-term effect can be fatal. I occasionally hear people bragging (at least they think they are bragging) that they use drywall screws on everything. Drywall screws are great fasteners—for drywall. They are lousy substitutes for wood screws, and they have no business being used outdoors.

The two principal concerns with exterior-grade fasteners are corrosion resistance and holding power. Fasteners that rust will discolor a fence and eventually disintegrate, and those that are too weak for the job will jeopardize the standing (literally speaking!) of the fence you build.

Corrosion Resistance

Hot-dipped (HD) galvanized nails are the best choice for most fences, offering good durability at a decent cost. Electroplated or zinc-plated fasteners also have a galvanized finish, but they do not stand up to the elements as well as HD galvanized products.

Stainless-steel nails are pricey, but they can be worth their weight in gold when you are building in a particularly damp or salty environment (along the coast, for example). Stainless steel is highly recommended for use with redwood or cedar, which contain tannic acids that will, if they come in contact with galvanized fasteners, result in ugly stains developing.

It is also a good candidate for use with pressure-treated wood, which can degrade galvanized coatings over time. If I was planning to paint the fence, and especially if I was painting it white, I would opt for stainless steel just to eliminate the possibility of rust stains appearing on the surface someday. Manufacturers offer two grades of stainless steel: Type 304 is the standard product, while type 316 has a bit more nickel to better stand up to a salty environment. For every $10 you spend on galvanized fasteners, you would need about $25 for a comparable supply in stainless steel.

When choosing screws, look for stainless steel or those with a galvanized finish and a weatherproof resin. I have not had great success using those low-cost "decking screws" with a yellow zinc coating, which seems to chip off or wear through as soon as the screw is driven into the wood.

Holding Power

Especially in exterior settings, wood swells and shrinks with the changing climate, and this unavoidable movement can result in a steady weakening of the bond between two pieces of wood. That is why holding power is so important on structures such as fences. It's hard to beat screws for being able to pull the components of a joint tightly together and hold them that way. Screws are also ideal fasteners if and when you decide to disassemble your fence, or parts of it. On the other hand, driving screws takes quite a bit more time than driving nails.

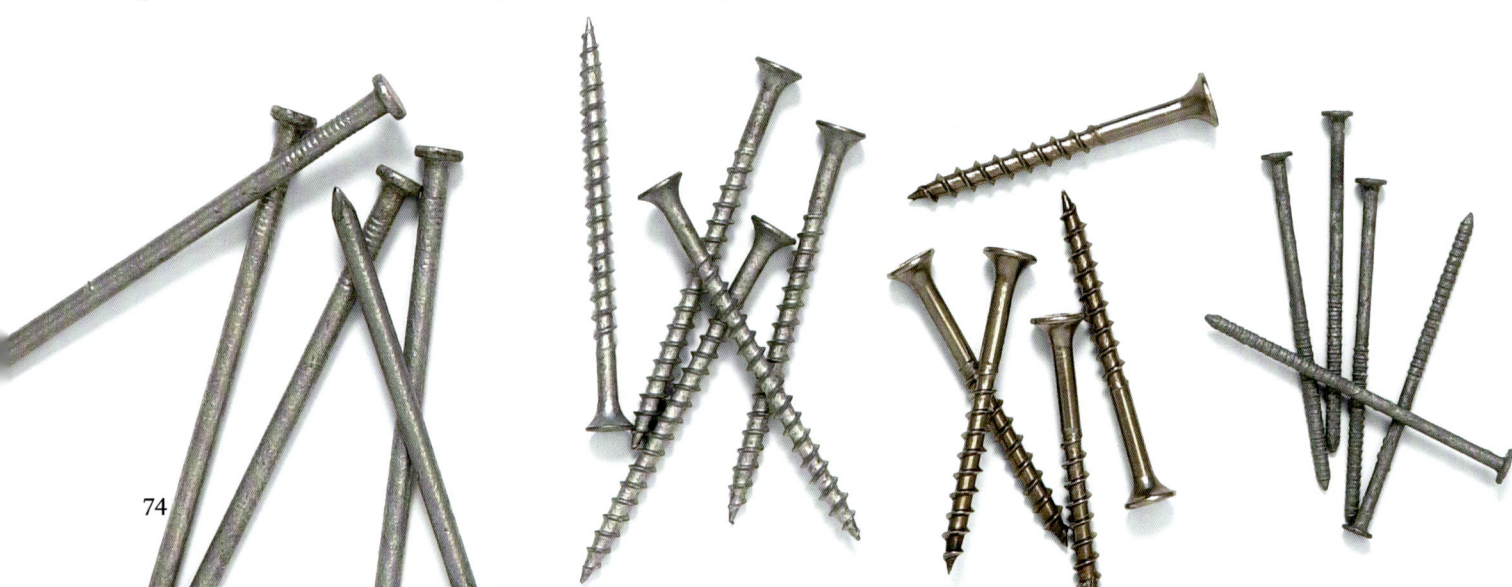

On the flip side, nails with smooth shanks have the least amount of holding power. Ring-shank or spiral-shank nails are a much better choice. Their only downside is that they can be tough to pull out, but of course this may not be a worry to most fence builders. For fastening pickets and fence boards, I most often use top-quality HD galvanized ring-shank wood siding nails that have been double dipped in zinc. But I would certainly use stainless steel if I was working with redwood or cedar, or if I was not planning to paint the wood.

Other Hardware

Some fences require heavy-duty brackets or fasteners, such as carriage bolts or lag screws. Carriage bolts are set in a predrilled hole and secured with a nut. The head has a smooth, round finish that should be located on the most visible side of the connection. Lag screws do not require a nut and are not quite as strong as carriage bolts. But they are often a better choice than normal screws for attaching large rails to posts. The best way to attach wire-mesh or barbed-wire fencing to wood posts is with galvanized fence staples.

When shopping for these larger fasteners and any other specialized hardware, you should still pay close attention to the corrosion resistance they provide. In my experience, stainless steel and hot-dipped galvanized carriage bolts and lag screws can be hard to find. Because of this, I have used electroplated bolts (you can identify them by their shiny surface) on outdoor structures, only to see rust start developing within weeks. In short, it pays to shop around.

Gate hardware is a subject unto itself and will be addressed thoroughly in Chapter 7.

New Wood Requires New Hardware?

In the early 2000s, pressure-treated wood containing arsenic was phased out and replaced with a new preservative formulation—alkaline copper quaternary (ACQ)—that was more environmentally friendly. The arsenic that was removed was replaced with a higher concentration of copper. This move raised prices but also posed less threat to plants and animals.

One unfortunate consequence of this change was that the higher levels of copper in the wood resulted in greater corrosion in metal screws and nails. Fastener manufacturers had to scramble to change the coatings applied to their exterior grade hardware.

Today, when you buy galvanized or stainless-steel fasteners for your fence project, you should feel confident that there will be no surprising failures a few years down the road. To stay on the safe side, however, it can't hurt to check with the manufacturers of the wood and fasteners you plan to use to make sure there are no compatibility issues. ■

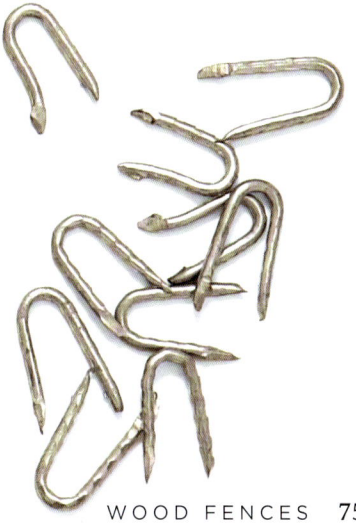

Setting Posts

POSTS ARE THE BASIC structural component of most fences. Wood posts are used with both wood fences and metal fences. In some fences, posts are a prominent visual element, while in others they are barely noticed. In either case, they serve a critical function. Set the posts right, and the rest of the fence-building work will be easier and more rewarding. Set the posts wrong, and you will regret the errors for a long time.

Setting the posts right means placing them in a straight line, spacing them evenly and properly, embedding them securely in the ground, and positioning them perfectly plumb (that is, straight up and down). The previous chapter provided general instructions for marking a fence line, determining square corners, and coping with sloped sites and other irregular geographical elements. I suggest you review that material along with the following discussion.

Size and Spacing

The size of posts and the spacing between them are often aesthetic decisions, but first and foremost they are structural factors. Your local building code may have minimal requirements that should be met regardless of what is said here.

There are two categories of fence posts. Terminal posts are those located on both sides of the gate, at all corners, and at the ends. Because these posts generally carry the heaviest load and receive the most abuse, they should be set deeper and often are sized larger than the intermediate posts. The illustration offers guidelines on size and spacing as applied to intermediate posts, which usually make up the majority.

Footings

Fences are subject to a lot of lateral pressure. When strong winds blow against the side of your house, the whole foundation and framing help brace the structure. When those same winds blow against a fence, it is only the strength of the buried posts that keeps it upright. And when you take into account the forces of snow drifts, leaning bodies, bumps from heavy animals and heavy equipment, and the numerous assaults that your fence may face from rambunctious kids, you can appreciate the importance of well-anchored footings.

The term *footing* refers to that vital connection of the posts to the earth. Since the world is a complicated place, there are of course several ways to form that footing, each with its pros and cons, and none with an iron-clad guarantee of perfection. Everybody seems to know somebody who smoked cigarettes, ate fatty foods, and drank every day through a long and healthful life, ending in death from old age. Those of us who also smoke, eat fatty foods, and/or drink regularly, ever eager for some bright light to shine on our vices, readily point to these rare birds as proof that we are doing ourselves no harm. Professional fence builders can be a bit like this when the subject turns to selecting the "best" footing technique: They all know

POST SIZE AND SPACING. **A** Tight spacing, nicely proportioned for short and tall fencing. Can handle heavy infill. **B** Most common post size and spacing. A good choice for rail fences and other light-infill styles. **C** Strong size and spacing combination. A good choice for heavy vertical-board fences or other styles that showcase the framing. **D** Strong but graceful spacing. Suitable for nearly any fence style.

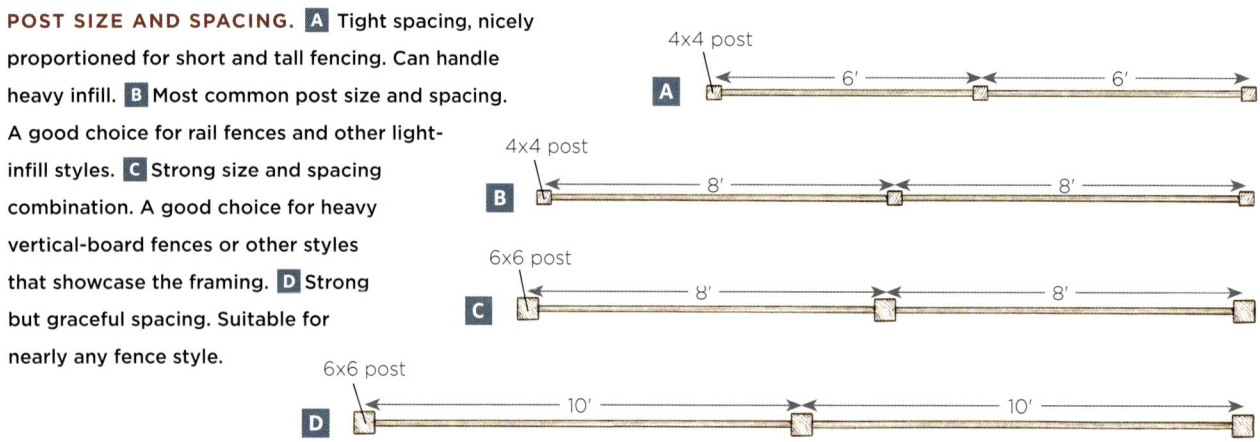

Fence-Post Rules of Thumb

■ Use only pressure-treated wood that has been rated for ground contact.

■ If the budget permits, buy "premium" pressure-treated posts that have received a water-repellent finish at the factory. For wood that is going into the ground, I like to have this extra protection.

■ Choose the posts yourself. Find the straightest ones you can by eyeballing each one lengthwise. Avoid posts with excessive splitting, checking, or knotholes.

■ Buy posts a little longer than you need so that they can be cut to length later. Believe me, there is almost no task more frustrating than trying to set a line of posts in perfectly dug holes so that their tops align vertically.

■ For reasons described in the previous point, avoid buying posts with decorative tops already cut into them. These look good only if the tops are aligned vertically, and that is tough to accomplish. If you want decorated post tops, you can easily tend to that chore after the posts have been set and the tops cut straight.

■ When in doubt about which size of post to use, use the larger size. There are no long-term penalties for erring on the side of strength, while there are plenty of potential headaches and regrets for erring in the other direction.

■ Bury the posts as deeply in the ground as you reasonably can.

of fence posts that were set their own preferred way that have withstood the test of time. Unlike the smokers and drinkers, however, they probably have solid proof to support their cases. So I am not going to try to convince you to set your posts one way or the other. I will, however, try to provide enough information that you can make a good decision.

First, the posts need to be buried in the ground—and the deeper, the better. For building a deck, I strongly recommend setting posts in metal bases attached to concrete piers. But this is not a good way to handle fence posts; they need the better lateral support that deep burial affords. Likewise, I would not set fence posts in those metal brackets that are hammered into the ground. I have my mailbox installed using one of these, but it is not what anyone would call a solid connection.

So, holes must be dug, and posts must be set in them. With that decision made, the next round of questions concerns how deep to dig the holes, how wide, and what to backfill them with. The traditional choices in backfill are alternating layers of tamped earth and gravel (see page 85), concrete (see pages 86–88), or a combination of the two (see page 89), as shown in the

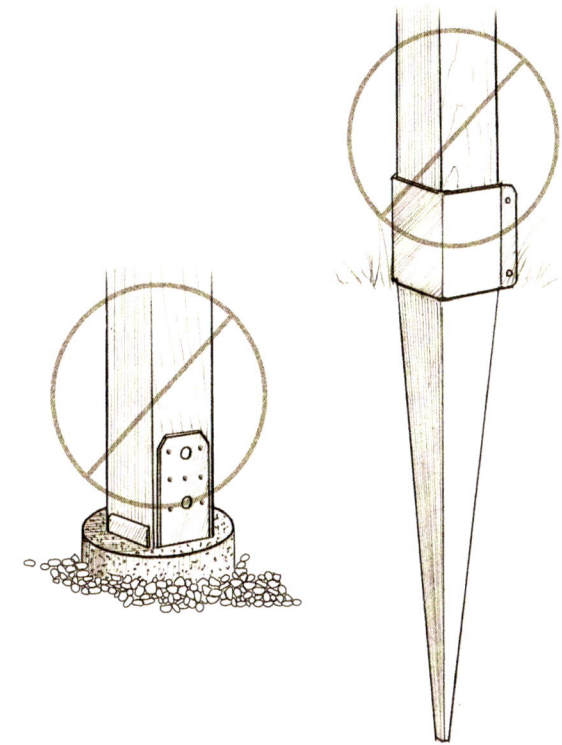

HOW NOT TO SET FENCE POSTS. Posts anchored to concrete piers [LEFT] or metal brackets driven into the ground [RIGHT] do not provide adequate support against the lateral pressure a fence post will face.

illustrations. Depending on the circumstances, I think that any of these methods can be used successfully. The depth and width of the holes will be affected by your choice of footing material.

Comparing Backfill Choices

Earth-and-gravel footings (or earth alone) have been used to anchor fence posts for centuries. They require relatively slim holes (roughly twice the width of the posts), which reduces the labor involved in digging holes. The only expense involved is for a bit of gravel, as the earth component can come from the dirt that you dig out of the hole. If and when the time comes to remove or alter the fence, or to repair or replace a section of the fence, these posts can be dug up easily.

Unfortunately, an earth-and-gravel footing will not work well in soils that are too light or too heavy. Soil containing a lot of sand or organic material, such as decaying plant material, or that has been "disturbed" (that is, dug up and backfilled before), may be too weak to provide sufficient lateral support. And heavy soil, containing a lot of clay, drains poorly and loses much of its strength and holding power when wet. In relatively dry, undisturbed soil that compacts well, the earth-and-gravel method will usually work just fine, especially if you are building a low, lightweight fence. And the detrimental effects of poor soil can often be offset by buying some good-draining soil along with the gravel to use as backfill (in which case, I would probably dig wider holes to hold more of the backfill).

Setting posts in concrete is a reasonable option under just about any conditions, but especially when the soil is less than ideal. It generally takes less time to set posts in concrete; all of that backfilling and tamping of earth and gravel can be time-consuming. But concrete footings do pose their own set of potential problems. First, the holes must be wider—for 4×4 posts, for example, you'll need 12-inch-diameter holes versus 8-inch-diameter holes for earth-and-gravel footings. That's a lot of extra digging, and I, for one, really hate digging postholes. Also, encasing wood, even pressure-treated wood, in concrete is an iffy proposition. Concrete (even cured concrete) is porous and effectively traps water around the post, which can lead to rot and infestation by wood-loving carpenter ants. Seasonal wood shrinkage can allow even more water in through the gap between the post and the footing at ground level (though this can be reduced by pouring the concrete so that it slopes away from the post to encourage drainage and by diligently caulking the joint to block water from entering).

In cold climates, concrete footings are also prone to the effects of frost heave, which is a truly powerful force. Most objects contract as they cool; ice is just the opposite. The colder it gets, the more it expands, and in doing so it can exert enough pressure underground to lift houses and roads, not to mention fence posts. Frost heave beneath a concrete footing can push the footing up and significantly weaken its connection with the ground. The simpler-said-than-done antidote

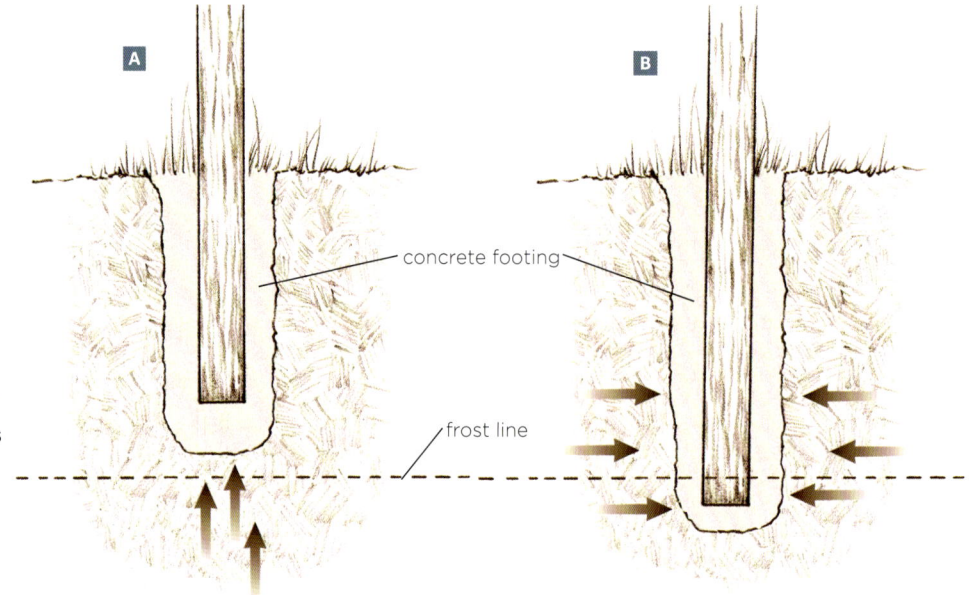

POST PRESSURES. If a post footing is not set below the frost line **A**, the ground can freeze and expand beneath the footing, pushing it up. Even when the bottom of the footing is below the frost line **B**, lateral soil pressure can loosen the footing unless it is buried very deep.

concrete footing

frost line

to frost heave is to be sure that the bottoms of each footing are below (and preferably well below) the frost line, which is the deepest point at which frost is likely to form in your area. (Your local building department should be able to tell you the local frost depth, and may even require you to set posts at or below this depth.) The only problem with this solution is that in much of North America, digging below the frost line requires holes that are 4, 5, or even 6 feet deep. Even in the best of conditions, that's a lot of work, and in many cases it is nearly impossible to dig that deep without running into hard soil and even harder boulders and bedrock.

Now, you would think that a natural, widespread phenomenon like frost heave that is so important to the construction industry would be the subject of lots of research and that we would have at our disposal top-notch scientific recommendations for dealing with it. In fact, there is a lot of murkiness in the literature. Some engineers argue that the only way to virtually guarantee that a post will not be heaved is to place one-third of the buried portion below the frost line; thus, if the frost line in your area was 3 feet, you would need to bury the posts 4½ feet in the ground. Another argument maintains that the posts are subjected to powerful lateral forces pushing against their sides as well as the pressure from below. Following this reasoning, the deeper the posts are buried, the greater the amount of surface exposed to this lateral force, and therefore the greater the chance that the post will be moved. Building codes that I am familiar with have not embraced the building strategies that are suggested by these views. I guess the only thing I can say is that there are no absolute guarantees; builders have followed every rule and regulation in setting posts, only to see them start heaving the following March, while others have ignored those same rules and regulations and can point to years of happy results.

Frost heave can affect any type of footing, but although I cannot cite conclusive proof for this hunch, my instincts tell me that earth-and-gravel footings would survive better than concrete footings when subjected to the same type of heave from below or lateral pressure. That is because posts embedded in concrete present a larger solid surface against which these forces can push. Even if I am wrong about this,

Think Before You Dig

Dirt and rocks and worms are not the only things you can encounter when digging holes. You may also run into buried pipes or wires, and breaking one of those can do more than ruin your day. Electrical and telecommunications cables, natural gas pipelines, water supply lines, and drainpipes can all be hidden belowground. If you are unsure about the location of these potential obstacles in your yard, contact your local utility companies well in advance. Once you learn the location, or even approximate location, of any buried lines, mark them on your site plan and dig very carefully when you get close to them. ■

at the very least, earth-and-gravel footings make it much easier to reset any dislocated posts.

Many books on fence building do not even mention frost heave, and quite possibly you are scratching your head right now, wishing I had followed their lead. While it is a vital consideration when planning foundations for houses, and even decks, frost heave does not need to be an obsession with fence construction unless you want absolute assurance that your fence will stay put through any type of weather. The 1892 manual *Fences, Gates, and Bridges* offers this bit of straightforward advice: "A fence post should be set two and a half or three feet in the ground, and the earth should be packed around it as firmly as possible." I can live with that, though with a little qualification. I suggest that you aim to bury at least one-third of the overall length of the posts. For a fence projected to be 6 feet tall, that would translate into posts buried 3 feet, which in turn would necessitate holes to be dug about 3½ feet deep to allow for a bottom layer of gravel. With good soil conditions and careful attention to detail, short picket fences 3 to 4 feet tall could be supported by posts buried no less than 2 feet deep in holes extending an additional 6 inches.

For terminal posts (posts at the gate, corners, and ends), the safest approach is to set them in concrete 6 to 12 inches below the frost line, regardless of which technique you use with the other footings.

Digging Holes

In the dozen or so years that I have been writing and editing in the construction and home-improvement fields, I have lost track of how often I have come across "how-to" instructions that covered the subject of post-hole digging in a brisk sentence or two. The truth is that for many do-it-yourselfers, there is no part of fence building (or deck building, for that matter) that is more difficult and discouraging than digging deep, narrow, straight, nicely aligned holes in the ground.

The first deep postholes that I can recall digging were to prepare footings for two posts to support a swing set I was building for my kids. Armed with a shovel, a brand-new clamshell digger, and a plan to have the posts buried in concrete before sunset, I nevertheless managed to get no more than a foot into the ground before the soil started turning as hard as a rock, and the rocks themselves started showing up every time I removed a little dirt. I wound up spending the better part of a weekend digging those holes, and most of that time was spent prying loose rocks with a large crowbar.

In many areas it is quite easy to dig holes, but even under ideal conditions, if you are building 50 feet of fence, you will normally need to dig holes for at least seven posts, and likely several more. A privacy fence around a moderately sized backyard could easily require a few dozen holes. This is why posthole digging is one of those chores I often recommend that do-it-yourselfers farm out to others.

A garden spade is the best manual tool for breaking through sod and getting a hole started. But once you get down a foot or so, it is best to shift to a clamshell digger. Hold the digger with the handles together and stab it into the hole. Spread the handles apart and carefully lift out the dirt. If you run into large rocks or hard soil, a digging bar is a better tool to have on hand than that crowbar I used to use. It is about 6 feet long with a 1-inch-thick heavy metal handle. One end has a chisel-like blade, and the other end has a tamping head that comes in real handy when you start backfilling the hole. A digging bar is also useful for loosening soil before you try removing it with a shovel or post-hole digger.

Some professional fence builders in New England, where the soil is famously full of stones, prefer to use

USING A CLAMSHELL DIGGER

a manual digger called the Nu Boston Digger. I have only seen photographs of this tool, which is not widely available but appears to be quite useful for digging deep holes manually. It is similar to a clamshell digger, except that one of the handles has a lever that operates a scoop on the bottom, which appears to offer far greater efficiency at dirt removal. (Clamshell diggers are notorious for spilling as much dirt back into the hole as they successfully carry to the top.) Word has it that this tool will dig 8-inch-diameter holes to a depth of 3½ feet (even deeper with 12-inch holes) before the lever mechanism starts interfering.

Another trick in dealing with hard soil is to pour water into the hole, then let it soak in for a few hours or even overnight. Scooping out mud is messy work, but it may also be easier than trying to hack through hard soil. Tree roots can be cut with a pruning saw, although I have often called upon my trusty and much-abused reciprocating saw (equipped with a special pruning blade) for such work.

If you run into bedrock (or ledge), a layer of solid rock near ground level, well, you have my sympathies. Bedrock has frustrated many building plans, and it

eter of the hole you need to dig. Often, the biggest bit you can find is 3 feet long and 8 inches in diameter. If you need to dig 12-inch-wide holes, you can rock the auger back and forth a little as it works itself deeper into the hole. (In truth, you may not have any choice on this.) There are both one- and two-person power augers available. One-person augers with the engine mounted on top of the tool scare me, and I would not use one. Two-person versions of the same design, as shown in the illustration, are much safer and easier to use. Still, I have seen strong, healthy adults nearly brought to their knees with fatigue after spending an afternoon using one of these tools. Safe, efficient operation requires a good grip, careful attention to what you are doing (that is, save the beer until after the tool is turned off), and the use of eye and ear protection.

If you must work alone, a flexible-shaft auger is worth looking for. The engine is separated from the auger, making it fairly easy to handle. But you may need a trailer hitch to tow one of these home. I have also seen, but never used, one-person augers in which the user stands on a platform at the rear of the device and safely guides the bit into the ground. These tools must be anchored to the ground with spikes and chains. With any tool you rent, be sure to get thorough instructions on its operation before taking it home.

can be a serious obstacle to fence construction. Sometimes you can chip away at the rock with a digging bar until you create a hole deep and wide enough to hold some concrete for the post. Another option is to drill a hole in the ledge, cement a rod or pin in the hole, and then secure the post to the rod or pin. I would highly recommend that you consult a professional contractor about this latter chore.

As you dig, pile the dirt on a piece of plastic or in a wheelbarrow or garden cart. This makes cleanup and backfilling much easier. Keep in mind that you do not want to dig holes any wider than necessary. If you are using earth-and-gravel footings, part of the strength comes from leaving the surrounding soil undisturbed. And with concrete footings, the wider the hole, the more concrete you need to buy, mix, and pour. Aim for holes that are as close to cylindrical as possible. And keep a close eye on your fence line: You want the centers of all holes to be aligned as closely as possible.

Power augers can produce postholes much faster than manual digging, which is not to imply that they make the job easy. Tool rental stores usually have at least one type of power auger available. Be sure to select an auger bit that matches the depth and diam-

USING A POWER AUGER

One of the keys to success with a power auger is to treat it as if you were drilling a deep hole in wood or metal. Advance into the hole a little, then withdraw the bit to clear the loose debris, then advance a little more. Also, do not try to clear big rocks with the auger; use a digging bar for that work. Finally, take regular breaks.

The last option for digging holes is my favorite. Hire someone with an auger accessory that they mount on a truck, tractor, or skid-steer loader, and let them dig the holes quickly while you sip lemonade. Yes, you will have to write a check to get these holes dug, but it may be smaller than you anticipate. This strategy will work only if the vehicle can get to the site, which is not always possible in backyards. And it may require that you hunt around for such a person; I suggest you contact fence-building contractors, talk to local farmers, or ask a building contractor for recommendations.

Preparing the Hole

Chances are good that if you have dug a dozen holes, no two of them will be exactly the same depth or diameter. Under ideal circumstances, every hole would be at least 30 inches deep, and preferably 12 inches or more below the frost line. In the real world, however, some of the holes may not even be as deep as you would like due to immovable rocks. It is not a catastrophe if you leave a hole or two short, although I would always strive for at least 2 feet of depth. If you run into major obstacles for holes that will be receiving terminal posts, however, I would not recommend fudging. Instead, make an extra effort to create an adequate hole, and if that proves too difficult, consider relocating that particular post.

When the holes have all been dug, it is a good idea to set up your string line or lines again to check the alignment. If necessary, dig a little more until you feel comfortable about setting the posts in a straight line. Under most circumstances, and regardless of what type of footing you are planning to use, I recommend putting a 6-inch layer of crushed rock into the bottom of the holes. Begin by putting about 4 inches of the rock into each hole. Tamp it well with a 2×4, then add another couple of inches after the post has been set in the hole and braced. The crushed rock allows water that reaches the bottom of the post to drain away.

I would think twice, however, about using crushed rock if the surrounding soil is regularly on the wet side. In this case, the drainage bed can actually attract water to the base of the post, which would be counterproductive to say the least.

Bracing and Plumbing Posts

I like to use my string lines to help align the posts as I set them. If I'm using 4×4 posts, I simply move the string lines 4 inches to one side or the other, then make sure that the face of each post remains ½ inch away from the string line through the setting process.

ALIGNING POSTHOLES

use string line to ensure that postholes are in a straight line

hole should be at least 30" deep or at least 12" below frost line

fill bottom of hole with 4" of gravel

I have found this to be a safer method than trying to set the posts directly against the string all down the fence line.

The posts need to be positioned perfectly plumb in the hole before you start backfilling or adding concrete, and they need to be braced well enough that they will stay put while you are filling the hole. There are two ways to do this, and I've found that each of them goes much faster if I am working with a partner.

Shim Bracing

The simplest bracing technique I've used requires that you make at least four bracing shims. The exact size of the shims may depend on the size of the holes relative to the size of the posts. For 8-inch-diameter holes and 4×4 posts, you can use a 30-inch length of 1×6 or a comparably sized piece of ½-inch plywood. Cut the piece of wood in half lengthwise, then lay out and cut two shims on each of the boards as shown in the illustration below.

To use the shims, set a post in the center of the hole as close to plumb as you can and offset the proper distance from the string line. Have your partner hold the post plumb with the use of a carpenter's level while you slip one shim into the hole directly under the level and another on the opposite side. Now shift the level to an adjacent side of the post and repeat the process. With all four shims wedged in place, the post will be able to stand plumb on its own. You can now begin shoveling in layers of rock and earth between the shims, tamping as you go. When the backfill nearly reaches the bottoms of the shims, carefully remove them to finish filling the hole. Then move on to the next hole.

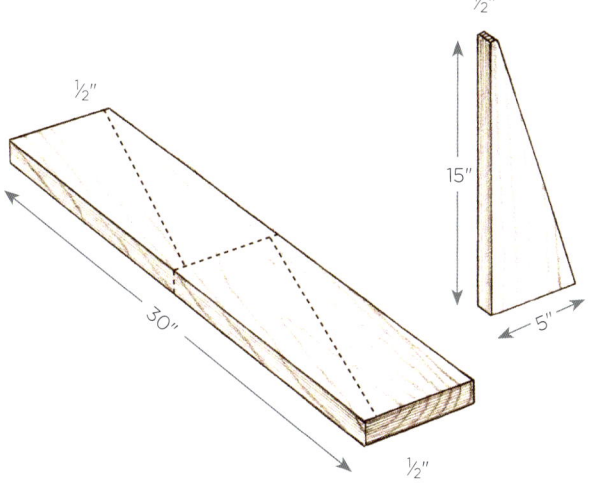

[ABOVE] One 30-inch-long 1×6 will yield two shims of an appropriate size for bracing 4×4 posts. You will need four shims altogether.

[RIGHT] **1** Set the first two shims on opposite sides of the post. With a level on one of the shimmed sides, push both shims snugly into the hole, keeping the post plumb. Then do the same thing on the other two sides of the post. **2** After all the shims have been placed, backfill the hole with crushed rock and earth, tamping as you go. Avoid disturbing the shims.

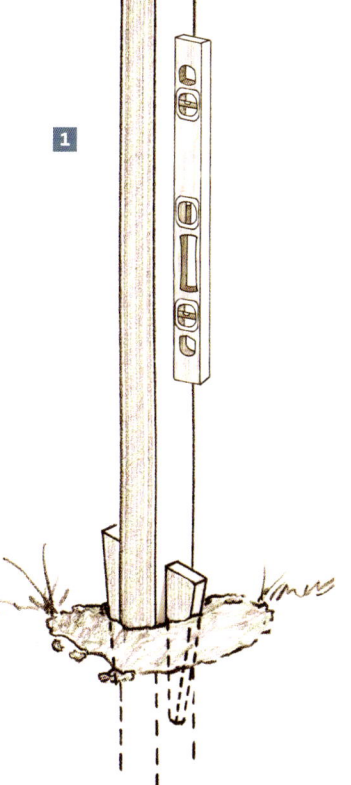

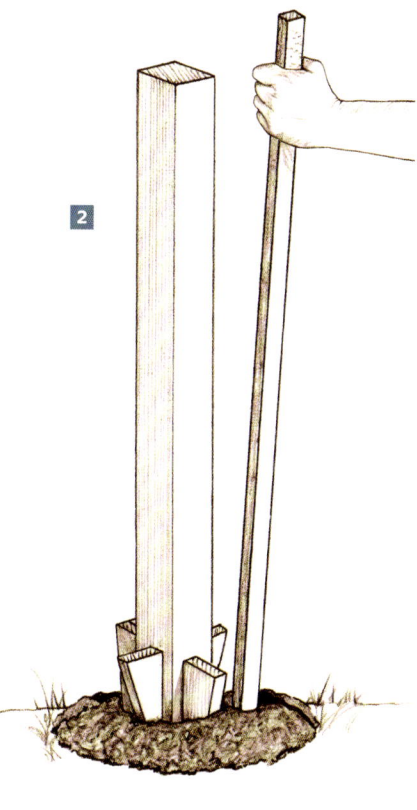

Lumber Bracing

The shimming method is tricky with wide holes and when using concrete. In that case you are usually better off using the more traditional lumber bracing method. I usually use 1×4s for braces. For posts that will be 6 feet or more above ground level, use 8-foot-long braces. You will need two braces at each post, along with stakes that must be hammered into the ground; 1×2s work fine for this chore, especially if you cut one end to form a sharp tip. Set the post in the center of the hole, holding it plumb and properly offset from the string line. Place one brace at an angle alongside the post. Drive a stake into the ground alongside the bottom of the brace, then drive a nail or screw through the brace into the stake. Check the post for plumb by setting a level on the side of the post facing the stake. When the bubbles are resting in the middle of the vials, drive a nail or screw through the brace and into the post. (I much prefer to use screws for this purpose, as that makes it easier to disassemble as well as to readjust. If you use a nail, drive it only partway in so that it will be easier to pull out.) Repeat this process with another brace placed at a right angle to the first one. The post will be plumb when the level demonstrates that two adjacent sides are plumb.

If you intend to use a concrete footing with the lumber bracing method, and you are having mixed concrete delivered, you will want to have all posts set in their holes and braced first. If, on the other hand, you are mixing concrete yourself one bag at a time, you will probably find it easier to brace one post at a time. That way, you can use the same braces over and over and you don't have to worry about braced posts being knocked out of plumb by a sudden storm or soccer game. Also, with this one-post-at-a-time approach, you can use clamps to hold the braces to the posts. Those clamps that can be tightened with one hand are particularly handy and allow for rapid readjustments.

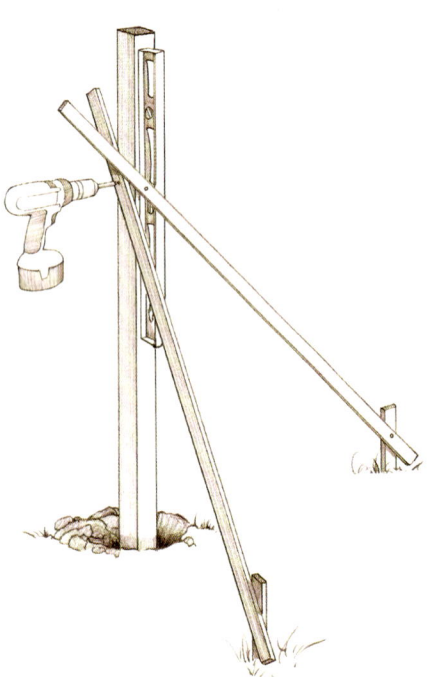

Braces are attached to two adjacent sides of the plumb post. This is much easier to do when working in pairs; one person can hold the post plumb while the other attaches the brace.

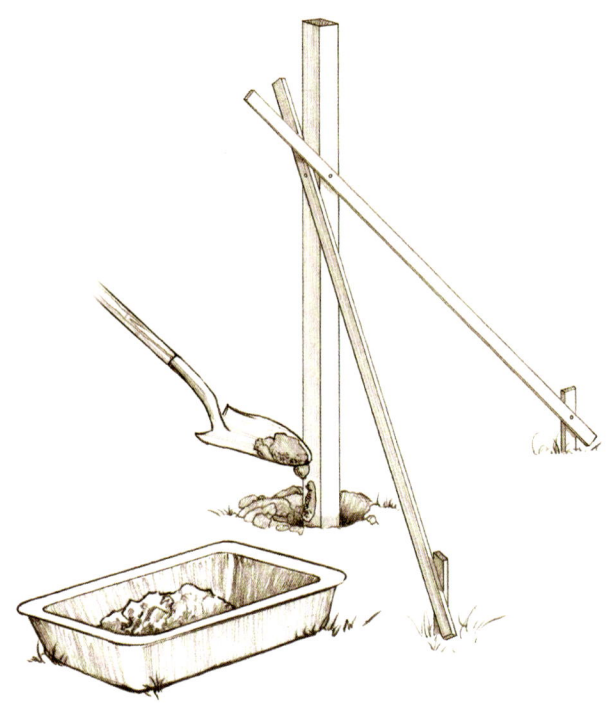

With the post securely braced, shovel concrete into the hole. Take care not to disturb the post itself. Overfill the hole so that the concrete can be smoothed and sloped to shed water.

Installing an Earth-and-Gravel Footing

Earth (that is, soil) and gravel work together to provide stability and drainage. The keys to success with this type of footing are alternating layers of each material and tamping each layer to ensure a well-compacted and snug bed for the post.

It also pays to use the right kind of gravel. The terminology in this wing of the construction business always manages to trip me up, so I will try to keep this discussion as simple as possible. Crushed rock (or crushed stone) has a self-descriptive name. It is created through a manufacturing process of crushing, which results in a product with jagged edges. Crushed rock is ideal for drainage, as the edges create large voids that allow water to pass through quickly. Crushed rock is often screened and graded for size; the larger the rocks, the better the drainage.

Crushed rock is fine for the bottom of the hole, as long as you tamp it well. But it is not the best choice for the earth-and-gravel backfill. The term *gravel* is often used generically (and I am following suit here) to cover artificially crushed materials as well as rocks formed from natural disintegration. Bank-run gravel, also simply called bank gravel, is a better choice for postholes, and it is a fine choice for the hole bottoms as well. It is a mix of rounded large and small rocks that compact nicely but also drain well. If bank-run gravel is hard to find, pea gravel (a graded product of fairly small gravel) is a suitable alternative and can usually be found in small bags at home centers, lumberyards, and garden suppliers.

With the post sitting on a level layer of gravel, braced and plumb, add another 2 inches of gravel in the hole. Tamp the gravel well, then shovel in about 4 inches of soil. Tamp that layer, then add more gravel. Continue this work until the hole is filled.

For tamping, I usually use a 2×4 and take well-aimed hard jabs at the material in the hole. The best tool I have used for tamping in the tight confines of a posthole, however, is a digging bar with a blade on one end and a tamping head on the other. If you didn't buy or borrow a digging bar while digging the holes, you may not want to shell out money for one now. But because it is made of heavy metal and is relatively easy to handle, this tamping device does a great job of compacting material around a post. Regardless of the tool

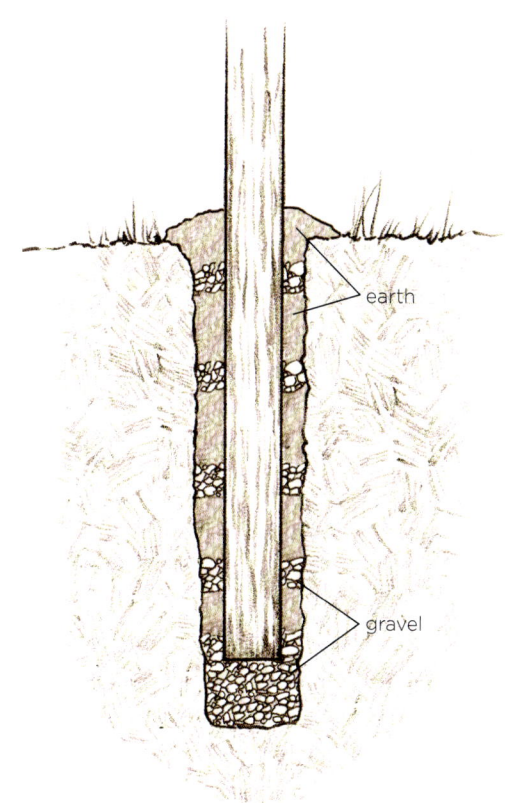

EARTH-AND-GRAVEL FOOTING

you use, try to avoid hitting the post. Work your way little by little around the post as you tamp, rather than trying to tamp one side completely before moving to another. Tamp each layer until it is as compact as you think you can make it.

This is tedious work, and it can take a toll on your arms and back—it certainly does on mine. To ease the burden, I suggest working with a partner and alternating the tamping duties.

I also recommend that you periodically check the post to make sure it stays plumb. If it shifts while you are backfilling and tamping, you may have to readjust your bracing and maybe even remove some dirt and gravel before getting back on track. The more the hole fills, the less likely it becomes that the post will move.

The top layer should be overfilled with dirt. Take care to compact this layer to create a slope away from the post to encourage water to drain away from both the post and the hole. This top layer may need to be refreshed from time to time with some fresh dirt and a renewed slope.

Installing a Concrete Footing

If you plan to bury your fence posts in concrete, one of the big decisions that confronts you is whether you want to mix your own concrete and, if you do, which products to use and how to do it.

There are circumstances under which it makes sense to have ready-mixed concrete delivered by truck. Under ideal conditions, it will save time and not prove much more costly than the alternatives. Ready-mixed concrete is ordered by the cubic yard. One cubic yard equals 27 cubic feet (or 46,656 cubic inches, if you need to do the math) and is typically the minimal amount you must order. Assuming that you dig 36-inch-deep holes that are 12 inches in diameter, dump 6 inches of gravel in the bottom of each hole, and plan to use 4×4 posts, you ought to be able to fill 15½ holes with 1 cubic yard of concrete (see the sidebar on the facing page for directions on estimating concrete needs). You should plan to overfill each hole, however, and you will almost certainly spill a bit, so a safer bet is that you could handle 13 or 14 holes with a minimal order.

However, before the concrete truck pulls up to your house, you need to have all the holes dug and all the posts placed and braced. You need to make sure that the truck can get fairly close to the holes (although you do not want a concrete truck driving through your yard), and you should have a small crew available as soon as the truck arrives, equipped with shovels and a couple of wheelbarrows, so that you can work quickly. If you keep the truck waiting to unload too long, you may have to pay extra.

The process normally involves sliding a wheelbarrow under the truck's chute, filling the wheelbarrow with no more concrete than you can safely handle, wheeling the concrete to the hole, shoveling the concrete into the hole, then returning to the truck for another load. Helpers can be handling another wheelbarrow as well as working on the filled holes to create a smooth surface that slopes away from the post (I usually use a margin trowel for this job; see the illustration on page 89).

If you have a large number of holes to fill but would rather work at a more leisurely pace, bracing and filling one or two holes at a time, then you would be wise to rent an electric or gas-powered concrete mixer. You

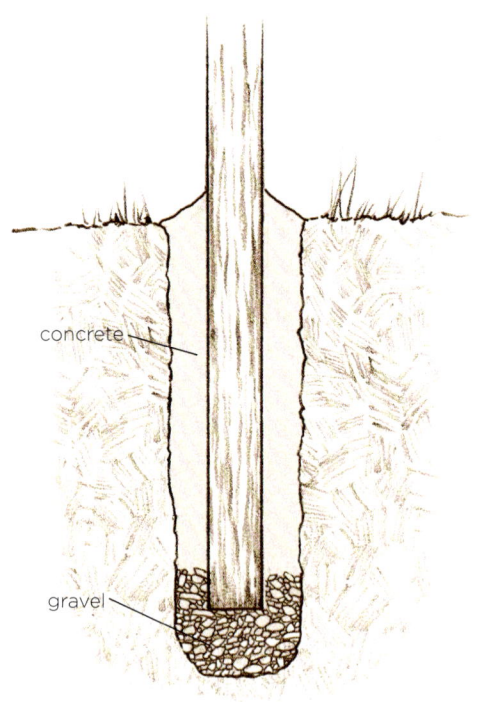

CONCRETE FOOTING

can throw bags of concrete mix (premixed concrete) into the power mixer, but it would be more cost effective (and only slightly more time-consuming) to mix your own dry ingredients: 1 part Portland cement, 2 parts sand, and 3 parts coarse gravel. All of these dry ingredients should be available at a good lumberyard or home center, but you may have to go to a rental store for the mixer itself. I have seen small power mixers that can be wheeled right up to and emptied into the hole. Use a shovel to stir the concrete mix (or the individual ingredients) in the mixer, then slowly add water until you reach the right consistency. When you can form a small pile of concrete that holds its shape, the mixture is ready to use. If water starts pooling on the surface of the mix, it is too wet.

For fewer holes, it is usually easiest to buy bags of concrete mix, empty the contents into a mortar tub or wheelbarrow, add a little water, and mix the ingredients with a hoe (a mortar hoe, which has two holes in its blade, is the best tool for this, but a standard garden hoe will work nearly as well). You can usually find bags of concrete mix (make sure that's what you buy; you don't want mortar mix or cement) in sizes ranging

How Much Concrete Do I Need?

When mixing or ordering concrete, you should know in advance how much you will need. That requires some work with a calculator. To determine the amount of concrete needed for a posthole (or any other cylinder), begin by determining the volume of the hole, using the following formula:

**radius² × depth of hole in inches × 3.14
= cubic inches in the hole**

From this result you need to subtract the amount of the hole that will be filled with the post itself, using the following formula:

**pole length × pole width × pole height (below ground)
= cubic inches pole occupies in the hole**

Assuming that the hole is a perfect cylinder, which it never is, and that your measurements are exact, the result would tell you how much concrete would be needed to fill the hole to ground level. In fact, you want to overfill each hole a bit, as it will settle some and you want to create a sloped top a bit above ground level. So you should add 5 or 10 percent to your estimates. Then, because mixed concrete is sold by the cubic foot rather than the cubic inch, divide your total by 1,728 (there are 1,728 cubic inches in one cubic foot). Multiply the quantity of concrete needed for one hole by the total number of holes to determine your total concrete needs.

The illustration gives details on how this extended math lesson would apply to a typical posthole.

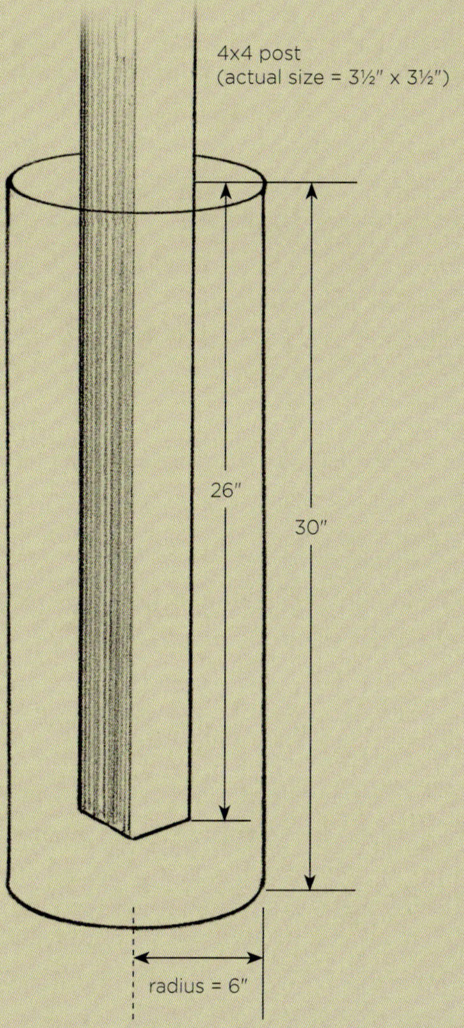

4x4 post
(actual size = 3½" x 3½")

26"

30"

radius = 6"

cubic inches in hole: 6² × 30 × 3.14 = 3,391.2 cubic inches

cubic inches pole occupies in the hole: 3.5 × 3.5 × 26 = 318.5 cubic inches

volume of post subtracted from volume of hole: 3,391.2 – 318.5 = 3,072.7 cubic inches

volume of concrete needed plus 10%: 3,072.7 + (3,072.7 × .1) = 3,072.7 + 307.27 = 3,379.97 cubic inches

volume of concrete needed in cubic feet: 3,379.97 ÷ 1,728 = 1.96 cubic feet per hole

Power mixers (electric or gas) can be rented from home centers and rental outlets. They are particularly useful if you want to mix your own concrete ingredients and have a large number of holes to fill.

Mortar tubs are usually large enough to mix one large bag of concrete mix with water. A mortar hoe, with two holes in its blade, speeds the mixing process, but a standard garden hoe will work nearly as well.

from 40 to 90 pounds. One large bag will yield about ⅔ cubic feet.

Concrete mix contains the proper proportions of Portland cement, sand, and gravel, but they aren't always mixed well in the bag. For that reason, I suggest you do not try to empty and mix only part of a bag. Use full bags only, and mix up the dry ingredients well before you start adding water. Read the instructions on the bag to determine how much water to add. Measure this amount of water, then mix about 90 percent of it with the dry mix. Add small additional amounts until you reach the right consistency (able to hold its own shape, without water pooling on the surface).

Keep in mind that the clock starts ticking as soon as you add water to the mix. Concrete will start hardening in 45 minutes or less in warm weather (in up to 90 minutes in cool weather). But even though the hardening (technically known as curing) process begins quickly, you should leave the posts undisturbed for at least two days after pouring the concrete. Once the concrete cures, apply a bead of clear silicone caulk at the joint between the post and concrete, and renew the caulk anytime you start to see a gap developing at the joint.

Yet another option is to use a fast-setting concrete mix. This stuff is sold primarily for setting posts and flagpoles and the like. It comes in 50-pound bags, and the manufacturers suggest that you dump the dry ingredients straight into the hole until they reach about 3 inches from the top. Then pour in the recommended amount of water and let it soak in. Under normal weather conditions, the concrete will cure in about 30 minutes. Top off the hole with some soil and tamp it to encourage water runoff. This type of concrete is a little weaker than normal mix, and from what I've seen a bit more expensive. For these reasons, I haven't used it, but I know of fence builders who swear by it. The "fast-setting" appeal of this product strikes me as more likely to appeal to a contractor trying to finish a job as quickly as possible, and I'm not sure most do-it-yourselfers would be too impressed by the feature. Some builders take the same approach with normal concrete mix (dump it dry in the hole, then add water). I have even heard of people dumping dry mix into a hole and letting the moisture in the soil take care of matters. With some luck and the right kind of soil, this may work. But I am not going to recommend it as a dependable technique for setting fence posts.

Preparing Hybrid Footings

Some fence builders have found that a footing that combines concrete with tamped earth and gravel stands up well to frost heave. Dig the holes about 12 inches below the frost line and shovel 6 inches of gravel in the hole. Brace the post in the hole and add a 2-inch layer of gravel. Now pour a 4-inch layer of concrete into the hole. When the concrete cures, fill the hole with alternating layers of earth and gravel, as described above.

To better embed the post in the concrete, drive some large nails into each side of the post before pouring the concrete. Be sure to use galvanized or stainless-steel nails (16d or larger). The same effect can be achieved by driving short pieces of ½-inch rebar through holes drilled in the post. (Some builders take this extra step with full concrete footings, thinking the protruding metal will help discourage uplift, but I've never bothered to do so.)

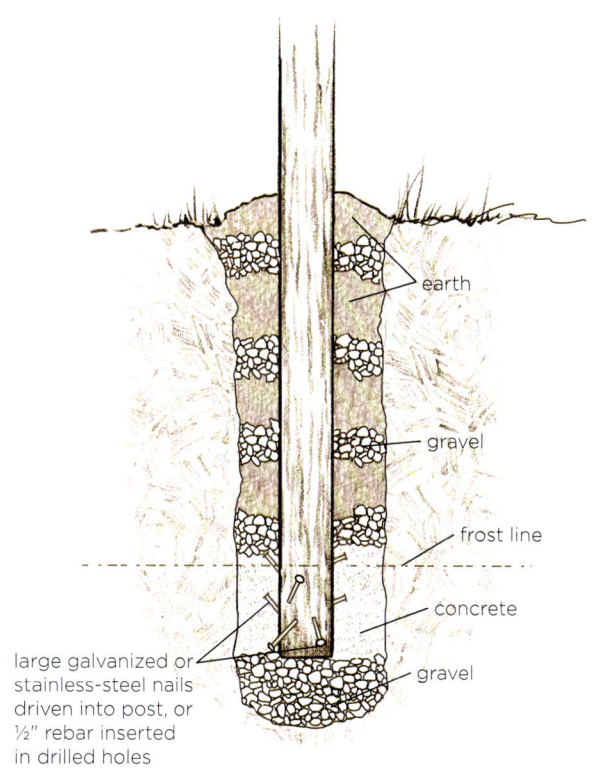

earth

gravel

frost line

concrete

large galvanized or
stainless-steel nails
driven into post, or
½" rebar inserted
in drilled holes

gravel

HYBRID FOOTING

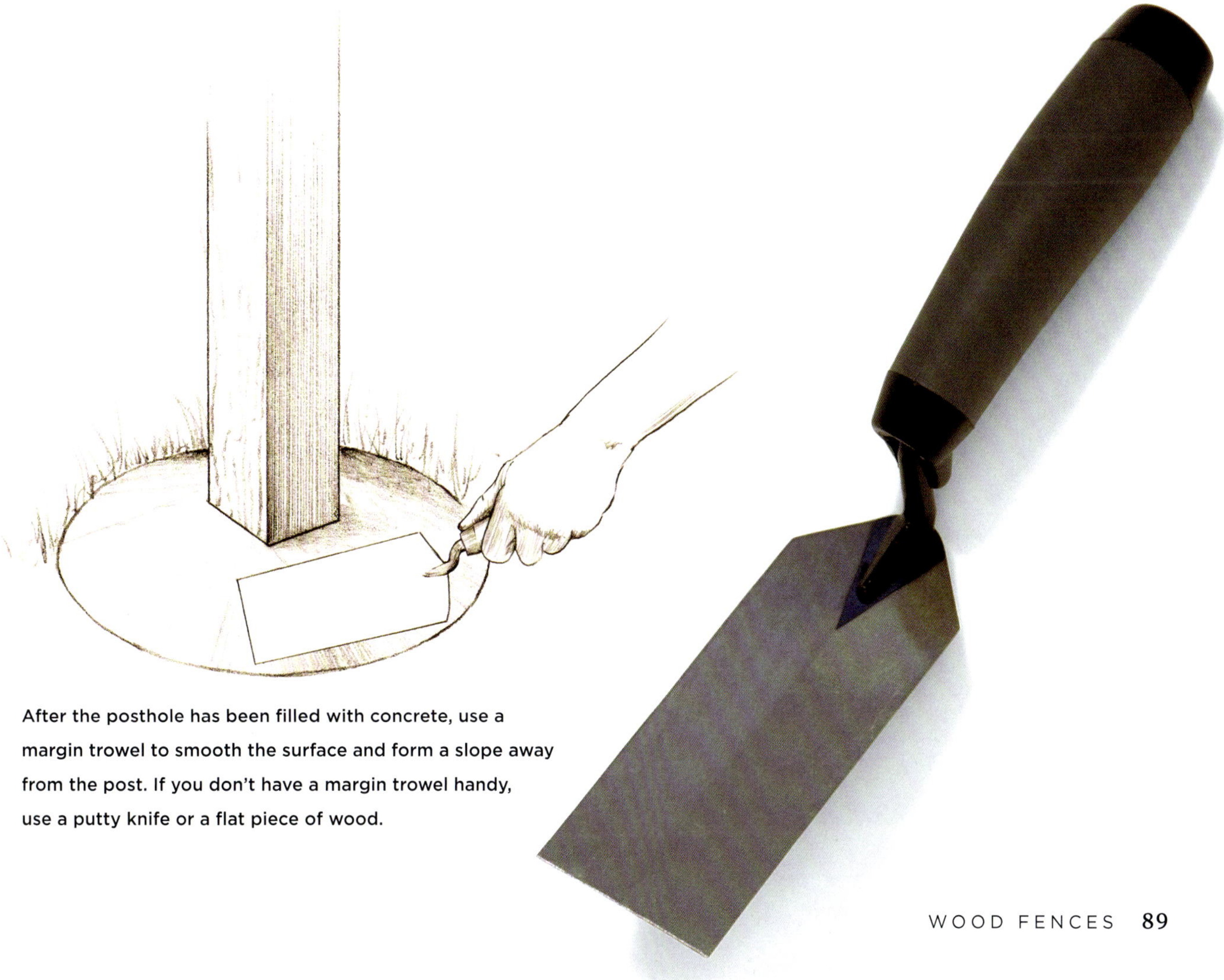

After the posthole has been filled with concrete, use a margin trowel to smooth the surface and form a slope away from the post. If you don't have a margin trowel handy, use a putty knife or a flat piece of wood.

Attaching Rails

Rails are the cross-members that span from post to post. In a basic post-and-rail fence, the rails constitute the entire infill and are therefore the principal visual element of the fence. In most other wood fence styles, the rails are primarily a structural element that stiffens the whole assembly and provides a surface upon which to attach boards or pickets.

With some fence designs, the posts must be trimmed to their final height before you start working with the rails. If that is the case with your fence, skip over to pages 98–99 for some hints on cutting posts before proceeding with this section. Whenever possible, though, I suggest getting the rails attached as soon as possible to provide stability to the posts before you start cutting them.

Over the centuries, wood rails have been attached to wood posts in more manners than I would care to try and count. Before metal fasteners were available, the strongest connections required that rails pass through holes or mortises cut into or through the posts. Today, most often, rails are nailed or screwed

Mortised post-and-rail fence materials can be bought ready to assemble. Once the posts are buried (with the mortises properly aligned!), the rails can be slipped into position without the use of fasteners.

onto posts. But if you are trying to duplicate an older style of fence, you will probably want to approach the rail-to-post joinery with that objective in mind.

Rail-Only Fences

The basic rail fence is a longtime fixture of rural life, and it remains a very functional style for controlling horses and other animals. But because of its simplicity, it also serves as a modest boundary and garden marker. Traditional rail fences had to be built one section at a time. You would set one post, cut mortises in it and another post, slip rails into the mortises, then set the other post with the rails already fit into its mortises. Today, it is more common to set all posts along the fence line, then nail the rails to the face of the posts.

The height of the fence is entirely up to you, although rail fences tend to be fairly low. Likewise, the number and size of the rails, and the spacing between them, are matters of personal choice. The most common type of rail fence used for animal enclosure has 4×4 posts set on 8-foot centers, with three or four 1×6 horizontal rails evenly spaced and nailed to the posts. For fences intended to be more decorative, two rails are often sufficient. You should be able to find boards that are 16 feet long, which can speed up installation.

If you would like to break up the horizontal lines of the fence a bit, consider adding a cross-rail feature to the design. Rails used for this purpose need to be cut at an angle, and this will slow down the installation just a bit. But the resulting fence is easier on the eyes than a long row of straight boards.

You can create a more contemporary look and greater privacy by spacing horizontal rails more closely together, perhaps by alternating boards of different widths, and by topping the fence with a continuous cap. In height and general purpose, this type of fence is reminiscent of a picket fence, but it creates a very different effect on the landscape.

Structural Rails

For rails that are intended to serve strictly a structural role, one of the biggest decisions you need to make is whether to install them flat or on edge. While I will grant that rails installed flat do have a certain symmetrical appeal and may simplify some types of fence construction, by and large I strongly suggest trying to install the rails on edge. Flat rails are almost guaranteed to sag at least a little over time, and sometimes a lot more than just a little. People lean against and sit on rails. Boards and pickets add weight to them. If you have any doubts about the relative strengths of these two positions, lay a 2×4 flat between a couple of sawhorses and start adding some weights in the middle. Very quickly you will see the 2×4 starting to sag. Now set the 2×4 on edge and repeat the exercise. The difference in strength is really enormous. The longer the span, the heavier the load, and the thinner the rail, the more significant the sagging effect is likely to be. If you just do not happen to like the appearance of rails installed on edge, I suggest you use 4×4s for at least the bottom rail, then install the top rail to the tops of the posts. Alternatively, plan to reduce the span between posts and set the rails flat.

Rails mounted on edge also offer more choices for mounting the infill boards or pickets. Often, the objective is to place the infill on the most visible side of the rails. With a board fence facing the street, for example, you would probably opt to place the infill on the street side of the rails. In your backyard, you might want to take just the opposite approach. The rails can be centered inside the posts, so that the faces of the infill are flush with the faces of the posts. With rails mounted on the edge or on the face of the post, infill can cover both posts and rails for a uniform look. For a typical picket fence, use two rails; for a high fence, especially if it will be filled with boards, use three rails.

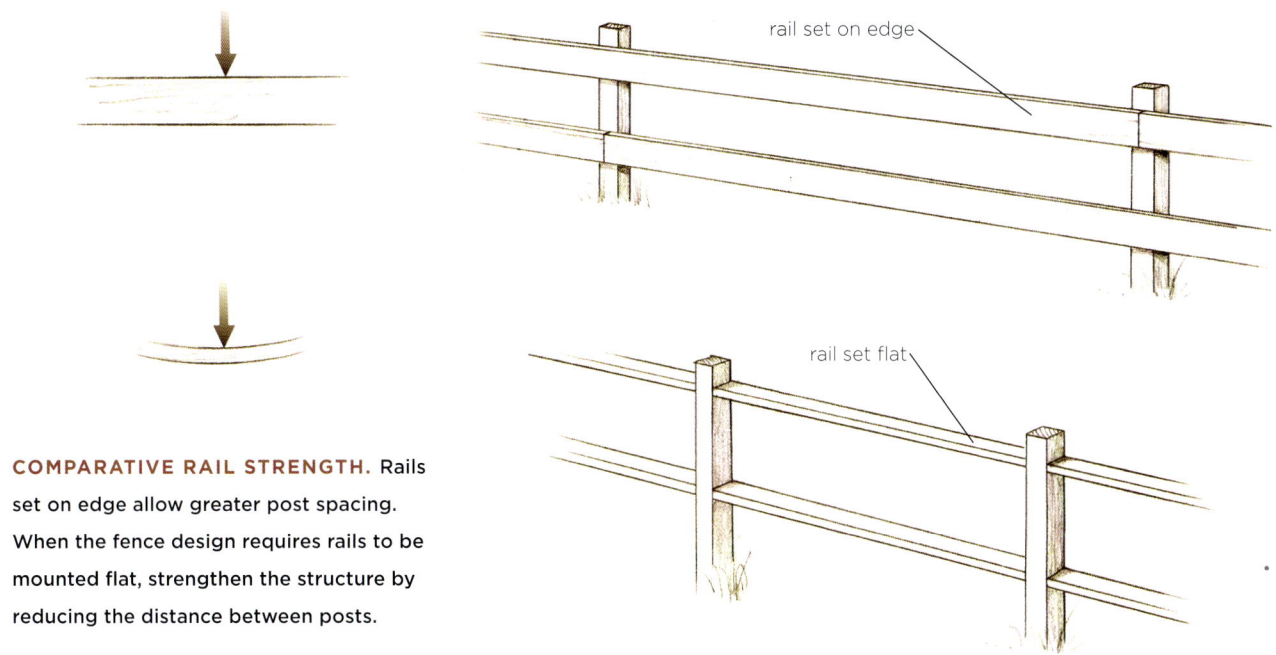

rail set on edge

rail set flat

COMPARATIVE RAIL STRENGTH. Rails set on edge allow greater post spacing. When the fence design requires rails to be mounted flat, strengthen the structure by reducing the distance between posts.

Rails A' Rolling Rail fences

combine a timeless, functional design with a practical and efficient construction technique. Whether you want to create a simple boundary line, enclose some horses, or add a bit of privacy to your yard, you can find a suitable fence style that uses only posts and rails.

Options in Joinery

The strongest joints between rails and posts are those that rely on rails sliding through holes or mortises in the posts, especially if the rails are installed on edge. The problem is that these joints require a lot of work. It takes time to cut the holes or mortises, and they must be aligned perfectly for the rails to fit right. I have seen fences built with rails installed flat in dadoes, but I cannot see why anyone would bother with this. Granted, dadoes can be cut fairly quickly, either with a router or by making several passes through the wood with a circular saw set to the depth of the intended dado, then cleaning out the notch with a chisel. But even when the rail is joined to the dado with a good-quality exterior-grade adhesive, the result is a joint that really is not much stronger than rails installed flat with metal brackets or fasteners, and it remains considerably weaker than almost any type of joint for rails set on edge.

Through-mortised posts can create a somewhat rustic fence that does not necessarily require any additional fasteners. I would probably consider using this type of joint only with 6×6 posts. That way I could cut through mortises sized for two 2×4 rails. To cut the mortises, I would mark the outline on the post, drill a fairly large hole through one corner of the layout, and use a long wood-cutting blade in a reciprocating saw to finish the cut. Alternatively, drill a series of closely spaced holes, then clean out the mortise with a chisel. Cut the rails about 2 feet longer than the distance between posts, and then slip them into the mortises. You can use this technique to set rails flat or on edge.

Butt joints are the most common means of joining rails to posts. Rails are cut to fit exactly between the posts and are then fastened with nails, screws, or special metal brackets made for just that purpose. I like using fence brackets because they are easy and inexpensive and create a strong connection. What I don't like about them is their appearance. But you can stain or paint the galvanized surface to mirror the color of the fence, if you like (prime them first, though, with a primer intended for metal surfaces). If you're going to use just nails or screws, they must be driven at an angle through the rail and into the post, and this can

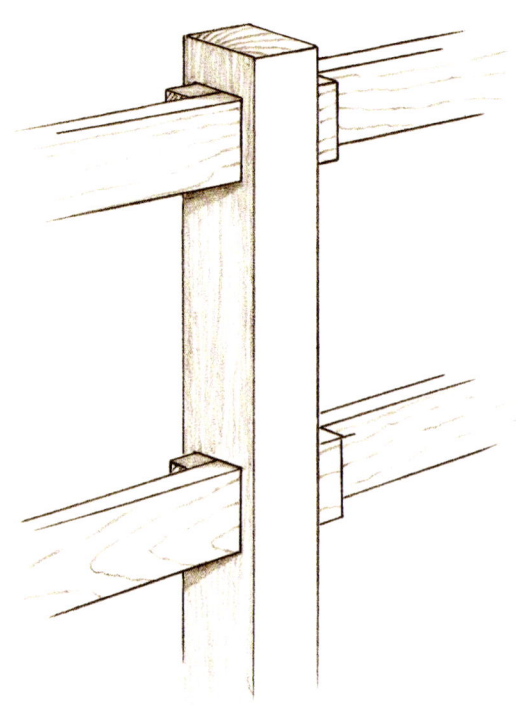

Through-mortised posts were used for rail fences long before metal fasteners became widely available. Though cutting mortises is time consuming, the results can be solid and attractive.

easily cause the end of the rail to split (if not right away, then later). To minimize the chances of this happening, drill a pilot hole through the rail before driving the nail or screw.

Face-nailing is probably the quickest approach, and it has the added advantage of installing the rails on edge. Normally, the rails would be attached to the back side of the posts, with boards or pickets going on the other side. I would still recommend drilling pilot holes before driving nails or screws.

For strength combined with neat appearance, I favor setting rails on edge in notches cut into the faces of the posts. The strength of this joint lies in the fact that the rail is secured by the post itself rather than simply by nails or screws holding it onto the post. Metal fasteners are needed with this solely to keep the rail in the post.

MAKING A MORTISE JOINT, OPTION ONE

1

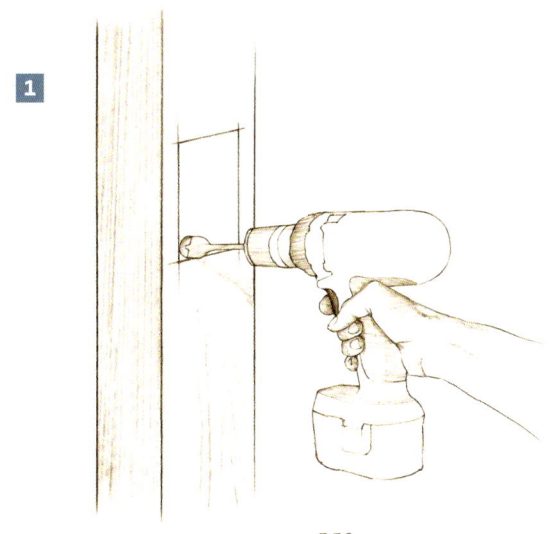

Mark the mortise layout on the post, then use a 1-inch spade bit to drill a hole through one corner of the layout.

2

Insert the blade of a reciprocating saw through the drilled hole and cut out the mortise along the layout lines.

. . . .

MAKING A MORTISE JOINT, OPTION TWO

1

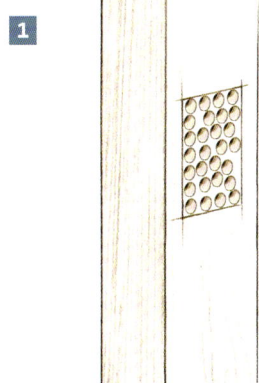

Mark the mortise layout on the post, then use a ½-inch bit to drill a series of holes through the post inside the layout.

2

Use a hammer and chisel to clean the wood out of the mortise.

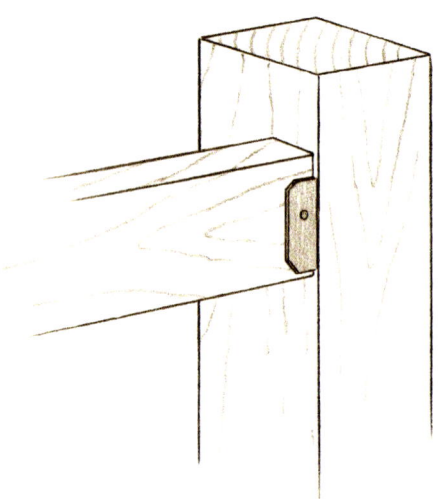

USING METAL BRACKETS. A metal bracket can be used to secure a butt joint, with the rail on edge, as here, or flat.

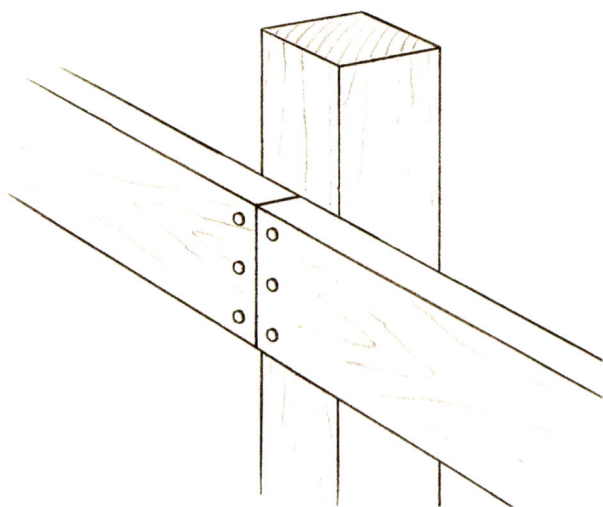

FACE-NAILING. Rails set flat can be face-nailed to posts. Be sure to stagger the joints in this arrangement, so that joints don't fall concurrently on the same post.

TOENAILING. Butt joints can also be secured by nails driven at an angle through the rail and into the post. To minimize the chances of splitting the wood at the rail, drill pilot holes before nailing.

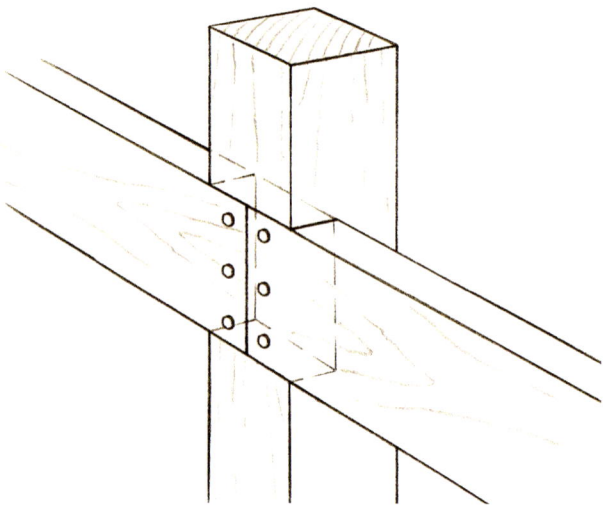

NOTCHING AND FACE NAILING. Setting rails in notches produces a stronger joint than when using butt joints or face-nailing. Rails set on edge and face-nailed in a notch make the strongest joint of all.

Notching a Post

Rail notches must be laid out and cut carefully for all the parts to fit together perfectly. Even if you set rails in only one notch per post (preferably, the top rail) and use a butt joint for the other, it will produce a stronger rail section than using butt joints or face-nailing exclusively. I think the safest way to approach this task (though, admittedly, not the most efficient way) is to finish one post before moving on to the next.

1 Begin by marking a layout on the first post. Hold a piece of rail stock in place perpendicular to and across the face of a post. Make sure it is level, then use a pencil to mark its top and bottom edges on the post.

2 With a circular saw set to cut exactly 1½ inches deep (or a tad less on the first post, just to play it safe), make a series of closely spaced passes between the lines.

3 Use a hammer to knock out pieces of cut wood.

4 Clean out the notch with a chisel and hammer. A flat wood rasp might come in handy on this chore as well.

5 Slide the end of the rail into the notch. Check the depth of the cut and adjust your circular saw if necessary. Now place the next rail section in the other side of the notch and use a level on the rail to guide the layout for the notch on the next post. Attach the rails with nails or screws. ■

1

2

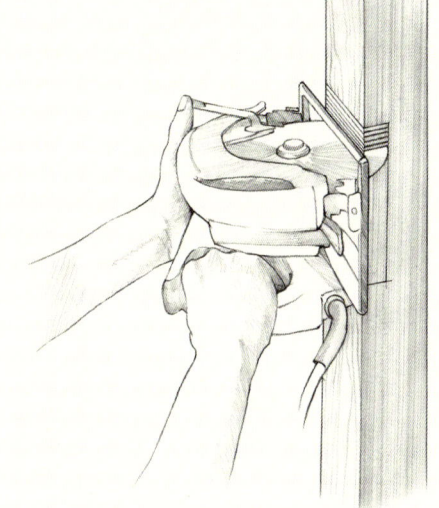

3

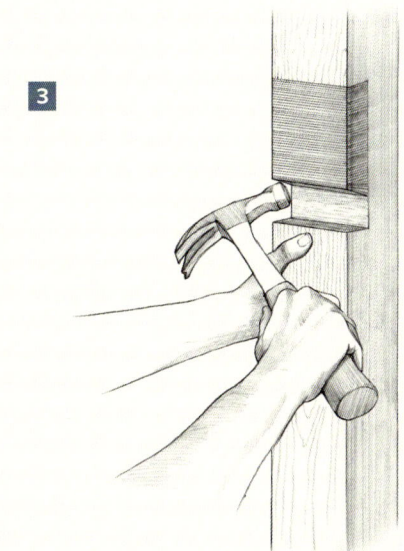

4

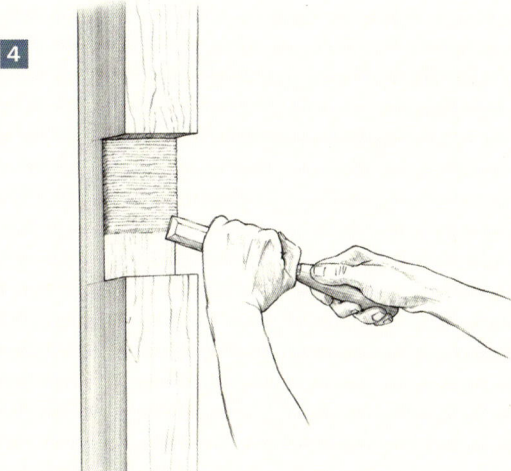

Post Top Details

Depending on the design of your fence, the posts may be virtually inconspicuous, moderately noticeable, or prominently displayed. Regardless of the final plan, the tops of the posts deserve a bit of special consideration. With lumber, the end grain (or cross-sectional surface) is the most vulnerable to water infiltration and damage. With rails or other components that are installed horizontally, the end grain is perpendicular to the ground, so water sheds easily. With posts, however, the end grain is fully exposed to the elements, and the large flat surface at the top of a 4×4 or 6×6 can soak up a lot of water over time.

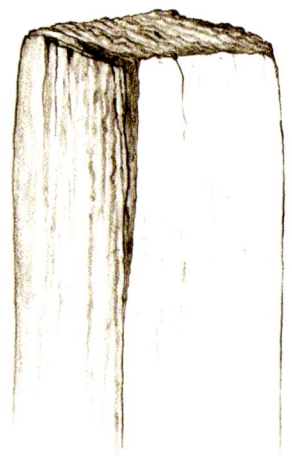

The end grain of posts and boards is the most susceptible to damage from the elements.

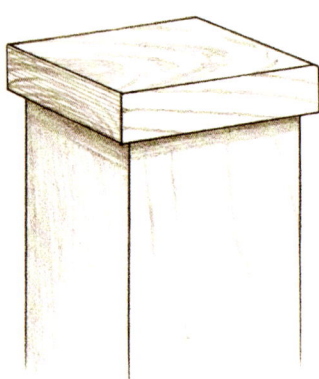

A simple square cap [ABOVE] or continuous cap [RIGHT] protects the end grain of posts and contributes to the posts' longevity.

Simple Post Tops

If your fence design largely ignores the posts as visual elements, I would recommend that you still make an effort to encourage water to drain away rather than congregate on the tops. You can do this either by placing some kind of cap on the top or by trimming the post at an angle.

With post caps, if all you care about is shedding water, most any type will do. A continuous cap rail is one solution; it both covers the tops of the posts and creates a mounting surface for fence boards or pickets. Another simple solution for 4×4 posts is to buy some 1×6 boards and crosscut them into squares. The resulting 5½-inch-square pieces could be stuck on top of the posts with a 1-inch overhang on all four sides. To attach the caps, I suggest applying an exterior-grade adhesive to the post top, then driving two galvanized ring-shank nails through the cap and into the post. Simple, effective, and just slightly decorative.

Cutting post tops, especially at an angle, can be a bit challenging. You could use either a circular saw or a reciprocating saw (or, for you hand-tool purists, a well-sharpened handsaw), but I often have better luck by using both tools. First, of course, you need to mark a layout on the posts. Use a level and a long 2×4 or a water level (see pages 62–63) to establish a level line from post to post. If you are cutting at an angle, use a combination square or Speed Square to mark a consistent 45-degree line on both sides of the level line.

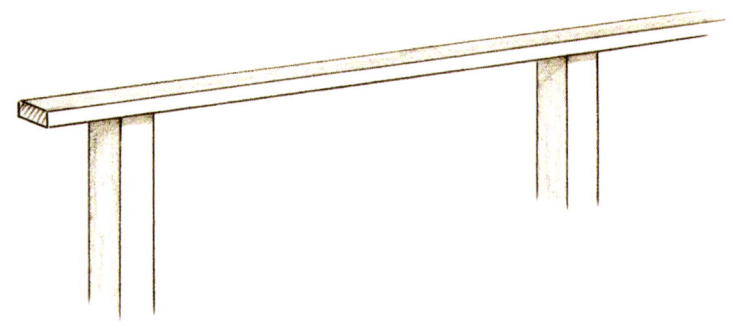

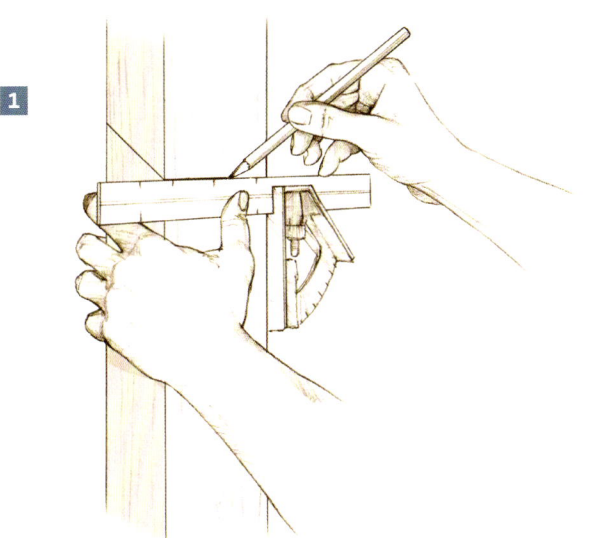

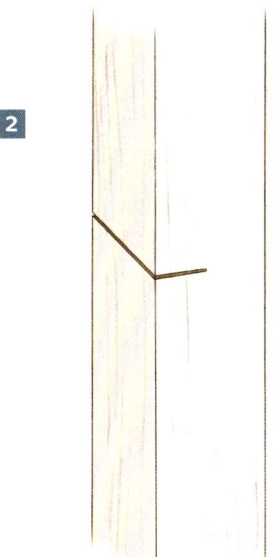

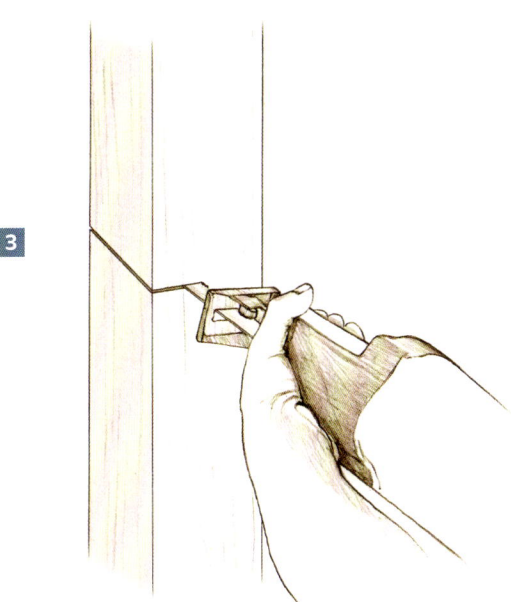

Using just a standard circular saw (with a 7¼-inch blade), set the blade depth to the deepest cut possible. Make one pass with the saw up one angled line, then move to the other side and cut down the angled line. If you are very careful, the two passes will produce a reasonably clean surface. If one side is a bit higher than the other, trim a little more off. (If you are not particularly experienced with a circular saw, practice these cuts on a 4×4 that is not being used on the fence, before you try cutting a post.)

With a reciprocating saw equipped with a long wood-cutting blade, you can make this cut with a single pass. But the blade may have a tendency to wander off-line a bit. Try to hold the saw at a consistent angle, keep the speed high, and move slowly through the post. (Again, when in doubt, practice on unimportant lumber first.)

For the dual-tool technique, make one deep pass with the circular saw, and then finish with the reciprocating saw. The first cut makes it easier to keep the second cut in line. When cutting 6×6 posts, you will not be able to cut to the center with a circular saw alone, so a reciprocating saw (or handsaw) will be mandatory.

CUTTING ANGLED POST TOPS. **1** Mark the layout of the cut on the post. **2** Use a circular saw to make the first cut, cutting at an angle. **3** Finish the cut with a reciprocating saw, keeping the blade at the same angle begun by the circular saw.

Decorative Post Tops

Decorative post tops perform a double duty. They cover and protect the end grain while also giving a special look to the fence. As I mentioned earlier, you can buy 4×4 posts with a range of decorative features milled into them, and if you see something you like, it is hard to resist the temptation to buy it. Even if you prefer to cut your own decorative post features, it is easier to do so before they have been set in the ground. The problem with both of these approaches, however, is that they force you to set the posts to their finished height before backfilling.

Now, this can be done. You need to dig the holes to a uniform depth, then add or subtract gravel in the bottoms of the holes to establish a uniform height for the posts. (You could also cut off the bottoms of the posts as needed to bring the tops into alignment, but I do not recommend placing the cut ends of pressure-treated posts in the ground, because the newly exposed surface will almost certainly not have the same level of preservative treatment as the discarded end. If you want to ignore this recommendation, be sure to coat the cut end liberally with a water-repellent wood preservative first.)

Find the center of the post top by drawing straight lines from corner to corner. The intersection of the lines marks the center.

I much prefer to set longer-than-needed posts in the ground, then worry about trimming them to finished height and adding any ornamental touches later. There is far more margin for error with this sequence, yet there remains a range of options for creating that decorative design statement.

Most lumberyards and home centers keep at least a few prefabricated post caps in stock, and you can probably special-order from a much broader selection. Local woodworkers and mill shops are other good sources. You can find pointed, rounded, flat, and other shapes and styles in solid wood or clad in copper. When looking for inspiration, study the trim and ornamentation on your house, especially the details on any porch posts.

Some prefabricated post caps have a screw in the end, so all you have to do is drill a pilot hole in the center of the post top and thread it in. (To be on the safe side, though, I would shoot a little exterior-grade adhesive into the joint before tightening the screw.) If your post caps do not have an attached fastener, use some glue and nails or screws to attach them to the post.

For an even greater selection of post caps, design and make your own. The simple squares cut from 1×6 boards described above can serve as the basis for any number of finished tops. With a power miter ("chop") saw or a table saw, you can produce uniformly sized pieces of wood trim very quickly. Cove or quarter-round molding adds a nice touch. Create a nice shadow line with a shallow cut using your circular saw or a router.

Tips and Tops

Like the colorful stocking cap you don on a cold snowy day, a fancy post top is both decorative and protective, preventing water damage to the vulnerable end grain on the post.

Boxed-In Posts

Making use of boxed-in posts is a great way to create large, visually dominant posts without having to struggle with the weight and expense of oversized solid timbers. They also offer a whole new range of possibilities for decorating.

Boxing in a post does not alter the structural capacity of the core post (that is, the buried post). Building up a 4×4 post to a size of, say, 8×8 does not suddenly create a post with the strength of an 8×8, but to the eye the added layer certainly appears stronger. Bulky posts are a particularly nice feature at gates, teamed with somewhat smaller and similarly styled posts used

elsewhere on the fence. By adding a layer of 1x spacers covered with 1x surface boards, you can increase a 4×4 post to a full 6½ × 6½, or a 6×6 post to 8½ × 8½. Thicker spacers or surface boards, or both, can create even bigger finished posts. Use pressure-treated wood for the spacers, and be sure to have a spacer at any potential rail or gate hardware locations to provide solid blocking for nails.

The surface boards can be made with good-quality pine boards, but I would recommend either redwood or cedar for this purpose. Alternatively, if you plan to paint the fence, you may want to look into one of the growing number of manufactured products being used increasingly for exterior trim on homes. Engineered wood and wood-plastic composites can now be found in primed boards that can be nailed on and look great under a coat of paint. I do not have any direct experience with these new products for this kind of application, but I see them showing up as trim on all kinds of new houses. Ask a reliable lumber dealer or some experienced builders or remodelers in your area for tips or recommendations. The cost, labor, and long-term maintenance obligations may compare very favorably with those of solid wood.

The corners of the box can be joined in any number of ways. The simplest is with butt joints, which require boards of two different widths. This is not a particularly attractive joint, in my view, but you can get away with it by placing the wider boards on the more visible sides (that is, front and back), which will make the long joint a bit less visible. Miter joints are neater looking, but they do require careful machining (preferably with a table saw). To keep the miters from opening up, I would use exterior glue and biscuit joinery, although newer exterior epoxies are probably strong enough to hold the bond all by themselves. A nice compromise is to use miters on the most visible side of the box, then enclose the less-visible side with a butted board. You can also dress up butt joints by chamfering the edges of the wide boards.

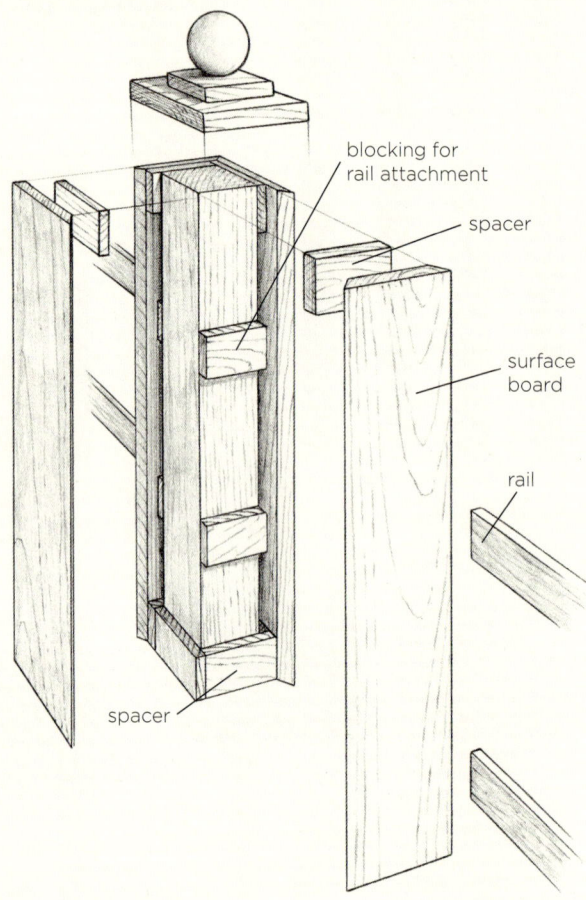

blocking for rail attachment

spacer

surface board

rail

spacer

The basic boxed-in post is composed of a central post surrounded by spacers, blocking, and surface boards and topped by a decorative cap.

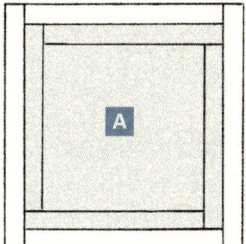
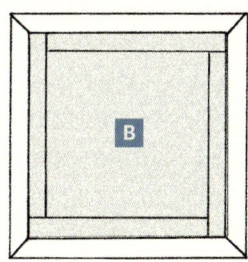

CORNER JOINERY OPTIONS. The surface boards of a boxed-in post can be joined in various ways, including butt joints **A** ; miter joints **B** ; miter joints on the most visible side, with butt joints on the back side **C** ; butt joints with chamfered edges **D** ; and no joints with the corners filled with quarter-round **E** or cove molding **F** .

Another approach is to cut all four surface boards to the exact width of the post and attach them, then fill in the corners with trim. Use a nice profile molding or some quarter-round or cove molding.

Fluted posts are a traditional feature that can really add to the attractiveness of a well-designed fence. To cut the flutes you will need a router, preferably a plunge router, along with a double-fluted core-box router bit. You can find these bits in diameters ranging from ⅛ inch to 1 inch or more. While machining the flutes is work that anyone with basic experience using a router could handle, it is not a job I would want to chance on solid posts, due to the human fallibility factor: One slip with the router, and you could really mess up the post. But I would have no qualms about routing the flutes on a surface board before it was installed. One mistake there and I can flip the board over and try again. ■

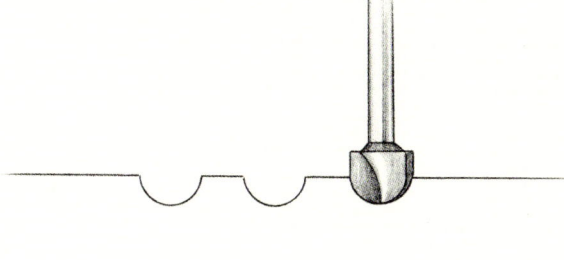

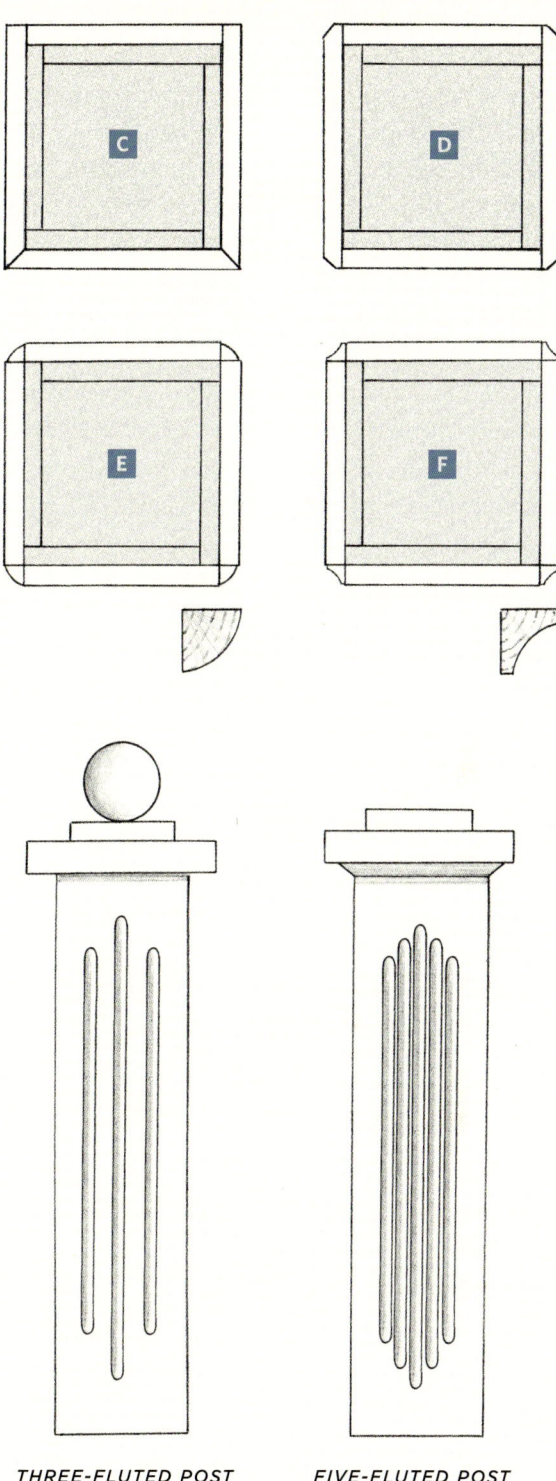

THREE-FLUTED POST   *FIVE-FLUTED POST*

MAKING FLUTED POSTS. Use a double-fluted core-box router bit [LEFT] to cut flutes into the surface boards of a boxed-in post. Three flutes or five flutes are the norm.

The Picket Fence

PICKET FENCES ARE, by and large, front-yard fences. They mark the boundary between the private residence and the public street. They are intended to look nice, keep people from wandering on to the yard, and perhaps offer a bit of protection to a garden. They do not offer much privacy or security, and they are not fences that are likely to serve as leaning posts for neighborly conversations. They are on the low side—generally in the 3- to 4-foot range. Although there are no legal or moral reasons that I know of to limit yourself to such height restrictions, I think thin pickets that exceed 4 feet start to lose their appeal.

Making Pickets

Pickets are normally made from 1×3 or 1×4 boards, although 1- to 1½-inch squares and round dowels are nice options in some situations. The tops can be flat, but they are usually cut in some decorative pattern both for looks and to help shed water from the exposed end grain of the wood. You can buy pickets in a limited number of shapes and standard lengths from lumberyards and home centers. A local woodworker or cabinetmaker could probably be convinced to crank out a large number of custom-designed pickets for a reasonable fee. But if you are building your own picket fence, why would you not want to design and make your own pickets?

Even if you decide on a standard style of pickets, making your own allows you to choose good-quality wood and to customize the lengths to suit the various ebbs and flows your fence is likely to have. It also offers you the opportunity to create a truly unique fence by opting for a less-common picket style or using an alternating-length design. Looking through old architecture texts at the library or wandering through neighborhoods of old and historic houses can generate dozens of potential ideas. You could even turn the design work into a family project, with everyone in the household submitting their own plans and then trying to find a consensus choice or compromise design.

Your choice of design should be based in part on the types of tools and tool skills you possess. Some styles can be mass-produced quickly with nothing more than a circular saw or jigsaw. Others make sense only if you have access to a power miter saw, radial-arm saw, or table saw (for straight and angled cuts) or a bandsaw (for rounded cuts).

The quickest way to produce pickets with pointed or angled tops at a consistent length is to set up a power miter saw with a stop block. Adjust the saw to the proper angle, then clamp or nail the stop block into place and start cutting. For pointed tops, cut one side of a board, flip it over, and cut the other. To create a pyramid-type shape on 1- to 1½-inch square pickets, make the same cut on all four sides.

When cutting a large number of pickets to the same length, measuring, marking, and cutting each one individually will consume a lot of time. Instead, take a few minutes to construct a stop block to use with a power miter saw.

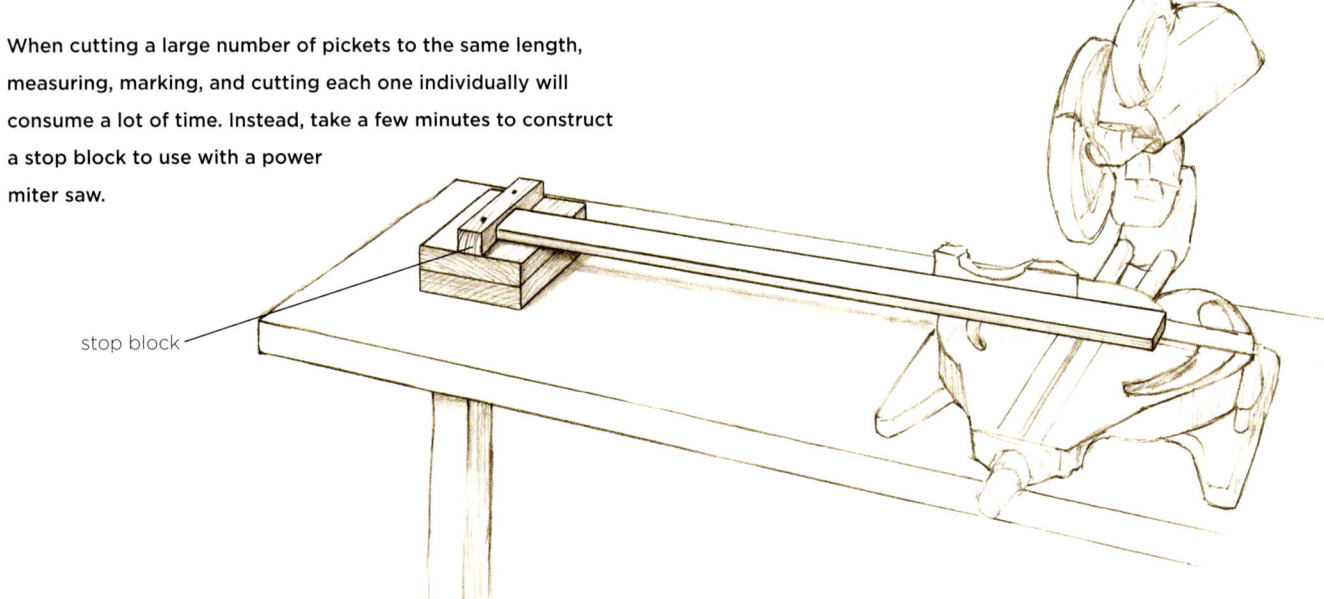

stop block

Picket Potpourri

Traditional picket fences tend to be simple in both design and construction. But they don't have to be. By adding a little variety to the pickets themselves, you can transform the fence into a one-of-a-kind structure. To create half-circle cutouts such as those in the bottom two photographs, drill the holes before ripping the pickets.

For more elaborate designs that require pickets to be cut one at a time, I like to make a template out of ¼-inch hardboard or plywood. Use the template to trace the design onto each board. This kind of work is almost always easier to do before the pickets are installed, and if you have a bandsaw, you may be able to cut three or four pickets at a time. But if you are installing the fence over irregular terrain, you may want to install overly long pickets first, then cut them to length and trim them into their final shape.

To create a curved top edge to a section of pickets, I think it is easier to keep the design fairly simple and to mark the layout and make the cuts after the pickets are installed. The best type of marking tool for curves is simply a thin piece of wood cut a little longer than the spacing between posts (the longer the piece, the more pronounced will be the curve). Drive nails temporarily into the posts to support the strip, then flex it up or down to form the shape you are after and trace the curve onto the pickets. Use this line to lay out the intended pattern on the pickets before cutting.

Attaching Pickets

If the pickets are being installed with their tops level, the easiest approach is to attach a string line from post to post and then align each picket with the string. With an alternating-height style, use more than one string line. Alternatively, make a spacer with a height gauge attached (see the facing page). If you plan to cut a shape along previously installed pickets, it is best to make sure that all of the pickets are a little longer than necessary so that each one will need a full cut.

Pickets may be all the same width and length, or they may have multiple widths and alternating lengths, depending on your design. But one place I look for uniformity is in the spacing between the pickets. If you want to use a variable spacing pattern, by all means go ahead. Just plan carefully, lest the pickets on your finished fence look like they were slapped on haphazardly.

The safest and quickest way to evenly space pickets is to use a spacer board that has been ripped to the width of the intended gap. For a traditional, relatively open approach to a picket fence, use a spare picket for the spacer. I think this spacing creates a nice balance. For a bit more privacy, however, use a thinner spacer. If you do not use a spare picket, make the spacer about the same length as the pickets you are installing. Attach a crosspiece near the top so that the spacer can rest on the rail by itself, freeing both of your hands to attach pickets. You can also attach a height gauge to the front of the spacer. This way, all you have to do is set the spacer on the rail, push a new picket tightly against the spacer, align the top of the picket with the height gauge, and drive your nails. I suggest that you

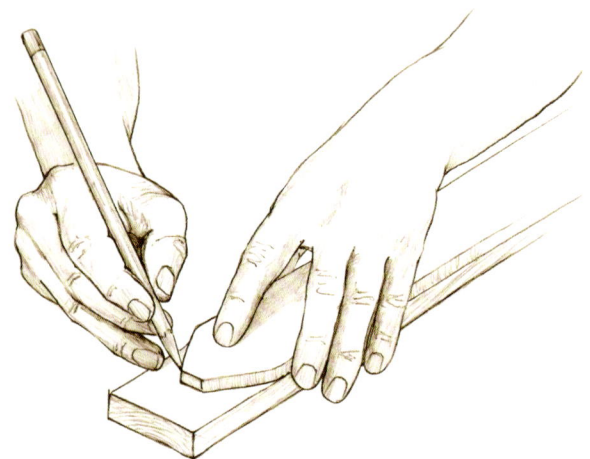

A simple template, made out of hardboard or plywood, greatly simplifies the task of uniformly marking pickets before they are cut.

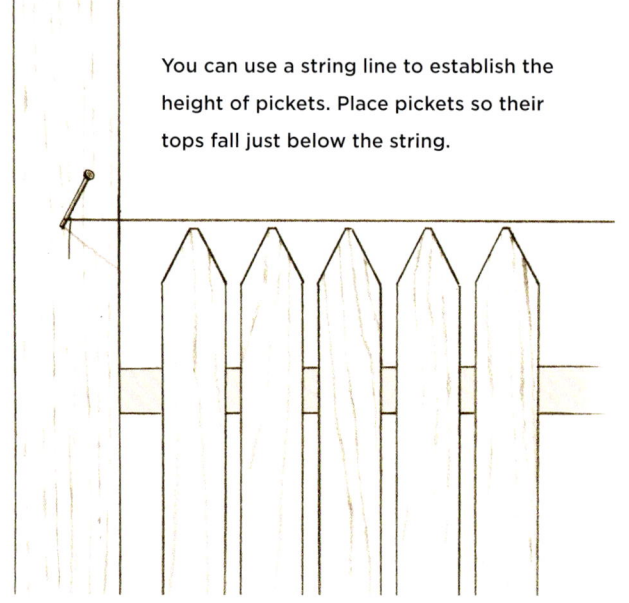

You can use a string line to establish the height of pickets. Place pickets so their tops fall just below the string.

The Perils of Pickets

Pickets are often made with sharp tips on their ends. It is possible that this is due to their lineage, as stockades of sharpened logs were used by early frontier settlers to protect their encampments. I cringe when I see fences like this along a sidewalk or pathway where pedestrians, bicyclists, and skateboarders are likely to pass. One thing a decorative fence should not be is a potential danger, but that is precisely what pointed pickets can present to someone who slips and falls into them. There are any number of picket designs that come to some type of point at the top, and I do not want to suggest that you ignore them in favor of pickets with more rounded tips. But you do not have to create sharp points to sustain the look. Rather than cutting the tops to form a sharp point, back the cut off just a bit to leave duller tips, as shown in the nearby illustration. You will not be able to notice the difference from even a short distance

away. Besides, sharp points in wood are not going to age well. They get banged up and dulled quite easily and inconsistently, compromising the uniform look they were intended to provide. ■

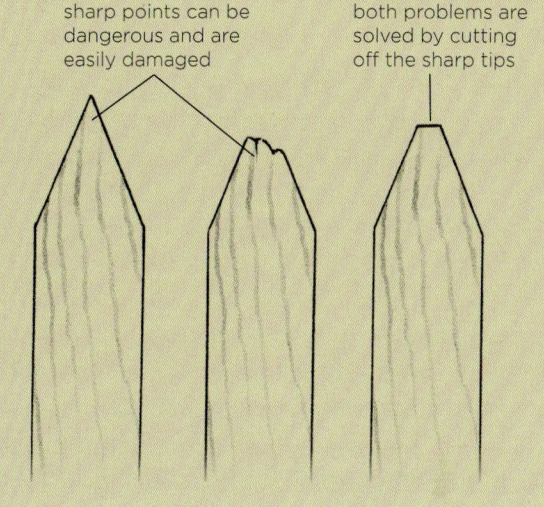

sharp points can be dangerous and are easily damaged

both problems are solved by cutting off the sharp tips

THE PICKET SPACER. Repetitive tasks can be performed much more quickly and more accurately with templates and spacers. This simple device, which can be made in minutes out of scrap lumber, ensures that pickets are installed plumb, properly spaced, and at the correct height.

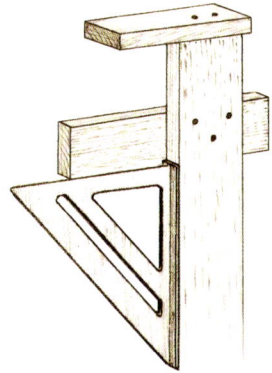

when building the picket spacer, use a square to ensure that the spacer hangs plumb

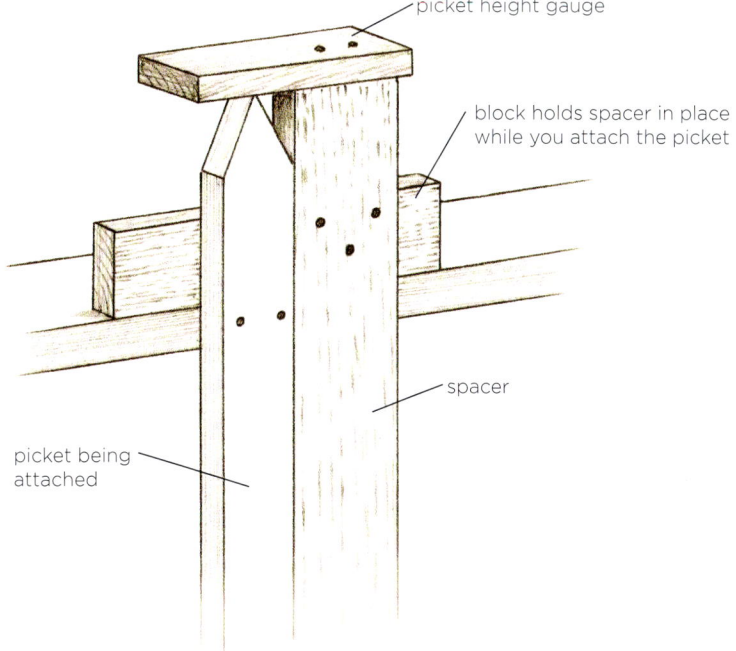

picket height gauge

block holds spacer in place while you attach the picket

spacer

picket being attached

use a framing square to carefully square up all pieces of the spacer as you nail them together; if the spacer does not hang perfectly plumb off the rail, you could wind up with a fence of leaning pickets.

Attach pickets to the rails with galvanized ring-shank nails. For 1× pickets and 2× rails, which would create a total thickness of 2¼ inches, use 6d nails, which are 2 inches long. Two nails at each rail junction should suffice for most pickets.

To be on the safe side, check the pickets with a level from time to time to ensure that they are plumb. I have actually used a level for a spacer on occasion, which streamlines this process a bit.

Vertical-Board Fences

VERTICAL-BOARD FENCES are typically higher than picket fences, with their boards more closely spaced. Thus, they are most often thought of as backyard fences, intended to ensure privacy, add some security, and keep kids and pets from wandering off. That said, there is no reason why boards cannot be treated like pickets, with decorative top cuts and with some spacing between boards to allow air currents to pass through and gazing eyes to see through. If that is your intention, see the above discussion on picket fences for suggestions on cutting and installing the boards.

The boards can be purchased in standard, uniform sizes (typically 1×4, 1×6, or 1×8); attached to rails (preferably three rails for a high fence, two for a picket fence); and cut to size, if necessary, quite quickly. But board fences lend themselves to a wide range of stylistic options. The tops can be cut into points, clipped (dog-eared) at the corners, or rounded. Alternating widths of boards in either random or organized sequences can create a nice effect. The board-on-board (or staggered) pattern is one I particularly like. Depending on how the boards are installed, it can completely block views or offer slim glimpses from the right angle. At the same time, it allows air to circulate and has a nice rhythmic look, with some contrast in light and shadow, unlike a wall of solid boards. It is also easy to build. Attach the boards with galvanized ring-shank nails or exterior-grade screws.

Clean Vertical Lines

Vertical-board fences tend to be high and solid. That makes them more visible than other fence styles, and thus worthy of careful and thoughtful design. Monotony can be tackled with color, shape, and fanciful fence tops.

Louvered Fences

Louvered fences are functionally much like the board-on-board style mentioned on page 108. They offer substantial privacy, yet still allow air to pass through. I think they look best around houses with a more modern architectural design. Louvered boards, typically either 1×4s or 1×6s, can be installed either horizontally or vertically. Either way, they do require careful layout and installation. You will use more wood with a louvered fence than with a flat style of board fence, and it will take longer to build. Unless you have a strong reason for doing otherwise, I suggest placing the boards at a 45-degree angle.

Horizontal Louvers

When installing 1×4 or 1×6 louvered boards horizontally, space the posts no more than 6 feet apart. Stay away from this style altogether, however, if you think it might inspire someone to try to climb over it.

You can find metal brackets made specifically for attaching louvered fence boards to posts, and I think they are the best way to handle horizontal boards. Start at the bottom, and make certain that the first louver is level. To speed up construction and ensure consistent spacing, determine how far apart you want to locate each row of louvers, then cut a short board to that dimension and use it as a spacer. With a level bottom louver and accurate spacer, you should be assured of an attractive result.

Vertical Louvers

You can also use those same metal brackets for a vertical installation, but another option would be to slip them between 1×4 spacers attached to the top and bottom rails. Cutting spacers on a power miter saw would be quick, and would save you from having to measure and mark the rails before installing boards.

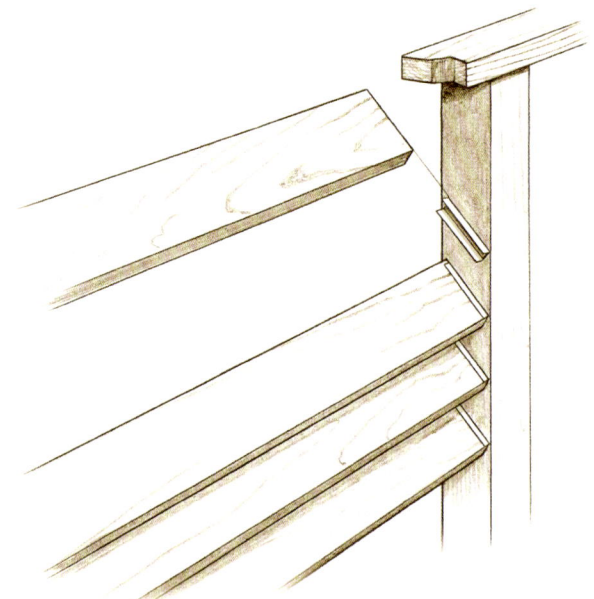

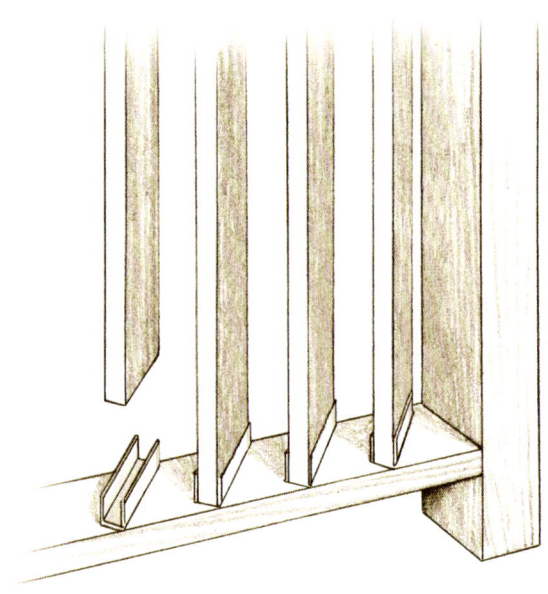

LOUVERS WITH METAL BRACKETS. Louvered fences can be quickly and conveniently assembled with metal fence brackets. For horizontal louvers, the brackets are attached to the posts [TOP]. For vertical louvers, they need to be fastened to rails installed flat [BOTTOM]. Careful layout and installation of the brackets will ensure properly spaced and aligned louvers.

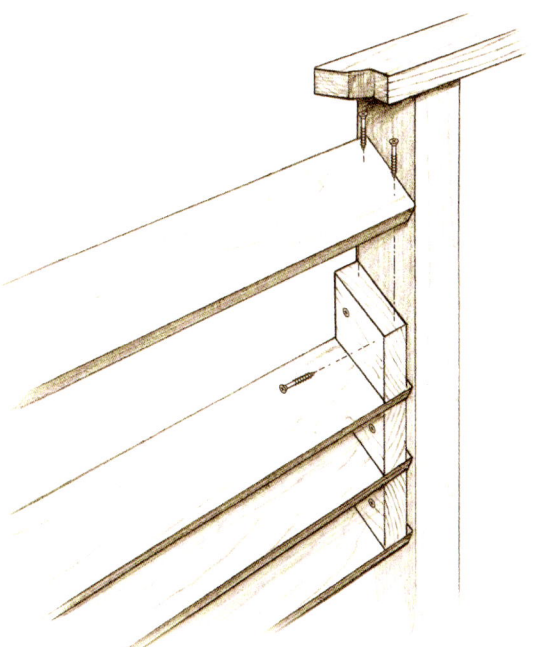
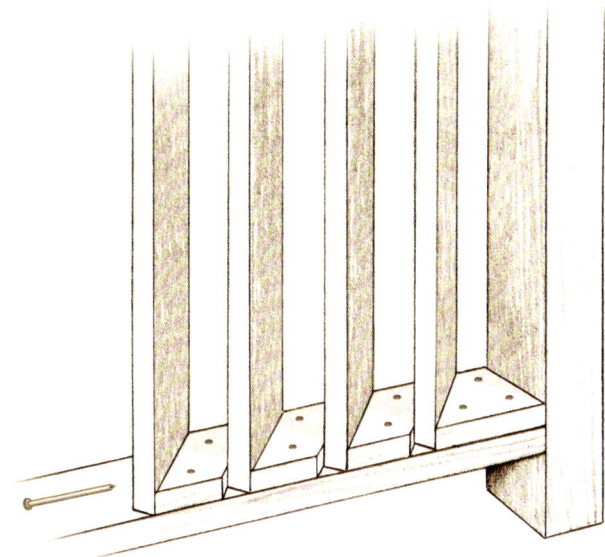

LOUVERS WITH WOODEN SPACERS. Wood spacers provide a relatively foolproof means of installing louvers for both horizontal and vertical applications. First cut all of the spacers, then install them one board at a time.

A Fence with a Sprinkler

If your fence will surround a garden or a row of plants, you might consider adding to it this handy and inexpensive product: a sprinkler kit that can be mounted directly to the fence. You can attach the main hose to the top or bottom rail, or even wind it through posts and pickets. Then you poke holes in the hose and insert the adjustable sprayers, attach the hose to a source of water, and turn it on. Turn the dials on the sprayers to create just the spray pattern you want. The sprinkler kit can be purchased from Lee Valley Tools (www.leevalley.com; 800-871-8158). ■

WATERING FENCE

Basketweave Fences

BASKETWEAVE IS ANOTHER TYPE of board fence that combines texture and pattern with privacy and security. You will need to use relatively thin boards (⅜ to ½ inch thick) about 4 or 5 inches wide in order to form this fence. Boards of that thickness are not commonly carried at lumber suppliers, so you may need to special-order yours from a sawmill.

Attach a 1×2 nailer down the center on the inside faces of each post and a vertical 1×3 spacer centered between posts. Place the centers of each board on alternating sides of the spacer, and then attach the ends to alternating sides of the nailers using galvanized ring-shank nails. I would not try to overdo the weaving effect by, for example, adding another spacer or two between pairs of posts.

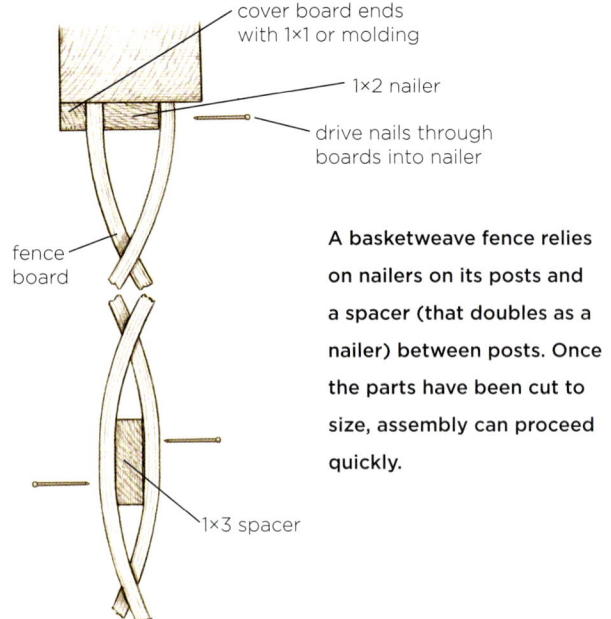

cover board ends with 1×1 or molding

1×2 nailer

drive nails through boards into nailer

fence board

1×3 spacer

A basketweave fence relies on nailers on its posts and a spacer (that doubles as a nailer) between posts. Once the parts have been cut to size, assembly can proceed quickly.

BASKETWEAVE FENCE

Adding a Kickboard

Ground level can be a rough spot for a fence, especially for fences made of relatively thin or vulnerable material such as lattice or basketweave boards. If you are worried about foot traffic kicking up a challenge to the bottom of the fence, consider adding a kickboard. Not only will a solid kickboard absorb kicks and bumps with little or no ill effect, it can also add a nice visual element to many fence styles. Also, dogs and other animals will find it more difficult to crawl under a fence with a low kickboard.

Use pressure-treated or decay-resistant wood for the kickboard. A 2x6 would be suitable for most situations. In areas that experience freezing temperatures, keep the bottom of the kickboard at least 2 inches above ground level. Fasten the kickboard to posts and, if possible, the bottom rail as well. Alternatively, you can inset the kickboard by attaching it to a nailer on the bottom of the rail. ■

KICKBOARD ATTACHED TO FACE OF POST

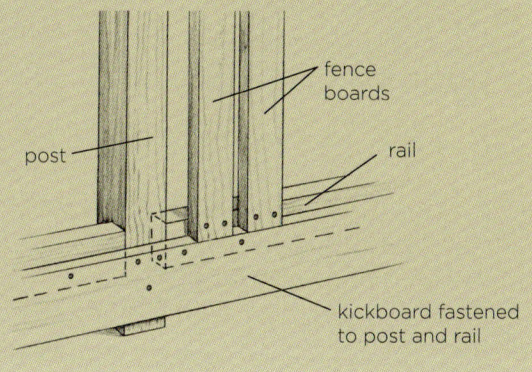

post

fence
boards

rail

kickboard fastened
to post and rail

KICKBOARD INSET BETWEEN POSTS

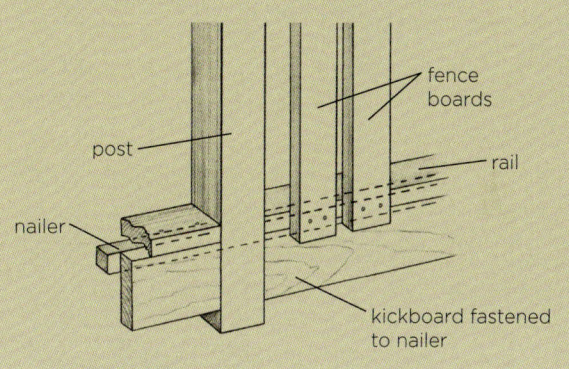

post

fence
boards

rail

nailer

kickboard fastened
to nailer

Lattice Fences

LATTICE PROVIDES A QUICK AND EASY method for screening an area. Standard lattice panels are not very strong, and I would recommend that they be used as infill in a regular fence rather than as freestanding structures. But lattice is ideal for helping enclose and conceal small eyesores, such as garbage cans, propane tanks, or air-conditioning equipment. Small sections of a fence can be filled with lattice, perhaps to break up the monotony of an otherwise solid board fence. You can use the open grid of the lattice as a support for some climbing vines or flowers, which can have a dramatic effect around your yard. And you can create a nice privacy screen by installing a panel of 4×6 lattice between posts spaced 4 feet apart.

Lattice is a particularly nice material for use as a fence topper. As mentioned in Chapter 2, you may be able to exceed the maximum height allowed by local codes for fences if you top the fence with an open material, such as lattice.

You can find lattice panels in various sizes. Lattice can be tricky to cut, so I suggest you try to develop a design that incorporates full panels if at all possible. If that is not possible, see page 115 for some lattice-cutting tips. The most commonly available lattice panels are ½ inch thick, which means that they are made with two layers of ¼-inch-thick slats. A much better choice for fencing is 1-inch-thick lattice, which is composed of ½-inch-thick slats. If you shop around long enough, you will discover that lattice can be found with both large and small openings, allowing you to control the level of privacy it provides. And, while diagonally oriented patterns are most common, lattice with a square orientation creates a very different and often more pleasing effect, at least to my eye. Vinyl lattice is also readily available in several colors, and might be a good choice for a fence topper.

If you cannot find lattice that is strong enough or constructed well enough for your tastes, you can build your own. I have seen homemade lattice created from a variety of approaches. The decision comes down to the size and type of the wood you want to use, the types of woodcutting machinery you have in your shop, and the amount of time you are willing to invest in the effort. You can rip 1x boards (or thinner boards, if you prefer) into strips about 2 inches wide, then fasten them together in a grid pattern of your own choosing. If you can find 2×4 cedar, you can rip those boards into ½-inch-thick pieces. Lumberyards and home improvement centers also usually carry thin pieces of rough-sawn 1×2 strapping that can be used for homemade lattice. Join the pieces with galvanized nails or staples and exterior glue.

The best way to install lattice is between wood stops attached to rails and posts. Install the stops on one side, attach the lattice, and then attach stops on the other side. You may also be able to find channel stock at your lumberyard, which has a groove already cut into it for holding lattice panels in place.

Lattice has been a popular garden accessory for centuries. With the widespread availability and low cost of lattice panels today, it can be used to create an attractive and quickly assembled fence.

Diagonally oriented lattice does not suit every need or taste. In that case, it is worth looking into lattice panels that form square or rectangular grids, which may seem a more appropriate match for the surrounding architecture.

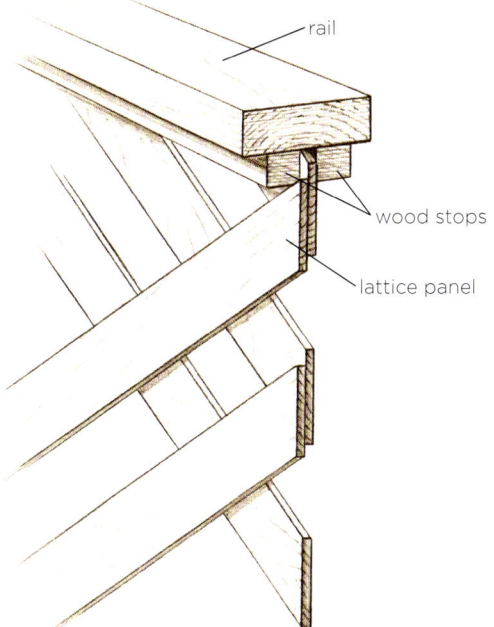

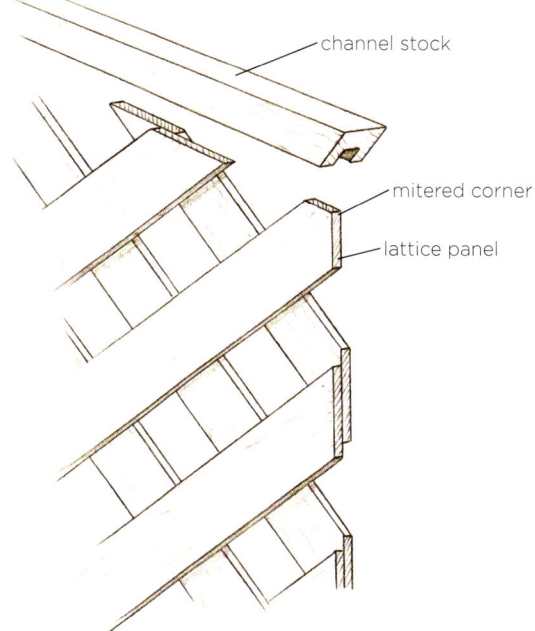

ATTACHING LATTICE WITH WOOD STOPS. Attach the first 1×1 wood stop to the rail, insert the lattice panel, and then attach the second 1×1 wood stop to the rail.

ATTACHING LATTICE WITH CHANNEL STOCK. If you can find it locally, preformed channel stock simplifies lattice installation. Cut the channel stock to fit the panel (mitering the corners), attach it with glue to the panel, then drive nails or screws through the channel stock and into the posts and rails.

Cutting Lattice

I have had decent success cutting lattice with a reciprocating saw and a jigsaw, with the lattice overhanging the edge of a deck or a sawhorse. But I have had the best success using a circular saw and supporting the panel on both sides of the cut line.

Set a sheet of plywood across two sawhorses. Then place the lattice panel on the plywood. Measure and mark your cut line. Then clamp 1×6 boards on top of and underneath the lattice, next to the cut line, so that the lattice is sandwiched in between. Set one or two 1×6 boards beneath the lattice nearby to keep the whole panel level. Set your blade to cut just deep enough to clear the bottom edge of the lattice when your saw is set on the top board, and then make the cut, pushing your saw along the smooth surface of the top board. The boards help keep the lattice from flapping around and separating while you are cutting. ■

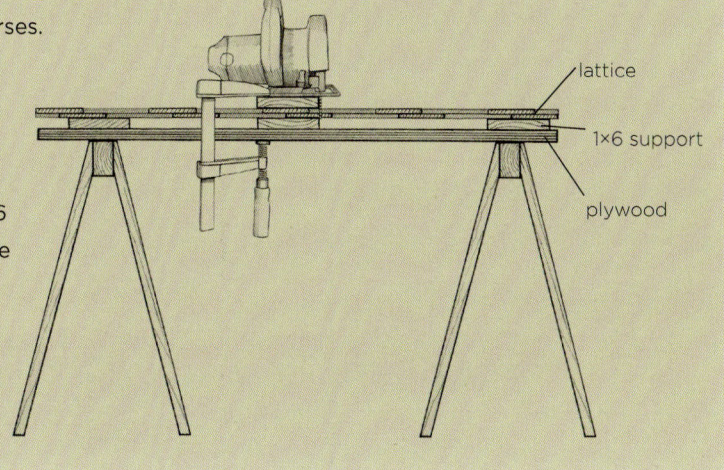

Prefab Panels

A VARIATION ON making your own fence panels is to buy prefabricated panels and attach them to posts you set yourself. You can find fence panels in just about any style you might want, including pickets, at most home improvement centers and many lumberyards. Stockade-type fencing seems to be the most common. However, before you invest your money in any kind of prefab panel, make sure that its rails are adequately sized for the weight of the fence and the span between posts, and make sure the fasteners are of good quality.

Good-quality prefab fence panels are available, but you may have to shop around to find them. Ask a reliable lumber dealer for some suggestions, and do some window-shopping on the internet. I have seen prefab picket and board fences made with nice wood and good fasteners, primed and even painted, that can be delivered to your door. The tricky part with these panels (aside from finding a good supplier) is making sure that they will fit adequately between your posts. If you run into a fit problem with a panel or two, one solution is to attach secondary rails between the posts and then attach the rails on the panel to these secondary rails.

The variety of prefabricated fence panels keeps growing. The quality is not always top-notch, but if you find a product that seems well constructed and fits your needs and budget, it can greatly simplify your construction project.

FASTENING PANELS TO POSTS.

Prefabricated fence panels are designed with extended rails that can be face-nailed to posts **A** or with rails cut flush to the edge of the panel, so that the panel can be slid into metal brackets installed on the posts **B**. If a panel doesn't fit the space you've allocated for it **C**, attach secondary rails to the posts and attach the panel rails to those secondary rails. Fill in any empty space with an extra fence board or picket.

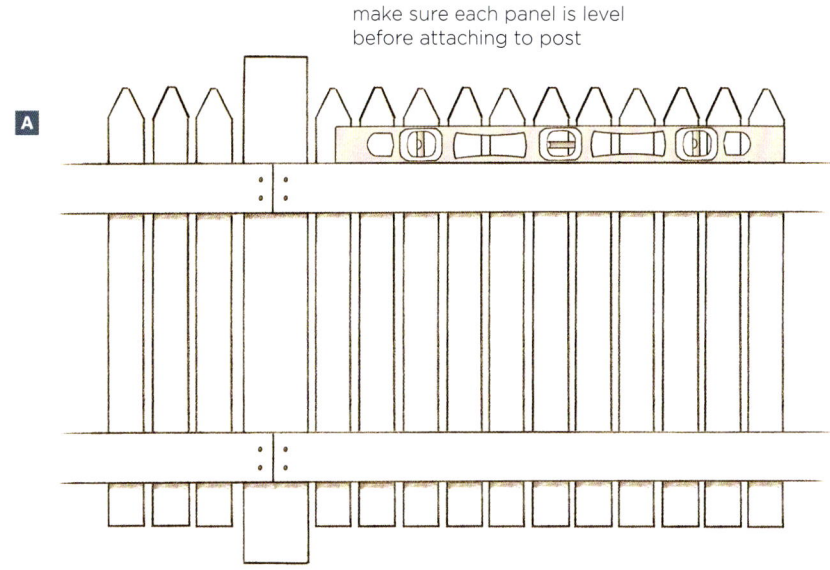

make sure each panel is level before attaching to post

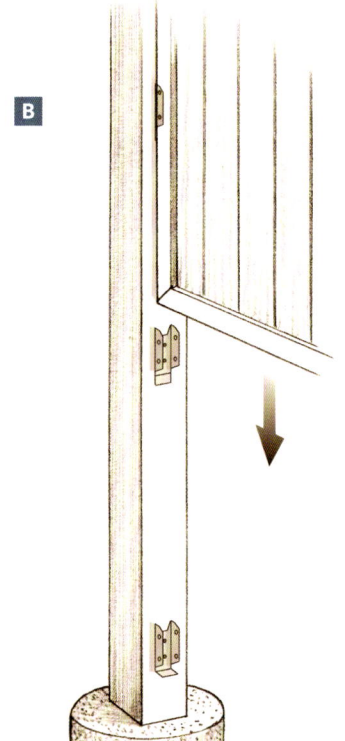

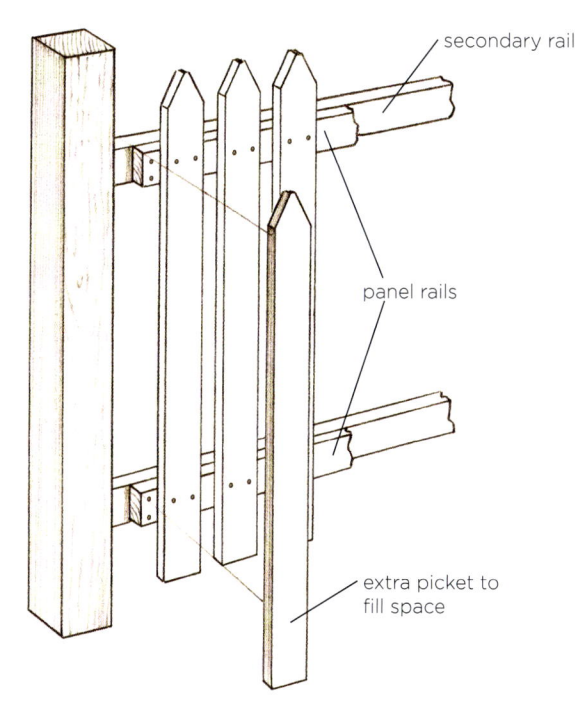

secondary rail

panel rails

extra picket to fill space

Building Fence Sections

In most cases, it is easiest and quickest to build a fence in place: Set all the posts, attach all the rails, and install the pickets or other infill. But there are circumstances when it makes sense to build fence sections—homemade prefab panels—one at a time, then attach whole sections to the posts. If you are building a long fence with evenly spaced posts, and you have a roomy workshop with easy access to and from the yard, you may find it quicker to prepare a jig and assemble the sections. Building fence panels also allows you to work on your fence during bad weather—even through the winter, so that you are ready to finish the project in the spring.

To prepare the jig, you will need a 4×8 sheet of ¾-inch plywood. Lay the plywood flat on a workbench or across two or three sawhorses. Mark the post locations on each end of the plywood; assuming the posts are set on 8-foot centers, you would need to mark only half of each post on each side. Attach a strip of ¼-inch plywood to the top edge of the jig to establish the tops of the pickets. Now mark the intended locations of the rails on the jig, and attach 2×4s to the jig above this line. Set two rails in place temporarily and attach some 2×4 blocks on the other side to hold the rails in place.

The next step is to set guide blocks on the jig to space the pickets. The blocks need to be the exact width of your intended spacing. In most cases, all you need to do is cut up a few spare pickets into blocks and nail the blocks to the jig.

Once the jig is complete, set rails in place, place the pickets on the rails and between the guide blocks, and nail the pickets to the rails. Once this is complete, carefully lift the fence section off the jig and carry it to the fence site. If the section is a bit longer than necessary, it is easy to trim it to size. But you do not want to wind up with a carefully built fence section that is too short. To avoid this potential problem, consider using 10-foot-long rails, with 1 foot overhanging each side of the jig. Once you test-fit the section in the fence, trim the rails as needed and add any extra pickets that are required. ■

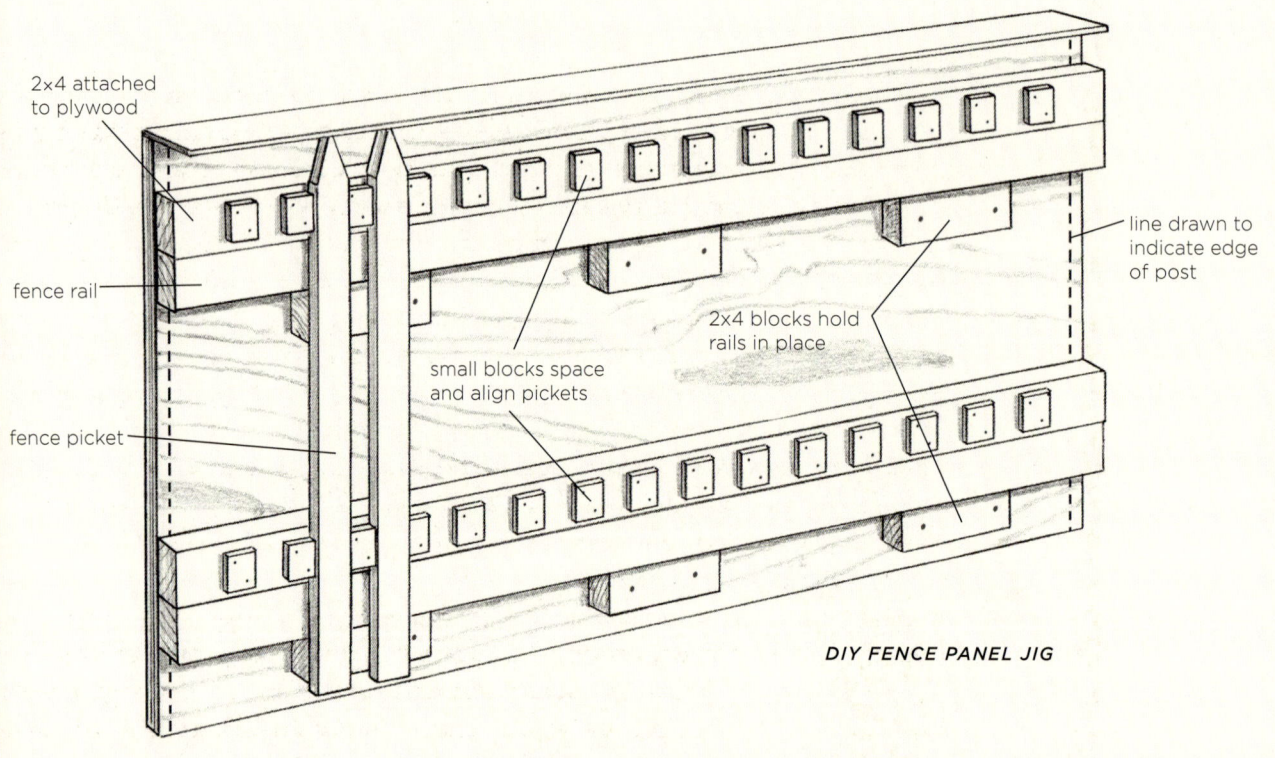

2×4 attached to plywood

fence rail

fence picket

small blocks space and align pickets

2×4 blocks hold rails in place

line drawn to indicate edge of post

DIY FENCE PANEL JIG

Vinyl Fences

Polyvinyl chloride, better known to most of us as PVC, poly, or vinyl, keeps showing up in an ever-growing number of products in and around our houses: flooring, siding, shower curtains, drainage pipes, roof membranes, window frames, and, increasingly, decks and fences. And whether or not vinyl is your cup of tea, in these uses one must admit that it offers strength, ease of installation, long life, low cost, and low maintenance.

Vinyl fences are available in just about any size, shape, or style you can imagine. I would suggest that you start your shopping early, however, as you may need to special-order the specific fence you want. You will almost certainly pay more initially for vinyl over a similar wood fence, but over the long term you may well save time and money in maintenance and repair. Vinyl fences do not have to be painted or stained, they will not rot or rust, and they can be effectively washed with a good rainfall. Finally, unlike with wood, you can expect to receive a warranty (and perhaps a very good warranty) with your vinyl fence. Note, however, that some warranties require that the fence be installed professionally.

White is the most common color choice with vinyl fencing, which is just fine with those fence styles that are often painted white. But a big, solid privacy fence in uninterrupted white can be an eyesore, as far as I am concerned. Although you can also find vinyl fences in gray, beige, and other light colors, dark colors are hard to come by. As I understand it, good-quality vinyl fencing contains titanium dioxide, which inhibits ultraviolet degradation of the vinyl. Because this additive is a white pigment, it limits colors of the finished product to lighter shades.

Vinyl fences will not splinter, and they have no exposed nails that could injure children and pets. Unlike with wood, horses do not care to chew on vinyl. I would have concerns, however, about horse kicks damaging a vinyl fence, especially in cold temperatures, when PVC is more prone to splitting upon impact. If you are looking at vinyl for a horse fence, I suggest that you ask very specific questions of the dealer and that you read the warranty carefully.

Vinyl fences are sold as complete kits that can include posts, post caps, rails, rail brackets, panel sections, gates, and gate hardware. You will receive step-by-step instructions with the kit, which I will not try to second-guess here. Typically, you must set the posts in concrete, allow the concrete to cure, attach brackets to the posts, then add the rails or panel sections. Post caps and other accessories are normally attached with PVC cement.

Though PVC can be made with recycled ingredients, it is itself a difficult product to recycle. And it is a toxic material. Pollutants are released during its manufacture, and toxins are released when it is burned (accidentally or during disposal). Furthermore, I enjoy working with wood; I can't say I find the same pleasure in working with plastic.

An alternative to PVC from within the general family of plastics is high-density polyethylene (HDPE). HDPE fences offer the same benefits as PVC, but they are generally stronger and more durable. The material performs better in extreme temperatures and stands up well to impact. It is available in a much wider range of colors, including dark colors. I have seen HDPE panels that have a woodlike texture, unlike the smooth, glossy appearance of PVC. And HDPE is nontoxic and easily recyclable.

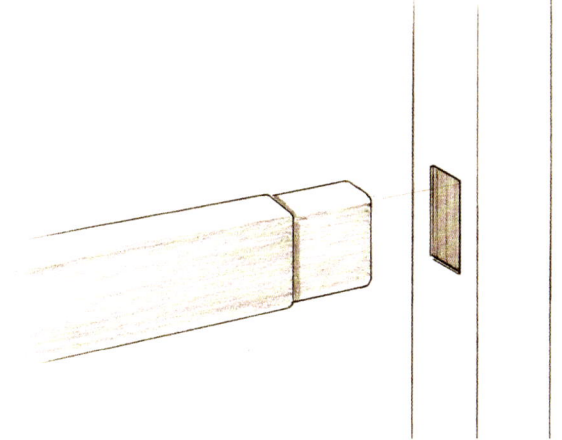

Vinyl fences are becoming increasingly popular as wood-fence look-alikes for homeowners who like their minimal maintenance needs. Constructing a vinyl fence requires only that you buy the right parts and assemble them according to the manufacturer's instructions.

Rustic Fences

THOUGH HARDLY WHAT ONE could call a technical term, *rustic* defines a type of structure that is inspired by age-old techniques and materials, utilizing few tools and little if any industrial-age hardware, and that is permitted to show its wrinkles through natural aging and weathering. Almost any type of fence can achieve a rustic look by being left alone to grow old without refinishing or repair, but the purpose of this section is to suggest fences that are intentionally built to such ends. These are fences that are more about looks and mood than they are about function, although some of them can serve very useful purposes. And, given their low-tech roots, they should not be thought of in the same manner as most other fences discussed in this book. That is, these tend to be structures with relatively short lifespans. I would not suggest one of these fences for someone looking to build a secure and long-lasting fence. With that limitation in mind, however, the following fence styles and materials may inspire you to design and build a perfectly suitable, unique, and perhaps even stunningly attractive fence.

If you have your own woodlot, or have access to someone else's, collecting the materials for a rustic fence can be simple work. Here, tree branches of varying sizes have been assembled into a fence section alongside of a stone pier.

Rusticity Abounds

Almost by definition, a rustic fence will be unique. Tree trunks, branches, and twigs can all be used, individually or collectively, to assemble fences in the same manner in which humans have been building them throughout the world for thousands of years.

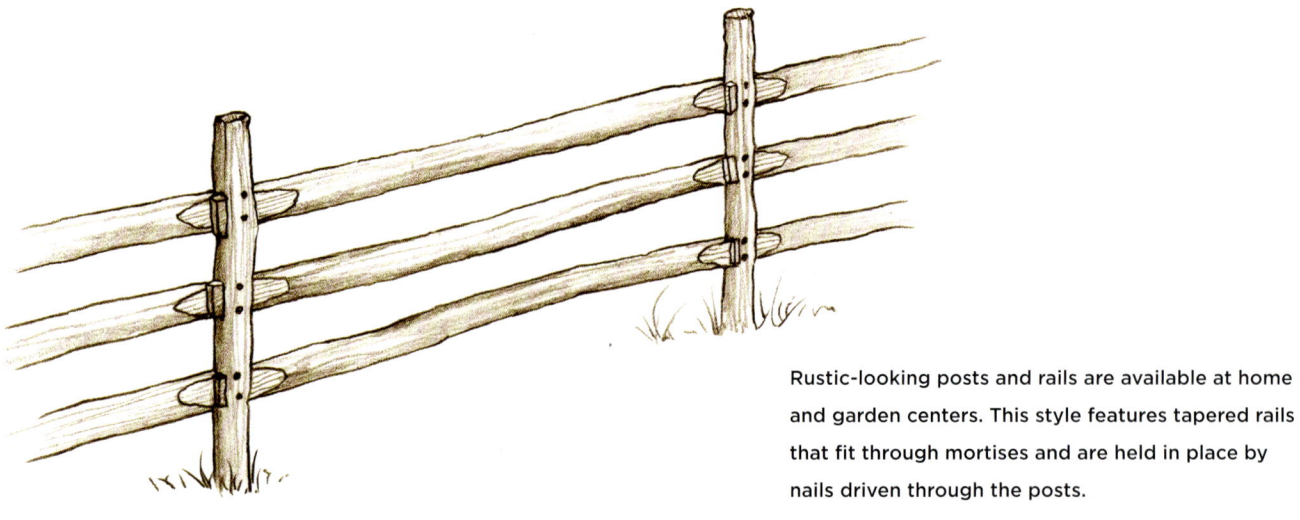

Rustic-looking posts and rails are available at home and garden centers. This style features tapered rails that fit through mortises and are held in place by nails driven through the posts.

Rough-Cut Round-Post Fences

You can achieve the rustic look by using the most unrustic method of all: Hop in your car and head over to the home improvement store. There you should be able to find rough-cut, round posts with mortises already drilled and rails with the ends tenoned or tapered to fit in them. You can even find these products available in pressure-treated lumber, which can help your rustic creation last a long time.

The round posts should be set just like square posts, as described on pages 76–89. If you are inserting rails into mortises, you need to assemble the fence one section at a time to ensure a tight fit for the rails. If the rails are tapered, they are intended to overlap inside of the mortises. For added strength and security, drive a couple of galvanized nails into the post and rails at each mortise.

If you cannot find premortised round posts, it is really not difficult to mortise them yourself. This is easier to do before the posts are set in the ground; however, if you follow that sequence, you do need to be very careful about setting the posts at just the right height to align all of the mortises horizontally. To cut a mortise, mark the outline on both sides of the post. With a spade bit or, preferably, a Forstner bit, drill holes in each side to remove as much wood as possible. Finally, clean out the mortise with a sharp chisel and a hammer.

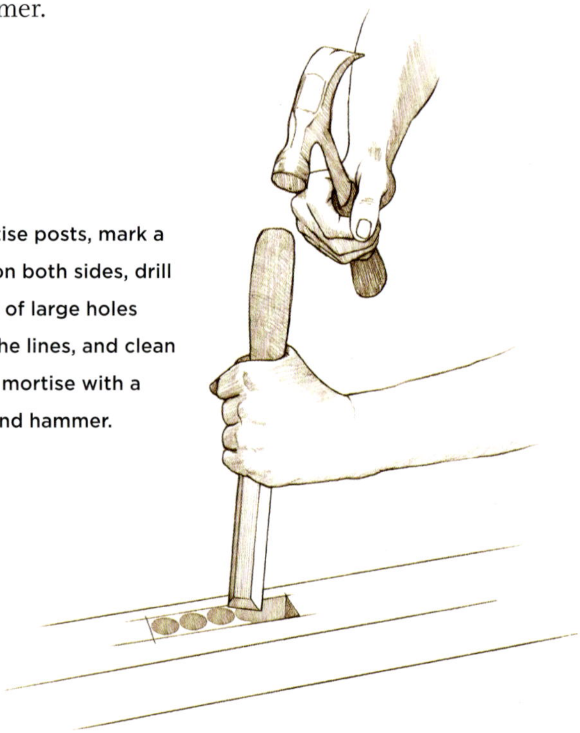

To mortise posts, mark a layout on both sides, drill a series of large holes inside the lines, and clean out the mortise with a chisel and hammer.

Bamboo

Bamboo is a woody grass that was being used as fence material in Japan by the twelfth century. Many of the roughly seven hundred existing species of bamboo are quite rot-resistant, which helps explain its popularity for a host of exterior construction projects. Houses, animal enclosures, and fences of all kinds have been built out of bamboo with no tool other than a machete. Unlike wood trees, bamboo has no bark to remove and its use generates virtually no waste.

Bamboo can be used to create a visually appealing fence, in the right circumstances. It looks great around a Japanese-style garden or enclosing a water garden or koi pond. In the wrong place, though, I think a long wall of bamboo can look silly and monotonous. A wide variety of bamboo products are available, although you may have to hunt the internet and special-order what you need. Many home improvement centers and gardening catalogs carry rolls of bamboo sections

As bamboo becomes easier to find, its use as a fencing material has spread. Whether designed in a familiar picket style [LEFT] or a more open latticelike configuration [RIGHT], bamboo is a great complement to gardens.

bound by wire or nylon cable. The bamboo may be split canes or thinner whole canes. I have seen these products sold in heights ranging from 3 feet to 6 feet.

These rolls of bamboo fencing are easy to carry and transport. They can be attached to wood posts with small nails or wire, and the joints can then be wrapped in twine to create a traditional-looking effect. If you're going to nail the bamboo to wood posts, drill pilot holes first to keep the bamboo canes from splitting. Large bamboo culms (that is, trunks) can be used for posts and rails, but be sure that you are using a rot-resistant variety. You can use the larger canes to create a frame that can be filled with fence sections or that can stand alone as a lattice or meshlike fence structure.

Bamboo fences are not especially given to long lives. A well-constructed fence made with top-quality materials may survive as long as 15 years, but I would not expect a fence made with sections of thin canes to last more than 5 to 7 years, depending on the climate.

Sticks and Twigs

Fences can also be built using twigs of willow and other trees, which can be handled much like thinner bamboo canes. You can even find prefabricated sections of twig fencing similar to those using bamboo, as described on page 123. Fences built with vertical sticks nailed to wood rails can create a charmingly rustic effect in the right situation, such as marking the boundary of a small garden or even as a background screen for a bed of flowers. For a full enclosure, space the sticks closely together and, if desired, attach a band of metal mesh along the bottom to keep animals out.

Some Native Americans built palisade fences by sharpening both ends of logs, driving one end into the ground, and relying on the sharpened tips of the closely spaced logs to discourage surprise attacks by enemies. You can mimic this fence style by using fairly narrow branches upright against a wood rail, joining the pieces with wire or nylon cord.

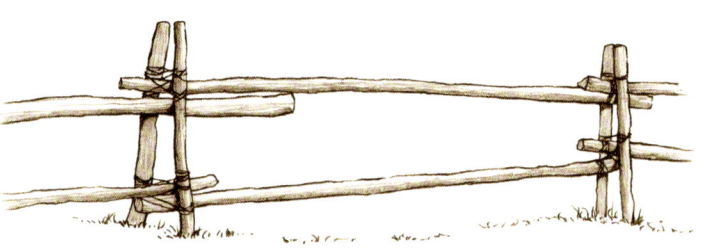

[ABOVE] Another "found wood" fence can be made from longer wood pieces to resemble a post-and-rail fence. In this case, however, the posts do not necessarily have to be buried, and the rails and posts can be tied together with twine or wire.

[BELOW] A random collection of twigs attached to a solid, rustic frame creates a truly unique fence.

Wattle

Rustic by definition, wattle is a style of fencing that has been used over the centuries just about everywhere that trees grow. Nomadic peoples of East Africa have long built portable wattle huts and animal enclosures with branches of acacia trees. More settled African communities built wattle houses that they plastered with clay to provide weatherproofing. Timber-framed houses in medieval Europe often used wattle as infill, woven through slender poles and then covered with mud daub.

The raw material for a wattle fence is composed of nothing more than thin, flexible saplings of willow, hazel, hickory, bamboo, and other plants and trees. Posts are driven or buried in the ground every few feet, then green (that is, recently cut) saplings are woven between the posts in layers. This does not produce a particularly strong fence, nor is it likely to last more than 5 or 10 years. Wattle was a popular style with colonial Americans, who used it as temporary fencing in new settlements until more permanent fences could be built. I think it makes a nice screen to conceal garbage cans or a compost pile (and it's inexpensive, if you have access to the saplings).

[ABOVE] Wattle fences have been built since the first humans met their first trees. Despite that long history, though, even today a well-designed wattle fence will draw envious approval.

[LEFT] Building wattle fences is a bit like combining weaving with construction. With posts buried in the ground a few feet apart, green saplings are woven through the posts in a pattern of your choosing.

Bentwood

Bending wood into decorative and even functional shapes has long been a staple pastime at summer camps and outdoor retreats. Many gardeners and low-tech landscapers use flexible green wood to create supports for tomatoes and climbing beans or to build trellises or decorative panels for the yard and garden. In turn, rustic fences and gates can be built from or accessorized with bentwood forms.

Several useful books have been published on the art of using bentwood techniques to build furniture and other fancy garden structures. I have seen bentwood benches constructed by individual artisans selling for thousands of dollars, which ought to demonstrate that there is plenty of room for acquiring skills and creating innovative designs from this seemingly simple, rustic craft. A brief but useful introduction to bending wood is a book by Jim Long, *Making Bentwood Trellises, Arbors, Gates, and Fences* (Storey, 1998). Long offers suggestions on the types of wood to use, where to find it, and how to shape and join it. The book also contains step-by-step instructions for building dozens of structures, including some fences and gates. If the subject intrigues you, I recommend that you take a look at Long's book.

One of the joys of bending wood is that once you learn the basics of wood selection and joinery, you can produce truly unique projects quickly and at little, if any, cost. Entire fence panels can be created with arches and elaborately interwoven limbs and saplings,

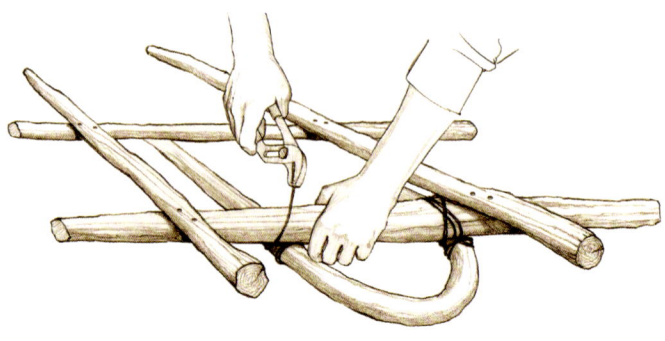

Use saplings that are at least 1 inch thick to build the frame for a bentwood fence. Then bend green saplings and attach them to the frame with galvanized wire.

For a stronger and more secure type of bentwood fence, attach wire mesh to buried fence posts. Then weave saplings through the mesh and tie them in place with galvanized wire.

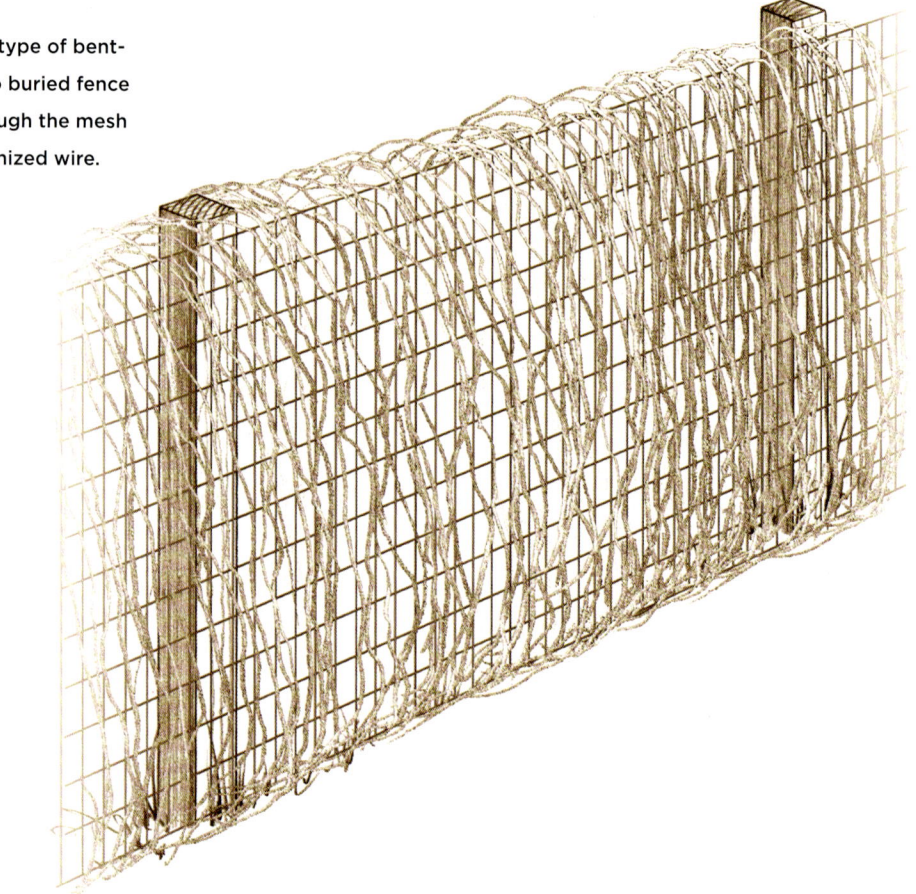

Modern Tools for Rustic Joints

Rustic structures are usually put together with nails or wire, or both. For a neater, and potentially more secure connection, however, you might want to invest in a tenon cutter or two, along with matching brad-point or Forstner drill bits. The tenon cutter is powered by an electric drill and can produce a large number of equally sized round tenons on limbs and saplings very quickly. The tenons can then be fit into holes drilled into mating posts or rails. Wood dowels

or brass pins are often driven through the joint to hold it together, although you can also use a slow-setting epoxy glue.

Lee Valley Tools (www.leevalley.com; 800-871-8158) sells Veritas Power Tenon Cutters that can cut tenons in diameters ranging from ¼ inch to 2 inches. The larger sizes are best suited to fence and gate construction, but note that they do require a drill with a ½-inch or larger chuck. ■

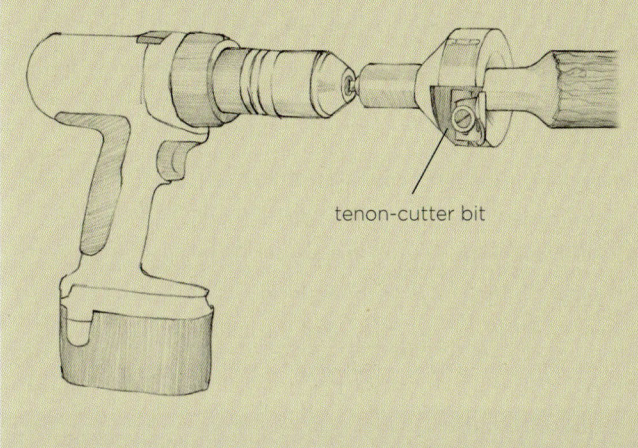

tenon-cutter bit

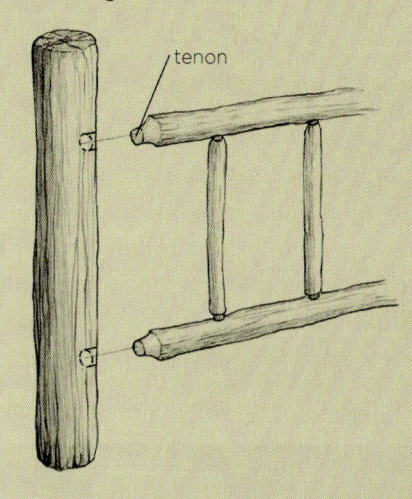

tenon

but a simpler approach is to build a barebones stick fence as described above and then add a few arches for effect. Or plan a fancy arched bentwood gate to complement a basic rustic fence. Use nails or wire to join the pieces.

Reading about the subject, however, is no substitute for experimentation. Start gathering flexible limbs and saplings and shaping them into designs you like, and you will likely accumulate all the knowledge you need for adapting the approach to your fencing needs. To construct a basic fence panel or gate, begin by building a rectangular frame using saplings of uniform thickness of an inch or more. With the frame joined by nails, use freshly cut branches (willow and dogwood are particularly good for this purpose) to form into shapes you like. Wood for bending needs to

be green and flexible. If you collect it too soon and allow it to sit around for a few days before use, it will be more likely to break or split.

Bentwood projects are frequently built with the bark left on the wood, but you can create even more unusual-looking (and longer-lasting) designs by removing the bark and staining the wood before you build the fence or gate. Removing bark is time consuming; it is easiest to peel in the spring.

I think bentwood and humdrum wire-mesh fencing are a natural combination. You can quickly and inexpensively build a wire-mesh fence to suit an immediate need, and then, as time permits, start adding bentwood to transform the look of the fence. Weave the wood through the mesh, then secure the connections with wire.

Choosing a Finish

THE FIRST OPTION IN FINISHING WOOD is to let it be. Outdoors, and left untreated, all wood will eventually turn a weathered gray. Dark wood will lighten, and light wood will darken, resulting in a uniform color over time. Ultraviolet radiation will cause the surface to deteriorate. Sand and dirt will scrape away at this weakened layer. Moisture in the wood promotes the growth of fungi; and termites, carpenter ants, and powder-post beetles can turn solid wood into a weakened material more closely resembling Swiss cheese. Mildew will likely appear. The grain of the wood will turn fuzzy. Boards will check, cup, warp, and split. And that, you may well say, is just as it should be. Nobody is going to force you to stop or slow down this natural aging process. I am certainly not going to tell you that your fence must be built and finished to ensure that it survives for generations to come. There are no moral or religious doctrines that I know of that should stop you from believing that aging is beautiful and that outdoor wood projects will not last forever.

On the other hand, one could find reasonable ethical arguments to be made for designing your fence and choosing materials with the goal of minimizing or eliminating the need for toxic treatments or finishes. You might wish to take some inspiration from the English "environmental artist" Andy Goldsworthy. Perhaps inspired by kids and their sandcastles on the beach, Goldsworthy is known for his works created with natural materials (stone, wood, and water) that are intended to wash away or melt shortly after they come to life. Not that I would expect anyone to build a fence with such an abbreviated life, but in many environments, untreated wood (even wood not known as decay-resistant) can last for many years.

To get the longest possible life out of your wood fence, apply a suitable finish immediately after it is built and renew the finish every few years. If you are willing to sacrifice some longevity for a more "natural" look [BELOW], don't bother with a finish, but do try to keep the unfinished sections of the fence away from direct contact with the ground.

There are houses, barns, and other buildings covered with naturally weathered wood that have survived for well over a century. If you want to follow this course of (in)action, keep an eye on developing problems, such as rotting posts, with the intention of making regular repairs, and you may sleep just fine knowing that your project is contributing to Mother Nature's divine composting plan rather than to the spread of toxins and foreign materials.

Most of us are a bit less altruistic, however, when we are dealing with significant investments in our time and money. Fence builders have long understood that they could substantially prolong the life of a fence with the application of coatings of one form or another. The 1892 manual *Fences, Gates, and Bridges* recommends that post bottoms be "given a good coat of boiled linseed oil and pulverized charcoal, mixed to the consistency of ordinary paint" before being set in the ground (p. 25). It further suggests that the finished fence be primed with a coat of crude petroleum, followed by paint.

Contrary to some misconceptions, using pressure-treated wood does not excuse you from needing to apply a finish. The standard chemical treatments used on the wood provide deep and effective protection against insects and fungi, but not against damage caused by sun and water. Even premium pressure-treated wood, with a water repellent already applied, needs to have its finish renewed from time to time, and would certainly benefit from a finish intended to provide protection from ultraviolet rays.

Finishes actually serve two separate functions: decoration and preservation. To serve the former, a finish must offer a color or colors that contribute to the landscape and to your design objectives. Preservation is accomplished when the finish addresses all of the physical and chemical abuses the wood is likely to encounter. Roughly speaking, exterior wood finishes fall into two categories: film-forming and penetrating. Penetrating finishes soak through the surface of the wood to provide deep protection, while film-forming finishes create a protective barrier on the surface to keep problems from even reaching the wood itself. Either type of finish can be used on a fence, with different visual effects and different levels of protection and longevity.

Penetrating Finishes

Penetrating finishes allow the wood to breathe and the wood grain to show through. Since they sink into the wood rather than forming a surface film, they do not peel or chip as they age. Wood preservatives, water repellents, and semi-transparent stains are the major types of penetrating finishes.

Clear penetrating finishes are widely available and affordable. They offer you the chance to protect the wood while retaining its original, natural color. Basic clear wood finishes provide protection against water damage only. They are the least durable choices and need to be reapplied every year or two. Water-repellent preservatives containing mildewcide, ultraviolet (UV) stabilizers, and other decay-fighting ingredients provide extra protection.

Due to the addition of pigments, semi-transparent stains provide significantly more protection against sun damage than clear finishes, including those with UV stabilizers. I think that oil-based semi-transparent stain mixed with a mildewcide is the best choice for a fence, unless you want to paint. A wide range of colors is available, and I have seen some wonderful wood fences that have been finished in several different colors. You can choose a stain that will give your fence a weathered gray appearance right away, or you can mimic the deep color of redwood with a stain applied to a less attractive type of wood. Shades of blue, green, and brown are common semi-transparent stain choices. Bleaching oils are similar to weathering stains except that they contain a bleaching ingredient that creates a silvery gray wood surface in a matter of months. You can usually count on getting two to five years of service out of a semi-transparent stain before it needs to be renewed.

Film-Forming Finishes

Both paints and solid-color (or opaque) stains are film-forming finishes that seal wood with a protective layer on its surface. Solid-color stains contain more pigment than semi-transparent stains but less than paint. Lacquer, urethane, and shellac are also film-forming finishes, but they should not be used on outdoor structures.

Because it does not allow the wood to breathe, the surface of a film-forming finish is prone to cracking and peeling. Unlike with penetrating finishes, when a film-forming finish begins to fail, it is immediately apparent and can be a very noticeable eyesore. This problem is particularly true of painted surfaces. When stained wood begins to fail, visible peeling is less common. And when it comes time to recoat, stained wood generally does not require as much preparation as paint: The surface needs to be cleaned, but usually not scraped or stripped. Solid-color stains can last anywhere from three to six years before needing to be recoated.

For longevity, nothing beats a properly applied paint job, which should be good for five to eight years before needing another coating. Note, however, that some types of wood do not really excel at holding paint. Douglas fir, ponderosa pine, and Southern pine are three such species, each of which is used for pres-sure treatment. Rough-sawn boards or posts of these woods will do a better job of holding paint than smooth-planed products. If you want to paint the latter, be sure that the moisture content of the wood is low (summer, following a long dry spell, is best) and that the surface has been lightly sanded to roughen it up a bit.

If you are going to paint, I recommend that you first apply a coat of paintable water-repellent preservative, followed by a coat of stain-blocking latex primer and then two coats of 100 percent acrylic latex paint. Although it takes more time, I like to apply the water-repellent preservative, primer, and first coat of paint before I even assemble the fence. That way, every surface is protected. The second coat of paint can be applied after the fence is completed. Read the labels carefully when shopping for a water-repellent preservative, however, as most products are not suitable for painting.

Bright-colored fences are lively additions to a flower garden. Do as much of the preparation and painting as possible before assembling the fence.

Applying the Finish

One of the worst pieces of advice you are likely to hear regarding finishes is to wait for several weeks or even months before applying the first finish to bare wood. The wait-to-paint rule applies in two circumstances: If you build with freshly cut, unseasoned (that is, green) lumber, then you would want to wait for the moisture content to decline. And if you used lumber with a factory-applied water repellent, you may need to wait a couple of months before finishing (check with the dealer or manufacturer for instructions). But with standard lumber, the minute you expose it to the sun it starts suffering from ultraviolet degradation. If allowed to weather unfinished for even a few weeks, the surface may be damaged enough to shorten the lifespan of a paint job. The most important coat of finish you will ever apply to your fence is the first coat, and you want to get it on as quickly as possible. As soon as the fence is built, apply the finish, as long as the wood is dry to the touch (if it has been rained on, let it dry out first). As a general rule, you can tell if the wood is ready to receive a finish if it quickly soaks up water sprinkled on its surface.

Finish can be applied to a solid board fence most quickly with a paint sprayer, roller, or pump sprayer. Brushing is more efficient, however, especially with that important first coat, as the brush more effectively works the finish into the wood. I would only use a paint sprayer on a solid fence; even if you tried to use one on a fence of thinly spaced pickets or boards, the amount of paint that would be wasted (and, as a consequence, be coating the ground beyond) would be substantial. You can buy pump sprayers at home centers and paint stores, but they are really only effective with relatively thin penetrating finishes. Even if you use a sprayer as your primary applicator, you will still need a brush for corners and edges and hard-to-reach areas. For oil-based stain or paint, use a natural-bristle brush for best results. For waterborne (latex) finishes, use a synthetic-bristle brush.

Smooth-planed lumber has a slight glaze on the surface that can inhibit the absorption of finish. So, before applying the first coat to smooth lumber, give it a light sanding. Rough-sawn lumber does not need sanding. All wood should be cleaned of any dirt and allowed to dry. You may want to go over the whole fence with a stiff brush to remove surface debris, and it wouldn't hurt to use a shop vacuum to clean out the nooks and crannies.

Apply a penetrating finish as directed on the label. Note that some products may strongly encourage you to apply two or three coats, while others (especially with some semi-transparent stains) may recommend only one coat.

When using solid-color stains, I recommend that you first prime the wood, and then apply two coats of stain. Take care to avoid creating lap marks, which can show up as dark splotches on the wood. Lap marks occur when wet stain is applied over a section of wood that already has a layer of dried stain on it. The best way to avoid this is to work in the shade as much as possible, and work on small sections at a time.

If you are painting the fence, and you want to make the paint job last as long as possible, apply a coat of paintable water-repellent preservative first. Let that coat dry in warm, sunny weather for a couple of days, and then prime with stain-blocking latex primer. If the wood contains a fair number of knots, make sure that your primer explicitly promises to block the extractives in the knots from bleeding through. If it doesn't, you may well find your nice painted finish disrupted by round, brown stains at the knot locations. Finish the job with two coats of 100 percent acrylic latex paint. That second coat of paint will probably buy you several more years of service before you need to repaint. Allow the paint to dry thoroughly before applying another coat.

Masonry Fences

UNLIKE WOOD FENCES, masonry fences do not need to be protected from the sun, moisture, and fungi. Whether built with stone, brick, or concrete, a typical masonry fence will have the durability of masonry walls used to envelop a house. When properly constructed, therefore, strength and security are givens, but masonry fences can really shine as landscaping elements when they assume more modest functions, as low boundary markers or garden enclosures.

hat's in a name? Build a 3-foot-high boundary marker out of wood, and it's called a fence. Build the same thing out of stone or brick, and it's more likely to be called a wall. This chapter is devoted to those structures that fall into the second category, and I assume that few if any readers will mind the terminological shift that is required in the process. "Stone fence" is not an unheard-of expression, but I can imagine some people asking themselves, "I wonder what the difference is between a stone fence and a stone wall?" To avoid such confusion, I will therefore call these fencelike structures walls.

[ABOVE] A uniform brick cap provides a neat top to this color-coordinated wall of mortared flagstone.

[OPPOSITE] Stone walls cannot be designed with the same predictability as brick walls. Each piece must be carefully selected to fill the available space, and this picking and choosing takes time. Few would argue, however, that the end result was not worth the effort.

Building with Stone

STONE HAS BEEN USED as a building and fencing material for as long as our species has roamed the earth. Think of those still-existing structures and monuments built thousands of years ago (Machu Picchu, Stonehenge, Egyptian pyramids, Greek Acropolis), and you will note that all of them are composed of stone. This may be a big reason why the ancient craft of stonemasonry is in such decline: When a good stonemason builds a stone wall, it will likely remain standing for centuries unless someone purposely removes it. There's just not much repeat business. Imagine if Detroit built cars with that attitude.

With stone fences or walls there are two basic approaches to construction: building with mortar, and building without mortar (dry stacking). Dry-stacked walls rely on gravity and friction alone to maintain their shape and integrity. But that is enough to ensure a very long life if they are built well. Unfortunately, it is the nature of a stone wall not to remain undisturbed. People, animals, soccer balls, and yard and farm equipment bump into them, sit on them, climb over them, walk across them, and pick away at them with unpredictable regularity. Children, being children, are prone to remove stones from an existing wall in order to create their own new structure nearby. Vibration from nearby traffic and frost heave will also cause problems. As a result, dry-stacked stone walls eventually fall apart. With those warnings aside, however, I must confess to great fondness for the look of a well-constructed fence built with native stone and no mortar. Dry-stacked walls also happen to be easier to build than mortared walls.

But mortared walls have their place. Along heavily traveled pathways they will certainly provide greater durability. In some areas, building codes require walls above a specified height to be mortared. But a mortared stone wall is much less flexible than a dry-stacked wall, which means that you need to be much more concerned about frost heave and a sturdy footing. Which is another way of saying that mortared walls require a lot more work.

Both types of walls require maintenance. The stones in a dry-stacked wall often have to be reset, and sections may have to be rearranged from time to time. But this work is generally much easier than having to repoint the joints in a mortared wall.

WET VERSUS DRY STONEWORK. Mortar initially creates a stronger structure, but as it ages it leaves behind a weaker—and uglier—wall [TOP] than a dry-stacked wall [BOTTOM].

Location, Location, Location

If you want your stone wall to last for generations, I suggest that you keep it at least 6 feet from any tree trunks. The roots of growing trees have a nasty habit of stretching out from the trunk, and if you build a stone wall over those roots, they are liable to do some serious damage over time. If you are concerned about roots of nearby trees, consider digging a deeper footing, which will allow you to find and cut out any potential intruders. ▪

Stone Hunting and Gathering

I think it's fair to guess that stone walls came into being not because people thought they looked good, but rather because there were so many loose stones lying about in fields intended for planting that something had to be done with them. Stacking stones alongside the fields was (and still is) one of the byproducts of growing crops. Ask farmers and gardeners in New England and other places where to find stones, and they are apt to tell you that they grow their own. Stones seem to pop out of cleared fields with great regularity in my neck of the woods. Those glaciers that passed through the neighborhood thousands of years ago carved up the landscape in my Finger Lakes homeland into gorgeous lakes and gorges, but they also left behind an abundance of stone.

Petrologists identify three primary categories of stone, which are defined by the manner in which they were formed. Igneous rock was formed by solidified molten material and is very hard and dense. Granite and basalt are two types of igneous rock used for construction. Sedimentary rock results from the accumulation of layer upon layer of the earth's debris (sand, mud, and small stones) that over time compresses into solid rock. This process creates a layered texture to the rock, as can be seen in such common varieties as sandstone, limestone, and slate. Metamorphic rock results from the transformation of one type of stone into another through pressure, heat, or chemical action. Marble, which is formed from limestone, is a good example.

Patterns in Stone

No two stone walls are alike, as the examples on this page can attest. From dry-stacked, thin layers of flagstone to weathered, round fieldstones joined with mortar, with multiple variations in between, the differences in pattern, color, and texture are instantly apparent.

The simple fact that so much stone exists, however, does not necessarily mean that it is easy for you to get your hands on a sufficient supply. The use of stone in construction and landscaping has grown enormously in recent decades, and many good sources have dried up. The stones that used to line farm fields or that were left lying around after the house or barn they supported rotted or burned away have been snatched up in many areas for new uses. You may still be able to find such sources in your area, but it may take some work. I am in the process of dismantling three old outbuildings on my property, each of which was built on substantial stone footings. Stones that were used in footings are usually flat and large, making them ideal for use in a wall. If you can find someone with dilapidated structures on their property, talk to them. Maybe they are planning to tear the structures down and would happily let you cart away the stones. New construction projects can also be a source of stone, but be sure to get permission from the owner or general contractor before you start poking around. If you gather stones in the wild, be aware that any number of biting and stinging creatures may already be using as shelter the stone you are about to grab. You may want to turn stones over with a long stick before bending down to pick them up. Also, do not assume that you have the legal right to remove stones from any and all public places, even if they are remote and rural.

You can also buy stone. A good dealer may carry a large supply of different types of stone in different colors and shapes and with varying characteristics. Although the price may be high, you can find good quality quickly and you should be able to arrange for easy delivery right to your building site. This latter option should not be underestimated, as moving stone is the hardest part of the construction process. It is also what drives up the cost of stone that your dealer imports from other areas. If you stick with native

FIELDSTONE

FLAGSTONE

ASHLAR

RUBBLE

stone, the price will be easier to handle and the stone itself will look at home in its natural surroundings.

Stone is usually sold by the ton, which makes it a little difficult to know how much to buy. A good stone dealer should be able to help you estimate the quantity you will need, as long as you have an idea of the size of wall you plan to build. For example, a 50-foot-long wall that is 3 feet high and 2 feet deep will consume a total of about 300 cubic feet of space. With a little experimentation at the stone yard, you should be able to estimate roughly how many cubic feet a certain amount of weight of the stone you are planning to buy will take up.

Stone dealers tend to have their own way of categorizing their merchandise, which has less to do with how it was formed than it does with how it is used. Rubble, as the name implies, is rough and irregular. It is a byproduct of cutting and blasting stone in quarries. This action tends to create at least one relatively finished side, which makes rubble useful in wall construction. Smaller rubble can be used as fill for the inside of a stone wall. Flagstone comes from quarries, where it is cut into flat slabs of a reasonably consistent thickness but with irregular edges. Limestone and sandstone are commonly cut into flagstone and used for walkways, paving, and decorative walls. Ashlar is also a quarried stone product, only it is cut into more consistent rectangular blocks. Mortared walls are best built with ashlar, because it gives the mortar a smooth, flat surface with which to mate. Fieldstone, technically speaking, is a form of rubble. Rather than originating in a quarry, however, it is gathered in fields and from dried-up riverbeds (which explains why it is sometimes also called river rock). Fieldstone tends to be weathered, irregularly shaped, and rounded. Because it is collected in nature, fieldstone is the best choice for building traditional, rustic stone walls. On the other hand, because of its lack of uniformity, it can also be the most difficult type of stone to build with.

I suggest that you not get too caught up in fussing over which type of stone or stones to use for your wall. While people do brag about their polished granite countertops, I never hear anyone doing the same about their granite walls. In most cases, simply select stones that are native to your locale. In my opinion, the best stones are those that fit, look good, cost little

or nothing, and don't have to be moved very far. Many stone walls are built with a combination of stone types and varieties. With a little luck, you may be able to find a source of free or nearly free fieldstone, which you could then supplement with a purchase of ashlar. Just try to match the type and color of stone. And also be sure to get a good mix of large and small stones.

Tools of the Trade

For cutting stone, you will need a few specialized hand tools. A heavy stone hammer with a blunt edge will be useful for large stones, while a smaller mason's hammer with a long blade and sharp edge will work better for small cuts and small stones. A mason's chisel with a 2½- to 3-inch-wide blade is needed to score, cut, and trim stone, brick, or concrete block. (Note that a brick set, or brick chisel, looks almost identical to a mason's chisel but does not have as strong of an edge.) A small sledgehammer (3 pounds or so) is needed to strike the chisel, while an 8- or 10-pound sledgehammer comes in very handy for quickly breaking very large stones into smaller, more manageable pieces.

Wear steel-toed boots on the job. Even if you do not want to invest in new footwear, do wear shoes that offer more resistance to falling objects than sneakers or sandals. And to protect your hands, wear leather work gloves while hoisting stone or cutting it with hand tools.

If you plan to use mortar, you will need tools and equipment for mixing and applying it. A mortar tub or wheelbarrow, mortar hoe, and shovel will take care of the former, while large and small mason's trowels, a margin trowel, pointing tool, and wire brush will suffice for the latter. Rubber gloves and eye protection are also advised.

Eye Protection

When cutting stone, brick, or concrete block with power tools or hand tools, always wear eye protection. Regular glasses are better than nothing, but a pair of good safety glasses or goggles is a better form of insurance against eye injury. ■

Cutting Stone

Most types of stone construction require some stone cutting. Careful stone selection as you build your wall may minimize the need, but you will almost certainly still have to do some trimming to keep the faces and edges of the wall aligned. Stone cutting can be an unpredictable part of the job, especially for a beginner. Each stone is different, with veins and weak spots often lurking in hard-to-determine locations.

Different stone-cutting tools produce different results, and you will want to experiment and practice before trying to execute any precision cuts. Likewise, different types of stone cut differently. The most common types of stone for building walls (limestone, sandstone, and granite) can all be trimmed and split with basic hand tools. Limestone is the easiest of the group to shape, while granite is the hardest. Slate is very easy to split into thin layers, but it has a tendency to break apart when you try to trim it.

Stonemasons often use a solid work table for cutting, but you may find the ground to be firm enough. If you want to build a table, frame it with 2×4s, and then attach a piece of ¾-inch plywood to the top. Attach 2×4 or 4×4 legs to the frame. The size and height of the table are up to you, but I suggest keeping it small (say, 2 feet by 3 feet) and low (2 to 3 feet high) so that it can be moved easily. I have also had good luck splitting stones on sand, which does a nice job of absorbing some of the impact. To keep it portable, pour some sand into a sturdy wheelbarrow or into a mortar tub set inside your garden cart.

Square-cut, mortared stone here creates a simple but unique garden background.

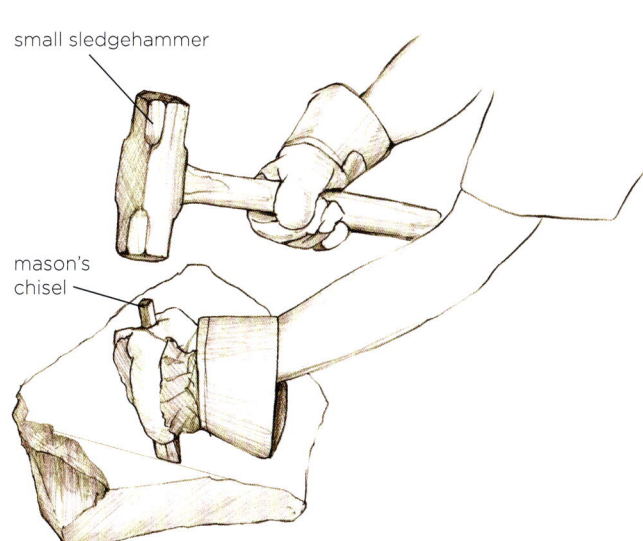

small sledgehammer

mason's chisel

stone hammer

SCORING STONE ALONG CUT LINE

SPLITTING STONE INTO THINNER PIECES

To trim a stone, use a mason's chisel and small sledgehammer. I recommend that you mark the cut line on all four sides of the stone with a piece of chalk. Begin by scoring (that is, making a surface fracture) all along the cut line. Hold the chisel upright, strike it with the hammer, then move the chisel an inch or so and repeat. Once you have created a groove around the stone, keep striking and moving the chisel. The biggest mistake is to try to force the cut prematurely. If you work patiently, the fracture will extend deeper and deeper into the stone until it meets the fracture from the opposing side. When the stone finally breaks, you may need to trim off some rough edges with a mason's hammer or a small chisel.

With some types of stone, you can speed up the process by scoring around the stone and then whacking the scored line with a sledgehammer. With practice, you will learn which approach works best for which types of stone.

Stones can be split into thinner pieces in much the same way, and sometimes much more easily. Align the chisel with a straight vein in the stone, and then work your way along the vein on all four sides until the stone splits. If you are splitting a fairly large stone, you may find it easier to use the pointed edge of a stone hammer instead of the chisel.

For even greater accuracy, you can use a circular saw to score the two long sides of the stone. For best results, however, you will need to buy a fairly expensive dry-cutting diamond blade. Less expensive abrasive blades are intended for softer masonry materials such as brick and concrete block. When cutting stone with a blade, you will create a lot of dust, so I suggest wearing a dust mask as well as eye protection. I do not recommend wearing gloves when using power tools, however. Make a couple of ¼-inch-deep cuts on one side along the cut line, then flip the stone over and do the same on the other side. Now use a chisel and hammer to finish the job, as described above.

Layout

String lines are almost mandatory to keep stone walls lined up both vertically and horizontally. Bright-colored mason's string is easy to see (for yourself and, more importantly, for anyone else who wanders nearby), and it can be tied to thin posts or stakes driven into the ground at the ends of the wall line.

For tapered walls, which are necessary for dry stacking, I suggest running two string lines on each side of the wall, one near the base to establish the bottom thickness, and another at the intended top, inset slightly from the bottom lines, to guide your tapering effort. You may find it easier to do what many professional stonemasons do, which is to build simple batter guides, one for each end of the wall (*batter* is a term masons sometimes use to define a wall's taper). Use 1×4s, or similar-sized lumber, and attach the horizontal boards at the intended thickness of the top and bottom of the wall. Fasten the guide to stakes or braces on each end, then run string lines between them. If you find that the top strings get in your way while working, remove them until the wall is about half of its intended height. But do visually check to see that you are following the proper taper by sighting down the wall toward the guide on each side.

For nontapered mortared walls, set up string lines for both faces of the wall and use a level to keep the faces lined up properly.

Dry-Stacked Walls

Anyone can stack stones, right? Well, yes and no. There is an art to building good dry stone walls, and most of that art is learned through on-the-job training.

For dry-stacked walls, the flatter the stones, the better. I do not recommend trying to dry stack round stones, as they are just too hard to keep in place. If all you have are round stones, shape them into having flat tops and bottoms before setting them in the wall.

Size and Shape

The thickness of your wall depends on its ultimate height. Experienced stonemasons tend to follow their own rules of thumb on these matters. Some use a 3-to-2 ratio: For every 3 feet in height, they make the wall 2 feet thick. Others try to make the bottom layer as thick as the wall will be high. Depending on which approach you follow, then, the base for a 4-foot-high wall would be either 32 inches or 48 inches thick. Either way, that's a pretty thick wall, and for beginners I would recommend that you aim more for the latter than the former. A thicker base will produce a stronger wall that is somewhat more forgiving of minor flaws in construction.

I emphasize the base here because a dry-stacked stone wall should have a slight taper to its profile of about 1 to 2 inches on each side for every 12 inches in height. Thus, a 4-foot-high wall would measure about

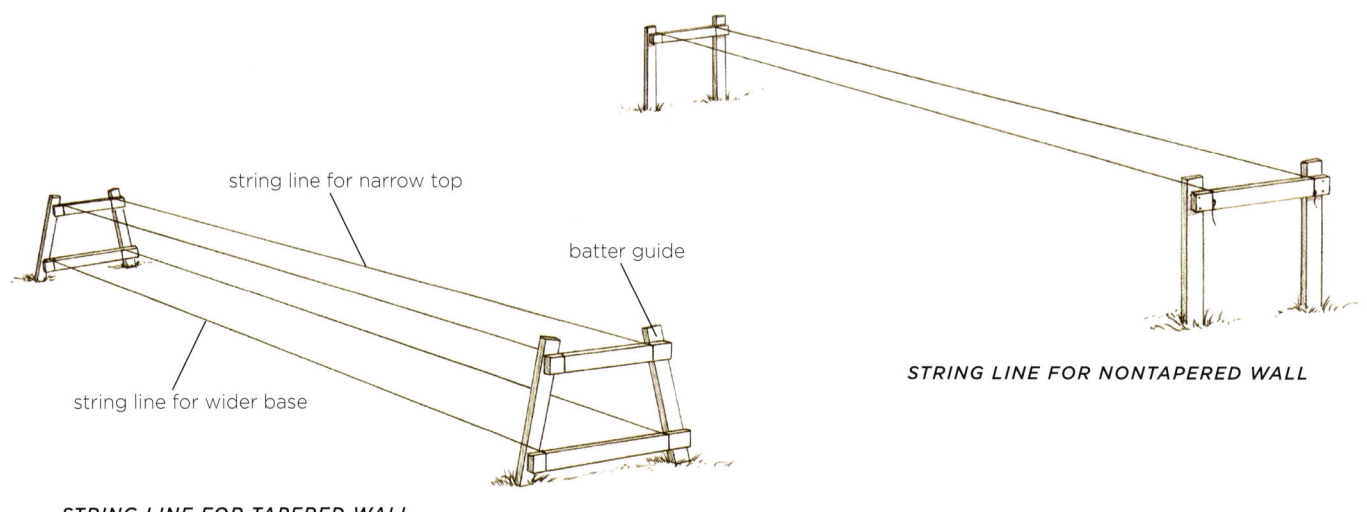

string line for narrow top

batter guide

string line for wider base

STRING LINE FOR TAPERED WALL

STRING LINE FOR NONTAPERED WALL

8 to 16 inches thinner at the top than at the base. The flatter the stones, the less you need to taper them. If you are looking to gain a little experience with your first stone wall, and you are using fieldstone, I would recommend that you build it 3 feet high, with a base of 30 to 36 inches, tapering to a thickness of at least 6 inches less at the top.

The Footing

Many stone walls have been built at ground level, with no footing at all. I think that's foolish. The work needed to construct a good footing for a normal dry-stacked wall is minimal, and by giving the wall a good footing you will add strength and years to it.

There is no harm in digging to the frost line for the footing, but it really is not necessary for most walls. For a typical 3- to 4-foot-high wall, I suggest that you dig a trench about 10 to 12 inches deep. Fill the trench with crushed rock until it is just a couple of inches shy of ground level, and then set the stones of the first course of the wall on the crushed rock.

If I were building a higher dry-stacked wall of 6 feet or more (which I wouldn't attempt without first gaining a lot of on-the-job training) in a cold climate, I would be more concerned about the damage frost heave could produce. The best insurance here would be to dig a trench about 3 feet deep and at least 4 feet wide. Fill the trench to within a few inches of grade, then start the first course on the crushed rock. This may sound like a lot of extra work, but a backhoe can dig the trench in very little time, and gravel is cheap.

Laying the Courses

A dry-stacked wall should have at least two rows of stones, one for each face or side of the wall. Set the stones with a slight inward slope, that is, with the thinner edge on the inside. Don't overdo the sloping effect, but do aim to steer any gravity-induced pressure toward the inside of the wall, with the force from one side counteracting that from the other. The ultimate goal is to pack the stones as tightly as you can, filling in small cavities whenever possible. You'll use smaller stones as filler to level out each layer.

Stone walls are built in horizontal courses, one course at a time along the entire wall, and not in sections. The first course (or layer) is the most important one, so take your time and do it right. Begin by setting the corners. Use large stones with a good 90-degree angle to them, and set two layers of stones at each corner. Next, set the largest and flattest stones in the pile along the line, leaving space between them to be filled with smaller stones. Then start setting stones into the voids, choosing each one carefully to fit an existing space. Try to place smooth, attractive faces on the outside of the wall, but with the thinner side facing inside. When you cannot find a stone to fit a particular spot, cut or trim one for the purpose.

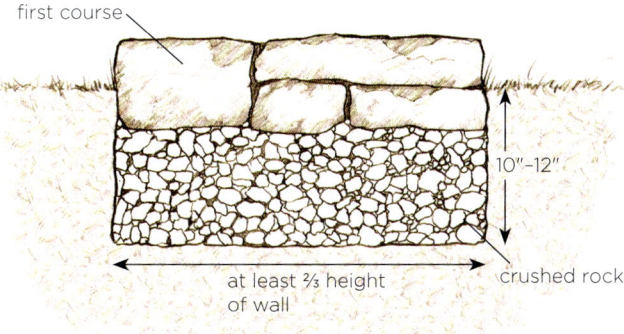

FOOTING FOR A NORMAL DRY-STACKED WALL

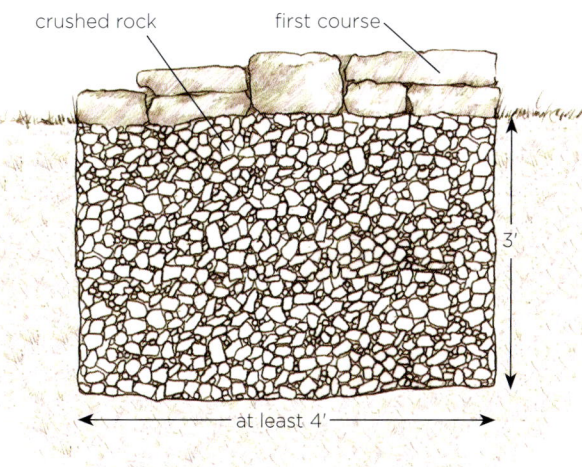

FOOTING FOR A HIGH, COLD-CLIMATE DRY-STACKED WALL

LAYING THE FIRST COURSES

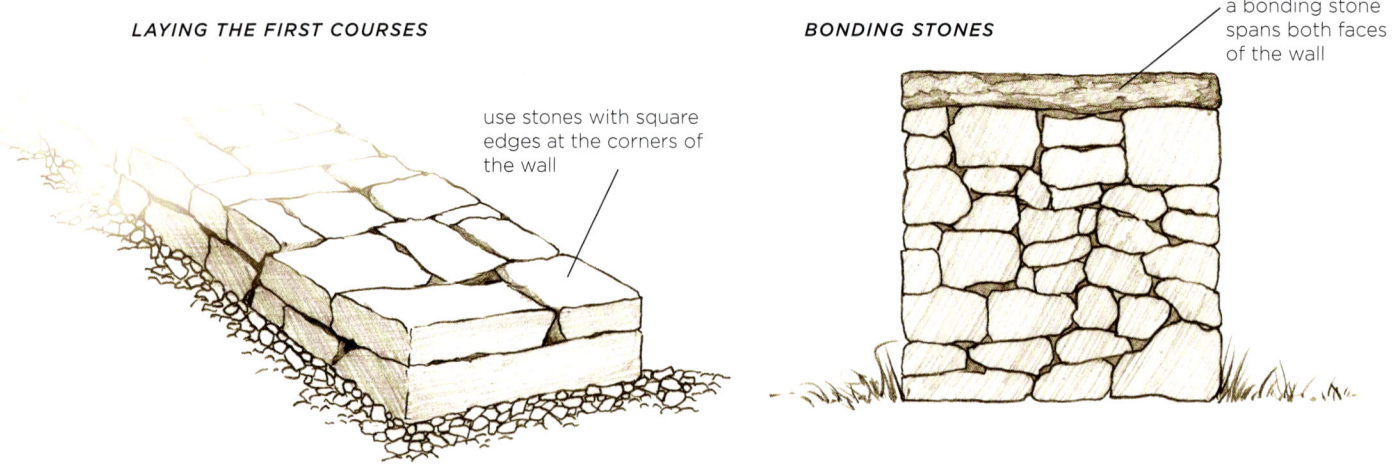

use stones with square edges at the corners of the wall

BONDING STONES

a bonding stone spans both faces of the wall

STONE WALL NO-NOS

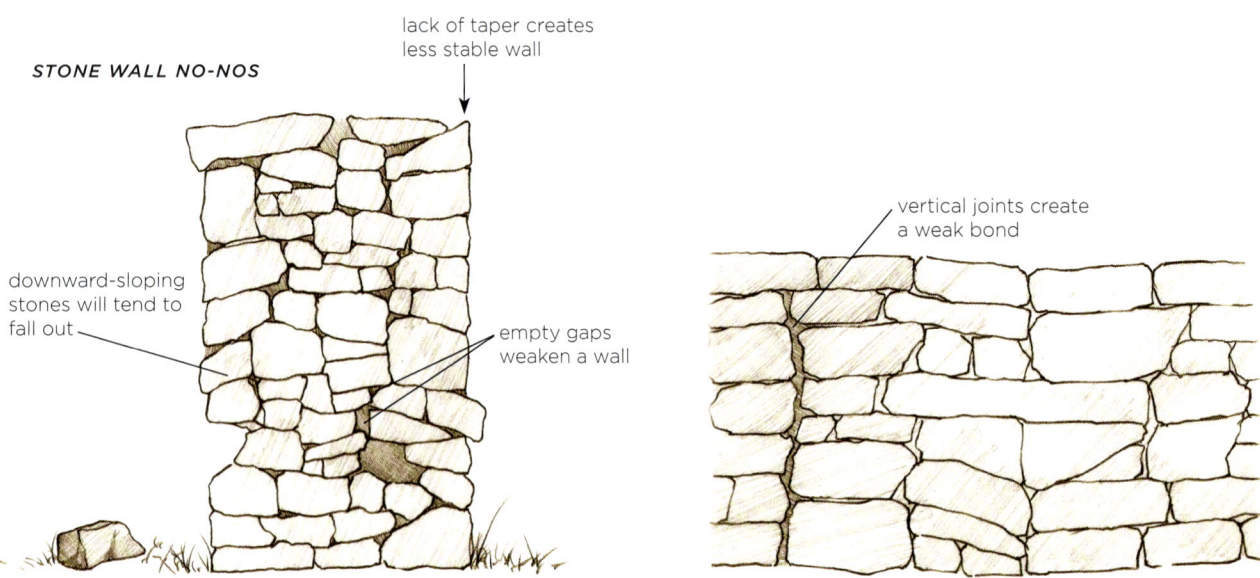

lack of taper creates less stable wall

downward-sloping stones will tend to fall out

empty gaps weaken a wall

vertical joints create a weak bond

STONE WALL ON A SLOPING SITE

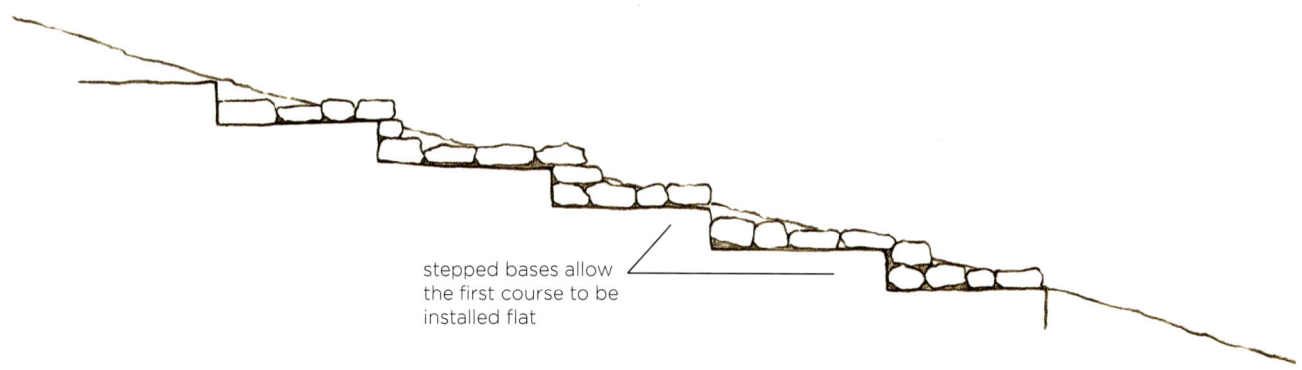

stepped bases allow the first course to be installed flat

Once you have stacked stones to the height of the corners, set new corner stones and repeat the process. Periodically, take a good look lengthwise along the wall to check the alignment and to make sure that it is tapering slightly toward the top.

Periodically along each course place a single stone that spans the full thickness of the wall. These "bonding" stones add strength to the wall and should be used on every other course at the ends. Some stonemasons like to create a shadowing effect by having the bonding stones overhang one or both faces of the wall a few inches.

Avoid vertical joints as much as possible. Not only do they look bad, they also create a weak spot in the wall. Instead, try to stagger the joints as much as possible by overlapping joints with stones above them. Make sure that each stone rests on a reasonably stable surface. In other words, stones should not rock back and forth. You may have to trim high spots off stones to allow for more stable surfaces.

Laying dry stones on a sloping site can be tricky. The key to success is to keep the layers oriented horizontally, and not allow them to follow the lines of the slope. That means that on a fairly steep slope, it is probably best to try to talk yourself into building a different type of fence. For a modest slope, however, use a terraced approach to the base of the wall, which allows the first layer to start off level.

Plants and Stones

Stone walls look great all by themselves, but they can assume a very different character with some plants growing through the gaps. While it is difficult to add plants after a wall has been built, it is very easy to do so during construction. The trick is to choose plants that not only are suited to your region but also will adapt well to the microclimate in and around the stone wall. Walls facing south and west are likely to be warm and receive a lot of sun, while those facing north and east will be cooler and more shaded. Talk with a knowledgeable person at your local nursery or garden center about your choices. Because the inside of a stone wall is pretty dry and difficult to reach with water, look for plants that are suited to an arid environment.

To add plants to the wall, line the large spaces and gaps with sphagnum moss every time you finish one course. Place the plant roots in suitable soil in the space, then tamp and water the soil. Keep watering the plant roots regularly until they are established. ■

Capstones and Coping

Capstones strengthen and protect masonry walls, but there are many ways to achieve those objectives. Flagstones can be dry-stacked upright [BELOW LEFT] or flat [BELOW RIGHT] to produce distinctively different looks, while a precast concrete cap [BOTTOM] mortared to the wall offers maximum durability.

Capstones

A capstone (also called the coping) adds strength to a stone wall. Good capstones should be long and thick, leaving as few joints along the top surface as possible. This helps to keep water out, which is important if you live in a cold climate. If water accumulates in low spots inside the wall and then freezes, the ice can push stones out of position and weaken the wall.

On the other hand, a heavy flat surface on top of your wall may encourage people to sit on it, which is usually not desirable. You can discourage this practice by lining the top of the fence with potted plants.

Once the cap is in place, gather a bunch of small wedge-shaped stones and fill in any remaining gaps in the wall as best as you can. Inspect the wall regularly and reset stones that have loosened.

Mortared Walls

I hope I have already convinced you that you do not need to use mortar to build a stone wall that will last longer than you do. There are times and circumstances, however, when a mortared job is preferable. Using mortar allows you to use a greater variety of stones, including round ones, since you no longer need to be so concerned about stones moving once the wall is built. A mortared wall also does not have to be tapered much, if at all, so you can build a thinner wall, using fewer stones than would be required with a tapered, dry-stacked wall. Finally, with mortar, you can safely build a higher stone wall.

It may sound somewhat contradictory, but the best mortar jobs are done by people who are experienced at dry-stacking. It is a big mistake to count on mortar to hold a wall together that might otherwise fall down. Think of the mortar as a supplement to a well-constructed wall, not a shortcut or an excuse for sloppy work.

The Footing

A mortared wall must be set on a concrete footing. That job, in and of itself, makes the task of building a mortared stone wall far more involved than dry-stacking. As strange as it may seem, the footing for a

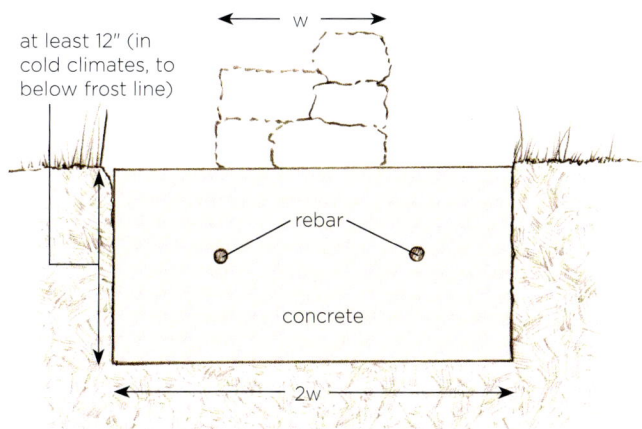

FOOTING FOR A MORTARED WALL

stone wall generally has to be thicker than the foundation walls needed to support a house. That's because the footing will support a structure that is much thicker than the 2×4 or 2×6 frame on the house. As a general rule, the width of the footing should be double the thickness of the wall. In cold climates, the footing must reach below the local frost line. In warm climates, make the footing depth at least 12 inches, and preferably as deep as the wall is thick.

It is important that the trench for the footing be dug very carefully, so as not to disturb soil on the sides and bottom. If you are lucky enough to have firm and compact soil, you can pour concrete directly in the trench. Otherwise, you have to dig a wider trench and set up forms for the concrete. Setting forms, pouring concrete footings, setting the rebar, and stripping the forms is a tough job, and not the kind of task that most do-it-yourselfers will care to tackle. If you are an exception to that hunch, I suggest that you study a few good books on concrete foundations before doing anything else. For the rest of you, find a contractor to handle the footings.

Allow the concrete footing to cure for several days before beginning to build on it.

With sloped sites, you will need to create a stepped footing so that all stones at the base rest on a level surface. See page 144.

Laying Courses

Use the same approach in selecting stones as you would with a dry-stacked wall: Save the biggest and flattest ones for the base, squared stones for ends and corners, and long, flat stones for the cap, and have plenty of small filler stones on hand. Also, like the dry-stacked wall, a mortared wall should have at least two rows of stones on each course, although you do not have to worry about tapering.

When you are ready to start building the wall, mix up a batch of mortar. Choose the stones for the first course, dry-fitting them if you feel the need. Use a trowel to spread a 1-inch-thick bed of mortar over the concrete footing. Then set the first course in the mortar, beginning with the corners. Press the stones firmly into the mortar.

For subsequent courses, dry-fit the stones first so that you know which stones go where. Remove the stones and spread a 1-inch-thick layer of mortar. Press each stone in place, rocking it back and forth while pressing some mortar out of the joint. Scrape away this excess with the point of a trowel or a piece of wood. Set the next stone to the side, not on top, and press it into a bed of mortar, filling the vertical joint between the stones with mortar as well. Use a pointing tool as needed to work mortar into thin cracks and to produce a clean edge. Fill large gaps with small stones surrounded by mortar, rather than mortar alone. Clean mortar off of the face of the wall with a wet sponge before it hardens.

Use bonding stones (see page 144) spanning both faces of the wall periodically along each course and on every other course at the ends.

If you are building a short wall, you may be able to set more than one course of stones in a day. I suggest, however, that you do not lay more than two or three courses per day, as you want to avoid adding too much weight on top of uncured mortar in the lower course. Each day you are building, use a wire brush to clean out the joints from the previous day's work before setting more stones.

Spread a 1-inch-thick layer of mortar on the concrete footing using a large trowel. Then press the stones for the first course firmly into the mortar. Repeat this process on subsequent courses.

The key to building an attractive stone wall with mortar, in my opinion, is to conceal the mortar as much as possible. Like 2×4s and electrical cable, mortar and concrete are fabulous building materials that function best if they are not seen. I routinely drive by a low wall built with beautiful, colorful round stones that is utterly ruined visually by huge mortar joints that extend to the front face of the wall. Why bother? Mortar can do its job of holding stones in place without having to detract from the appeal of the stones themselves. Try to recess the mortar joints as much as possible, and keep them as thin as possible. Thin, recessed mortar joints create a nice shadowing effect on the wall. And thin mortar joints are stronger than thick ones. Keep in mind that the stones may last forever, but the mortar is bound to crack and loosen over the years. If you rely on large amounts of mortar, and it starts to degrade, the integrity of the entire wall may be at stake.

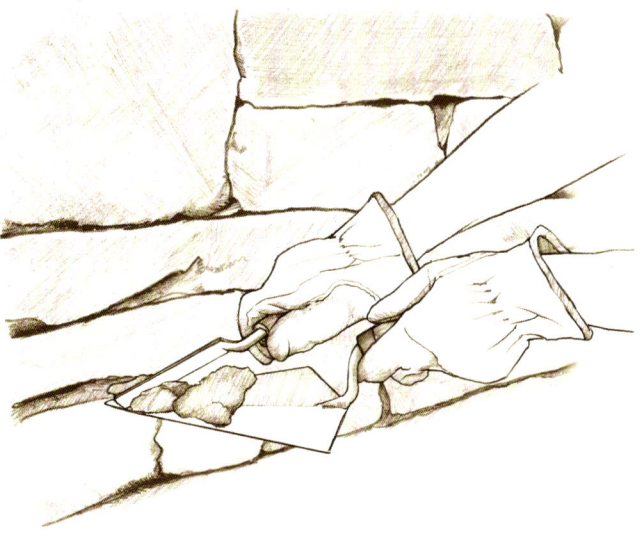

Before the mortar begins to harden, use a pointing tool to push the mortar firmly into the joints between the stones and to remove excess mortar.

Capstones

The capstones are particularly important on a mortared wall, as you want to keep any water from seeping into the wall, where it could freeze and undermine the mortar. Spread a flat bed of mortar over the entire surface of the top course before setting the capstones in place.

Curing

Try to encourage the mortar to cure slowly. This is best accomplished by keeping it damp and shaded from direct sunlight. At the end of the day, drape a plastic tarp over the wall, or use wet burlap bags if you have them. On those mortar joints that have had 24 hours to cure, give the joints a gentle shower with the hose before covering them. Keep up this routine for five or six days. And avoid putting undue stress on the wall for several weeks.

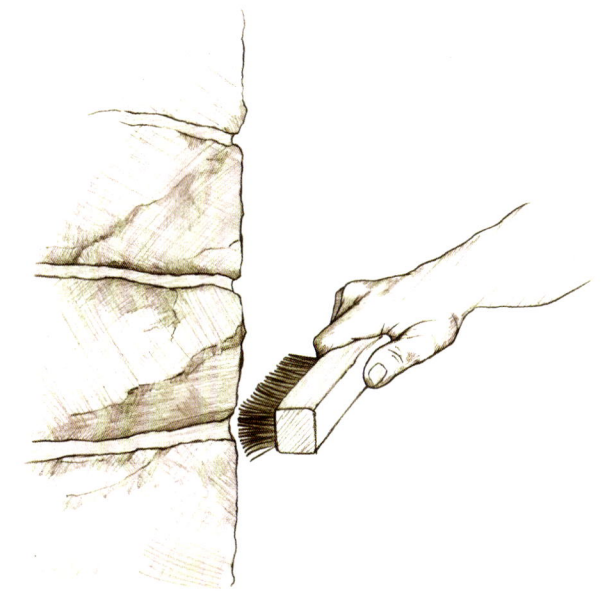

Allow the mortar to cure overnight, then use a wire brush to clean out the joints, recessing them slightly, before you begin setting more stones.

Building with Brick

Brick may not have the same romantic allure as stone when it comes to fencing purposes, but it is a wonderful material for do-it-yourselfers looking to build a solid, attractive wall. Brick walls must be mortared, and constructing one can be expensive and labor-intensive; but if the cost and the prospect of repetitive work don't bother you, it is a project that can be undertaken at the pace you choose, with the conviction that the finished project will remain standing for a long, long time.

The basic technique for making brick has not changed substantially for centuries. Dry, ground clay is mixed with water, formed into the desired shape, dried, and fired. Manufacturers today try to control the consistency and quality of the finished product by combining different types of clays and firing the bricks under carefully controlled conditions. But brick is still an old, enduring material that was created specifically as a building material that fit into the human hand.

If your house is built with brick, I would think that brick would be your first choice (though not your only choice, of course) for security and privacy fencing or a boundary marker. In this case, you should look for brick that closely resembles that on your house, and you should install it in a similar pattern. Regardless of your house style, however, brick today is manufactured in so many colors, textures, and sizes that you are sure to find a product that can fit into any surrounding.

If you live near a large city, you should be able to find a well-stocked brick supplier. Home improvement stores also carry, or can order, brick. Shop around to compare prices, colors, and delivery options.

Types of Brick

The brick industry wouldn't be a proper entity if it didn't have its own language that the normal citizen would be hard-pressed to comprehend. Fortunately, for the purposes of this book we can keep the terminological exegesis to a minimum.

Weathering Classification

Brick is made in three different weathering classifications. Brick that is graded SW (severe weathering) can withstand repeated exposure to freeze and thaw cycles and is best for cold climates and use below grade. MW (moderate weathering) brick, not too surprisingly, is made for more moderate climates, while NW (negligible weathering) is used only in warm climates and indoors.

There are many ways to break up the monotony of a solid brick wall. Here, bricks are staggered to create an infill that mimics lattice.

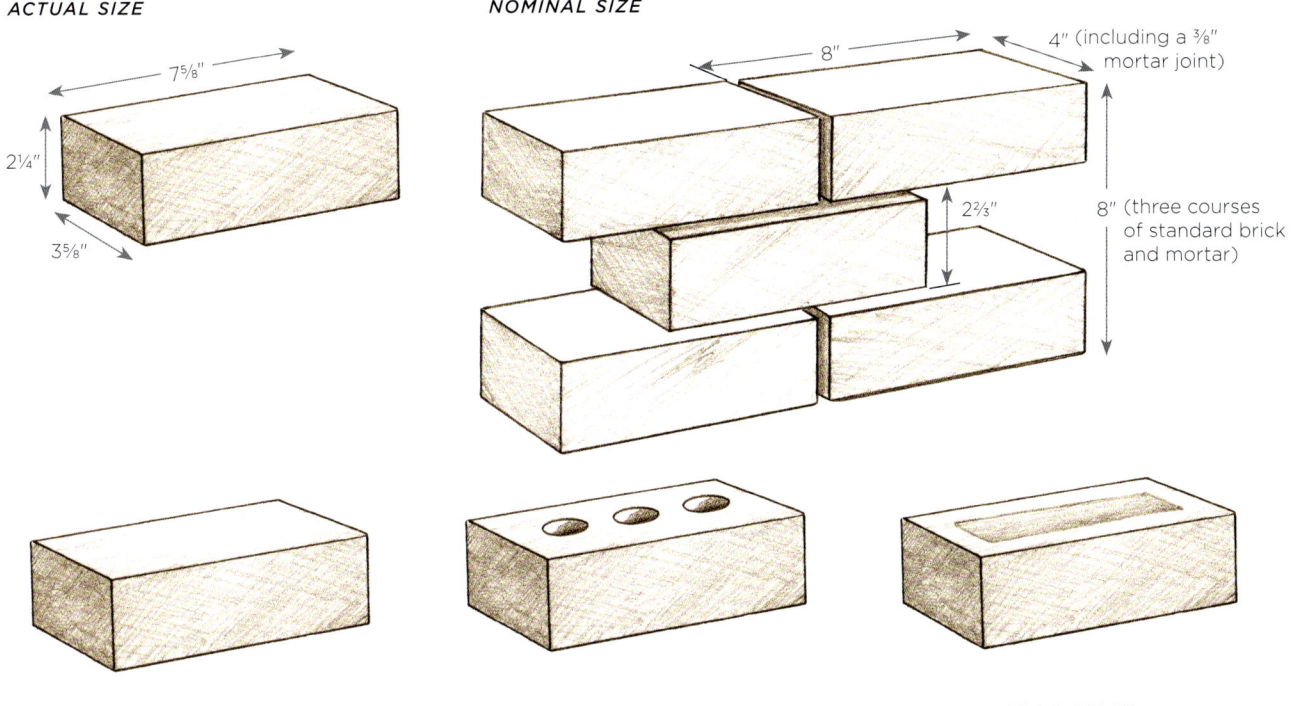

ACTUAL SIZE

7⅝"

2¼"

3⅝"

NOMINAL SIZE

8"

4" (including a ⅜" mortar joint)

2⅔"

8" (three courses of standard brick and mortar)

SOLID BRICK

CORE BRICK

FROG BRICK

Sizes

Most experienced do-it-yourselfers understand that a 2×4 does not actually measure 2 inches by 4 inches. Brick also has a conflict between actual and nominal size. The actual size refers to the measurements of the brick itself as it leaves the factory. Nominal size refers to the installed brick, taking into account the actual size plus the normal mortar joint. When drawing plans and placing an order for brick, it is easiest to work with nominal sizes.

The most common size is conveniently called "standard" brick. It is manufactured to an actual size of 3⅝ inches × 7⅝ inches × 2¼ inches. For planning and ordering, however, standard brick is more often referred to in its nominal size of 4 × 8 × 2⅔. The planning assumption here is that three courses of brick plus mortar joints will be 8 inches high.

Styles

Standard bricks can be purchased solid, with cores, or with what are called frogs. Cores are holes through the brick, while frogs are modest indentations. Both of these devices reduce the weight of the brick while adding greater surface area for the mortar, and thus

create a stronger wall. Frogged bricks are solid on one side and depressed on the other, and the frog should always be installed facing down.

Bricks often have an identifiable front and back side if you look carefully enough. Plan to install the bricks with the straightest, smoothest side facing out. For large jobs, bricks can be purchased and delivered in metal-strapped cubes containing about five hundred standard bricks each, weighing approximately one ton.

Colors and Textures

Brick can now be found in dozens of colors, including various shades of black, white, gray, pink, red, and brown. Antiqued brick that looks as though it has been standing for decades, or more, can be bought right off the rack. You can choose faces that are smooth, tumbled (rough), stippled, scored, and pressed. Throw all of these options into the mix, and you can come up with literally hundreds of potential looks for your brick wall. And, if that is not enough, consider also that mortar can be tinted in several different colors to complement the color of the brick you use. Finally, the type of finish you apply to the mortar joints can affect the look of your wall.

Wall Patterns

The overlapping orientation of bricks is both a structural and a decorative consideration. When bricks are laid flat and lengthwise, they are called stretchers. Stretchers provide longitudinal strength to the wall. When bricks are laid on end and vertically, they are called soldiers. When a brick is set flat to span front to back, it is called a header. Headers tie two vertical layers, or wythes, together. A rowlock header is formed when the header brick is set on its side.

In the brick-setting trade, patterns are most often called bonds. The most popular is the running bond, in which all bricks are stretchers. Bricks for each course lap over half of the bricks below. Running bond is the pattern of choice for brick veneer on houses because it can be built in a single, 4-inch-thick wythe. When used on a freestanding, double-wythe brick wall, the wythes are tied together with metal ties.

Bricklaying terminology is a language unto itself. The terms illustrated below are used to identify the functions of bricks as they are laid in a wall.

Common bond, also sometimes called American bond, uses full courses of headers at regular intervals, separated by five or six courses of stretchers. It is a good choice for a freestanding wall, as it is nearly as easy to build as a running bond wall, but provides additional structural support with the use of headers. Note that the corners of each course containing headers in this bond must contain a three-quarter (that is, 6-inch) brick.

English bond consists of alternating courses of headers and stretchers, with the headers centered over the stretchers and the mortar joints between them. Flemish bond has stretchers and headers alternating along each course, again with the headers perfectly centered over stretchers below.

Additional variety can be obtained by using the Flemish bond pattern to create a latticelike wall. This look can be achieved on a single-wythe wall by leaving out the headers, either in a regular pattern (as in the illustration on the facing page) or randomly. Or you can add some texture to the wall by recessing or projecting bricks along the face, again either in a pattern or randomly.

STRETCHERS

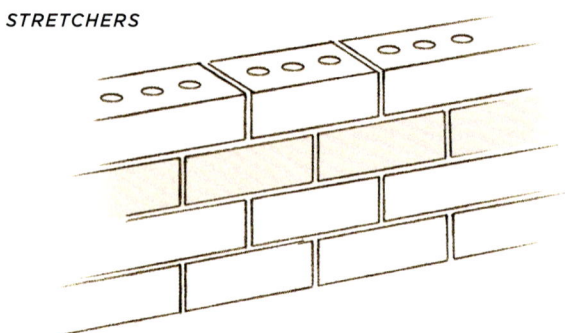

SOLDIERS

HEADERS

ROWLOCKS

This wall is a good example of common bond, in which rows of bricks laid as stretchers are interrupted by single rows of bricks laid as headers.

Brick Patterns

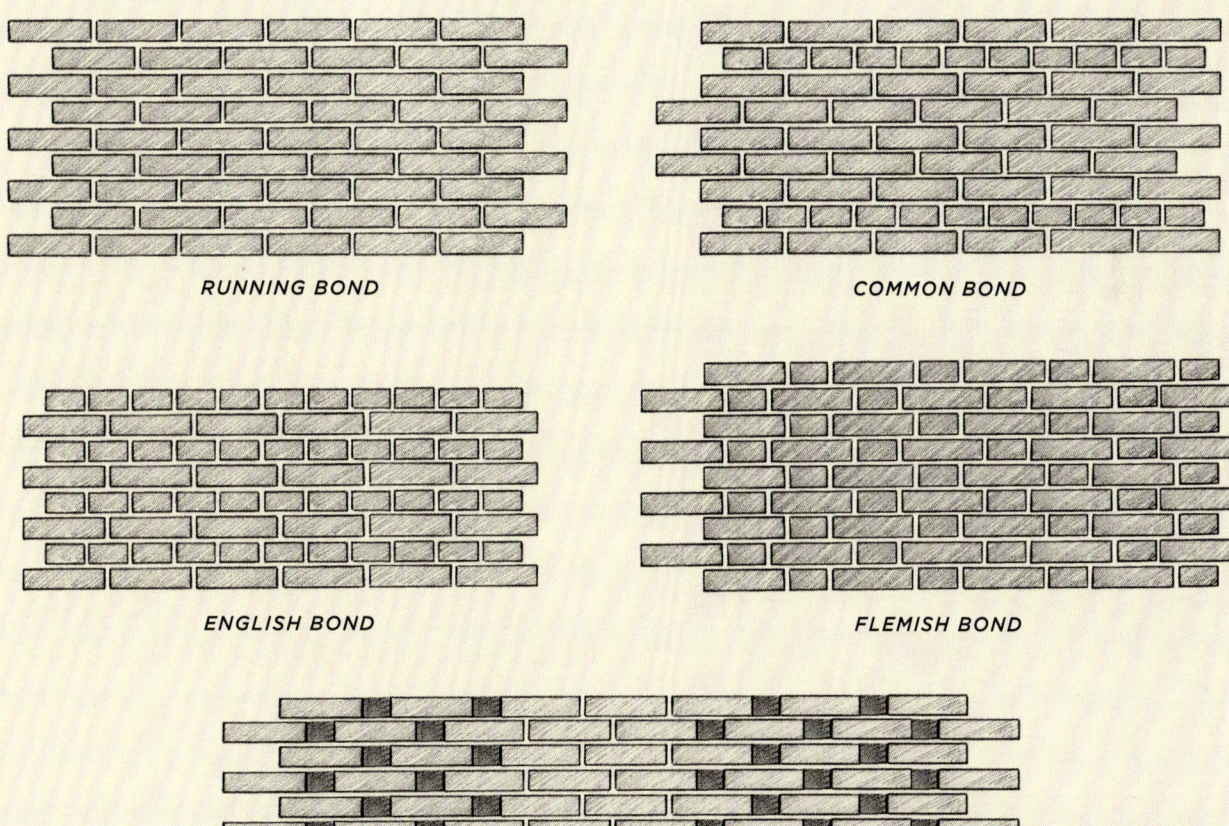

RUNNING BOND

COMMON BOND

ENGLISH BOND

FLEMISH BOND

LATTICE EFFECT

Tools of the Trade

To build a basic brick wall, you really do not need many tools at all. In fact, you may not even have to cut any bricks. If you do, you will want to have a brick chisel and hammer (and the requisite eye protection). For the mortar you will need a mortar tub or wheelbarrow, a mortar hoe, a trowel, and a finishing tool for the joints (see page 157). A chalk line and a 4-foot level will be useful for getting you started in a straight line, and keeping you there. A stiff brush will be helpful for cleaning the brick.

The Footing

A brick wall 3 feet high or more should be built on concrete footing that is twice as wide as the thickness of the wall. In areas that experience freezing temperatures, the footing should extend below the frost line. In warmer areas where freezing is not a concern, make the footing at least as deep as the width of the planned wall, and twice as wide. Plan to lay the first course below grade, so that the footing itself will not be visible once the wall is built.

The construction of the footing for a brick wall is the same as for a mortared stone wall. See the discussion on page 147. Allow the concrete footing to cure for at least a couple of days before beginning construction of the wall.

FOOTING FOR A BRICK WALL

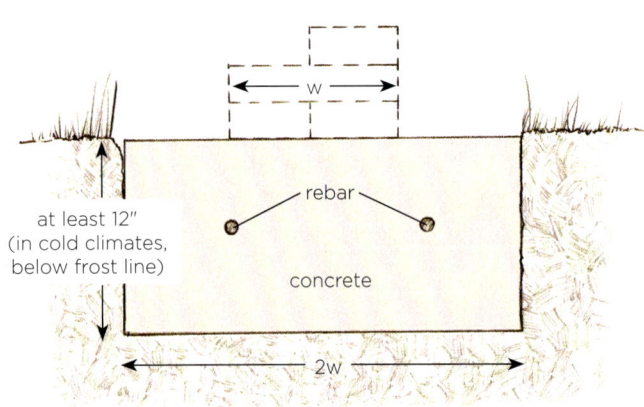

at least 12"
(in cold climates, below frost line)

rebar

concrete

2w

w

The Mortar

Bags of dry mortar mix are your best choice for brick work. Pour the entire bag into a mortar tub or wheelbarrow, add water, mix, and you are ready to go. I do not recommend trying to mix partial bags. Since the contents of the bag tend to settle and separate, you may not wind up with proper proportions. If you do need to mix only a small amount, empty the entire bag into the tub or wheelbarrow, mix it all up, then return all that you do not need to the bag. If you want to tint the mortar, this is also the time to add some colorant.

Mortar mix is sold in 60- and 80-pound bags, and at least one brand can also be found in 40-pound bags. I suggest that you start with small bags, mixing one at a time, until you build up enough experience and speed to be able to handle the mixed contents of a larger bag before it starts drying out on you. Figure on laying 30 to 40 bricks for each 60-pound bag of mix.

Laying Bricks

A few hours before you begin building the wall, give the bricks a good watering with the hose. Dry bricks absorb water from the mortar too quickly, which results in a weaker bond. Snap chalk lines on the concrete footing to mark the outlines of the wall and guide the first course. For a normal double-wythe wall, leave a gap of ⅜ inch between wythes. This will allow you to use standard bricks as headers. If you do

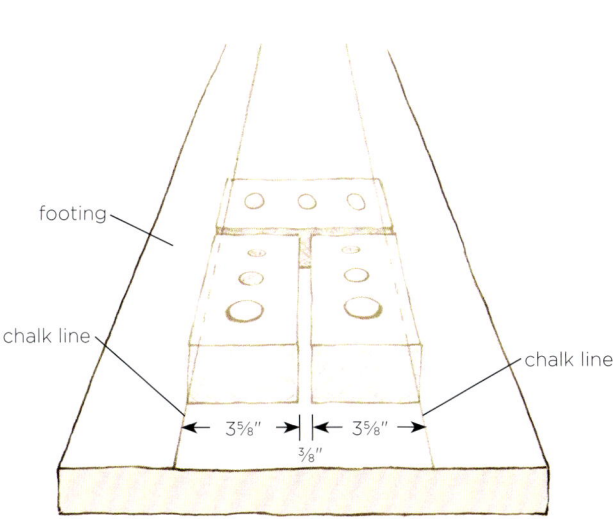

footing

chalk line

chalk line

$3^{5/8}''$ $3^{5/8}''$

$3/8''$

Snap chalk lines on the footing to serve as a guide in laying the first course of bricks. For a double-wythe wall using standard bricks, the chalk lines should be $7^{5/8}$ inches apart.

not plan to use headers, however, you can create a bigger cavity between wythes. The center of the wall should be aligned with the center of the footing.

The best way to lay bricks for a wall is to build the opposing ends first, then work your way into the middle from each side. Scoop up an evenly shaped line of mortar with the side of your trowel (a trowel with a 10- to 11-inch blade is best for this job), and "throw the mortar" (that's the appropriate jargon) along the chalk line by twisting your wrist and giving the trowel a little snap downward. With some practice, you should be able to master this move as well as the right quantity of mortar needed.

Apply a 4-foot-long line of mortar. Now take the tip of the trowel and form a furrow in the mortar. The goal is to shape the mortar to be about 1 inch deep and as wide as one brick, and it should be consistent in size and shape from end to end. Try not to cover up the chalk line.

Lay the first corner brick on the mortar and press it firmly into place. Beginning with the second brick, "butter" one end of the brick with mortar and press the brick both down into the mortar and sideways into the adjacent brick. With each brick you should see some mortar squeeze out of all joints; aim to create mortar joints that are consistently $3/8$ inch thick. Lay the first course along one line, then lay the six bricks

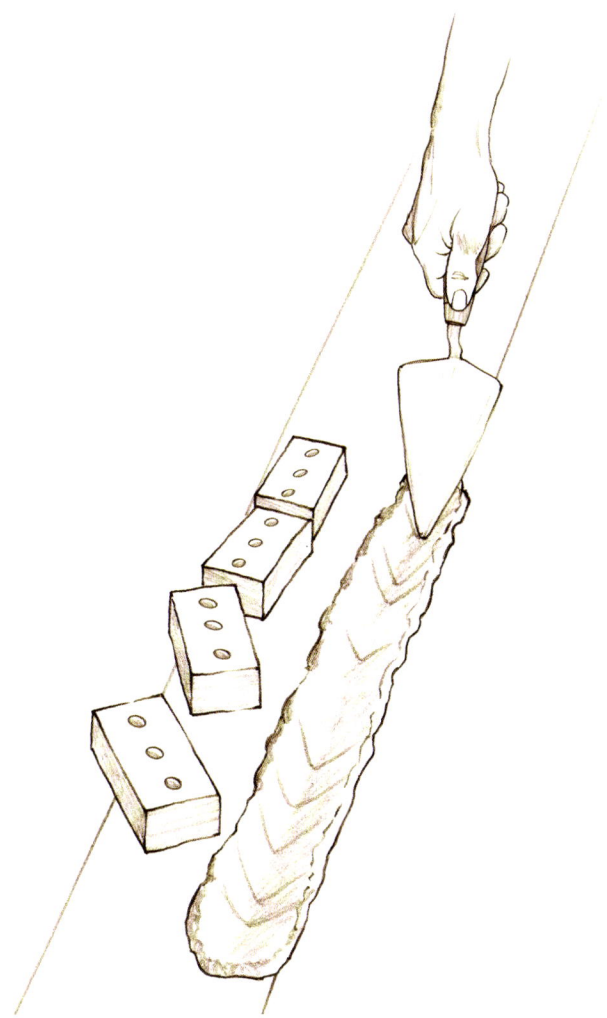

Carefully throw a 4-foot line of mortar along each chalk line. Then use the tip of the trowel to create a furrow in the mortar.

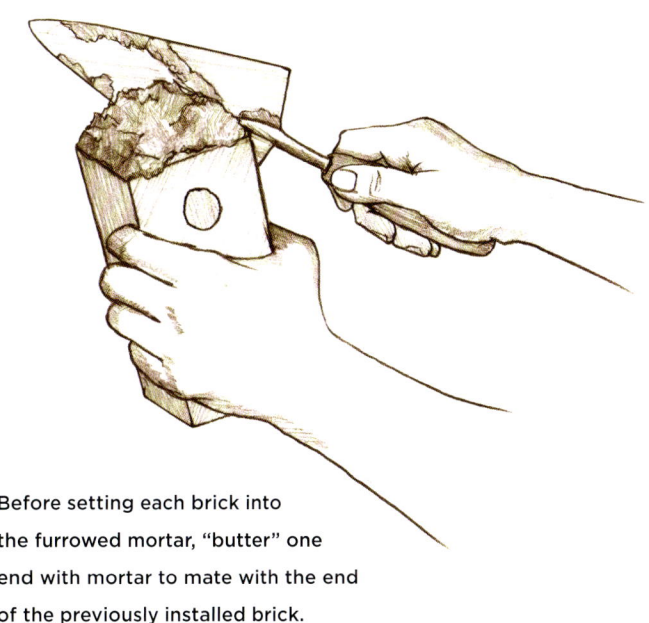

Before setting each brick into the furrowed mortar, "butter" one end with mortar to mate with the end of the previously installed brick.

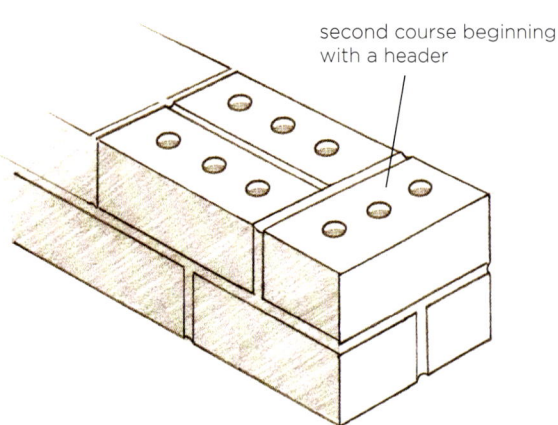

second course beginning with a header

The second course of a double-wythe wall, and of every other course after that, normally begins with a header.

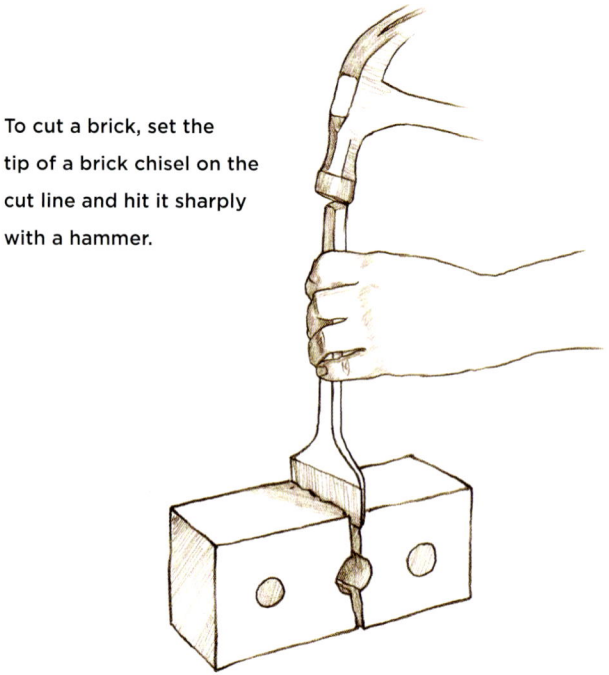

To cut a brick, set the tip of a brick chisel on the cut line and hit it sharply with a hammer.

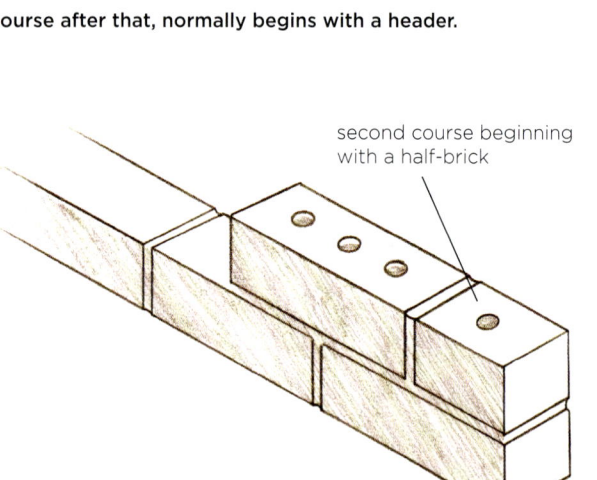

second course beginning with a half-brick

Single-wythe walls and some other styles use a half-brick to begin the second course.

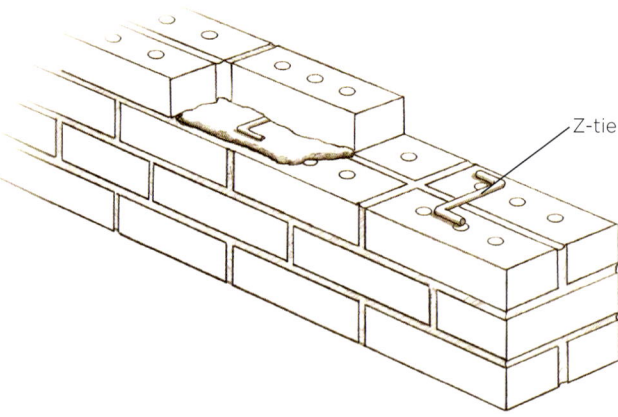

Z-tie

The wythes (rows) of this wall are tied together with metal Z-ties embedded in mortar.

for the second wythe alongside. As you work, use a 4-foot level regularly to check that the bricks are plumb and level (both along each wythe and across the wythes).

The second course will usually begin with a full-sized brick installed as a header or with bricks cut in half, depending on the pattern you are following. To cut a brick, mark a line on the brick and set the brick on its edge on a piece of wood. Set the brick chisel along the line and give it several good whacks with your hammer until it breaks. Use a mason's hammer to clean up the edge, if necessary.

Lay the remaining bricks so that they are centered over the mortar joints below. Keep checking for plumb and level, and keep the width of the mortar joints as consistent as you can. Use the side of your trowel to scrape away excess mortar and throw it back into the tub. Begin the third course just like the first, with full-length stretchers. Continue this alternating pattern as you work up the wall. If you are building a high wall, or are not including headers in the design to bond the wythes together, plan to set metal Z-ties (or another approved tie) in the mortar 16 inches apart in every third or fourth course.

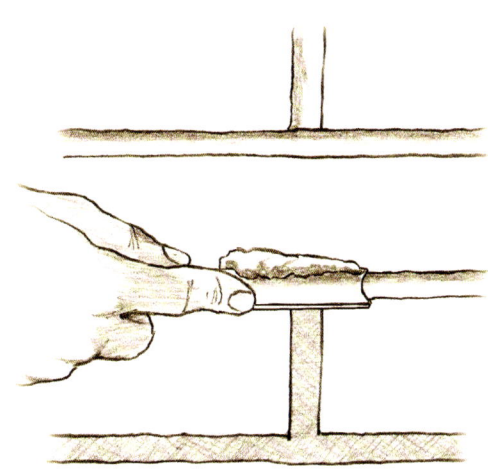

A jointing tool seals and fills the joint between bricks while simultaneously creating a clean, finished look. Here, a convex jointer is pulled along a horizontal joint.

Plan to stop laying bricks from time to time to finish the mortar joints. Finished joints look better and also are better sealed than unfinished joints. Just how often you stop to finish the joints will depend on how quickly you are working and on the climate. The mortar should not be too wet, which will make the job excessively messy, nor should it be too dry, which will make the job excessively difficult. Again, with a little experience you will find the right rhythm. Finish the vertical mortar joints first, and then work on the horizontal ones.

The most common tool for finishing brick mortar joints is a convex jointer, which creates a concave joint. Press the tool into the joint at a slight angle and pull, applying just enough pressure to form a consistent indentation. Rounded mortar joints are not your only option, but if you would prefer one of the other profiles shown in the illustration, you will need to buy the appropriate tool. V-joints, which are equal to rounded joints in terms of strength and weatherproofing, require a V-shaped jointing tool. You can use a flat pointing tool or even the trowel itself to create the weathered joint, but you will have better luck with a jointer made specifically for the purpose. The weathered joint gets its name from the downward slant of the joint, which encourages water to shed quickly. You can use a flat pointing tool for raked joints, but the preferred tool is a skatewheel jointer, which is not expensive but may be tough to find. Raked joints create an attractive shadow line, but they also create a resting place for water, ice, and snow. I would recommend using raked joints outdoors only in dry climates.

After the joints have been finished, use a fairly stiff brush to clean the surface of the brick.

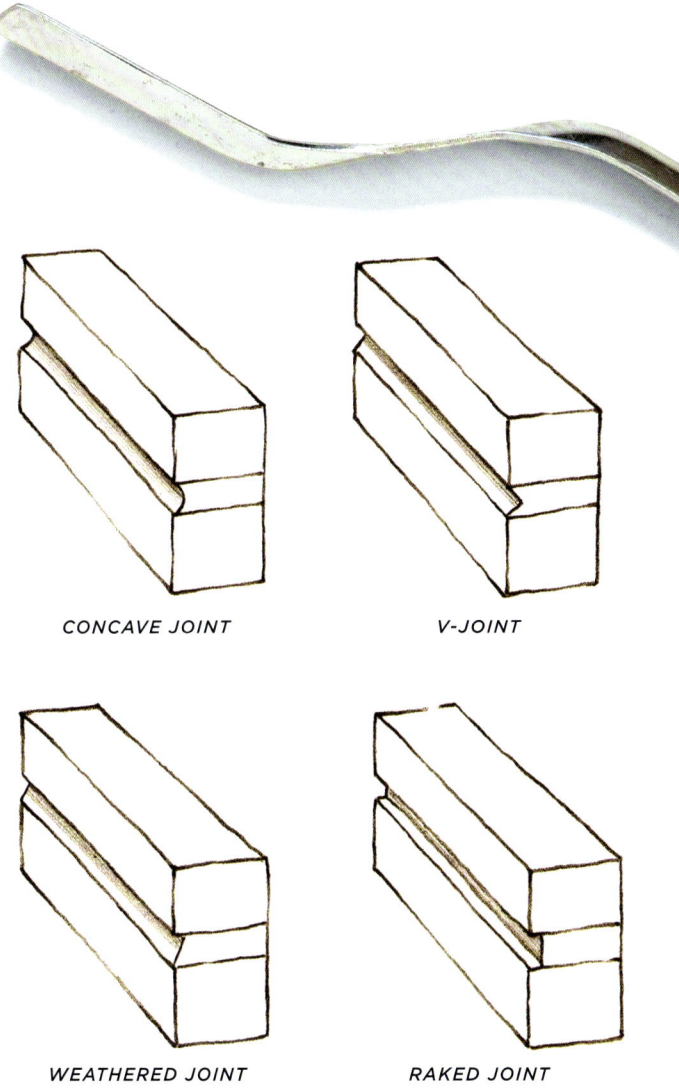

CONCAVE JOINT　　　　V-JOINT

WEATHERED JOINT　　　　RAKED JOINT

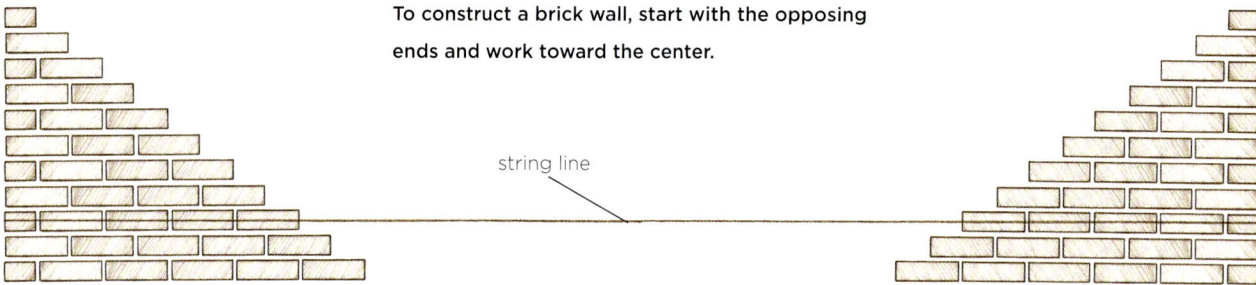

To construct a brick wall, start with the opposing ends and work toward the center.

string line

Once you have set the top brick at one end of the wall, move to the other end and repeat the same sequence. Once the two ends are completed, fasten a string line along the faces of the two sections to help guide you in laying the remaining bricks in the middle of the wall. Set the bricks flush with the string line, but continue to use your level to check your progress. Add additional courses on top, if necessary, to bring the wall to the desired height.

Top the wall with solid bricks set as headers, stretchers, or rowlocks, or use flat stones or precast concrete headers. To maximize weatherproofing, allow the coping of concrete headers to overhang the wall on all sides by at least ½ inch. This creates a drip edge that discourages water from running down the face of the wall. Slope the coping to one side or the other just a bit to help water drain off.

At the end of each workday, cover the wall with plastic tarp or wet burlap bags. After the mortar has cured for about a week, clean the brick, if necessary, with a stiff brush and a solution of 1 part muriatic acid to 10 parts water.

Brick piers can be particularly effective when combined with other materials. Here, offset piers mate beautifully with a mortared stone wall.

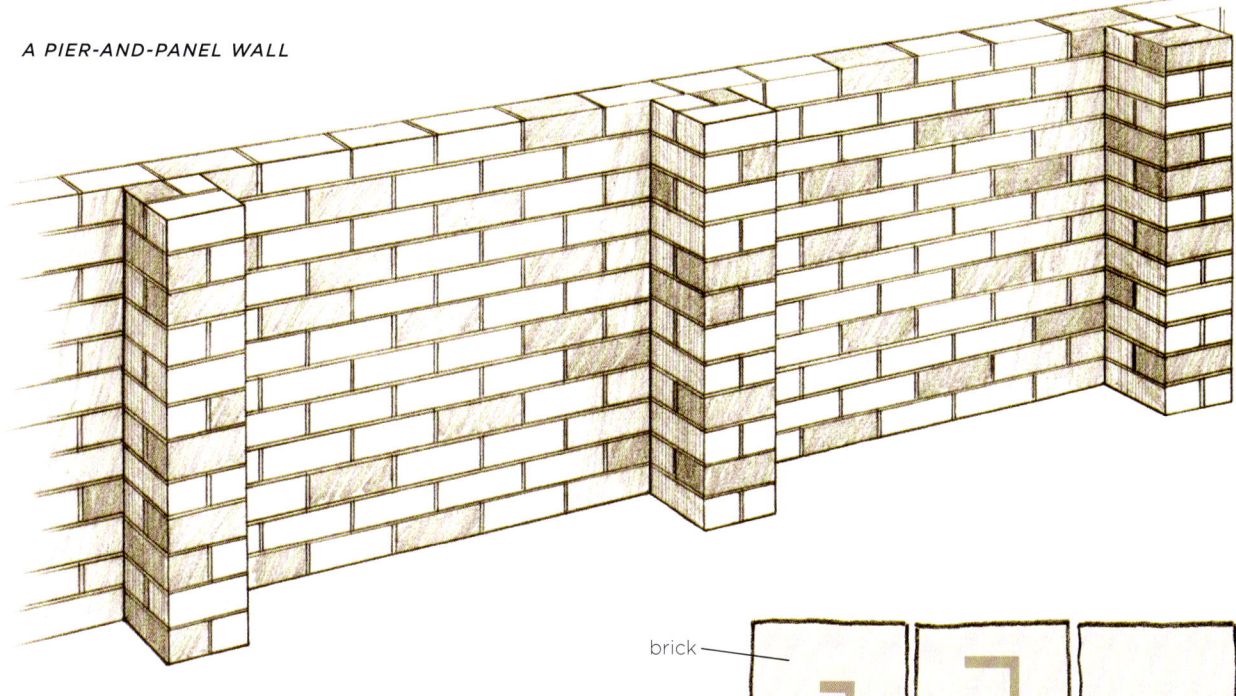

Pier-and-Panel Walls

An alternative to the double-wythe wall just described is a pier-and-panel wall. This approach to brick walls allows you to build relatively high walls that are only one wythe (that is, 4 inches) thick but that gain significant strength from the addition of brick piers every 8 to 12 feet.

Because these walls are thin, however, they are potentially more vulnerable to damage from high winds and other stresses. As such, they are often more carefully engineered than a standard double-wythe wall. The piers must be sized, spaced, and reinforced to provide adequate strength for the conditions of the specific site. Before proceeding to build a pier-and-panel wall, I suggest you talk to an architect or engineer, who can advise you on these technical matters. With the structural details in hand, however, the actual construction of such a wall can be relatively quick and affordable. The illustrations provide a general idea of how piers can be put together and incorporated into the wall.

Brick piers can be built to protrude from one side [TOP], to rest at the center of the wall [MIDDLE], or to allow for offset walls [BOTTOM].

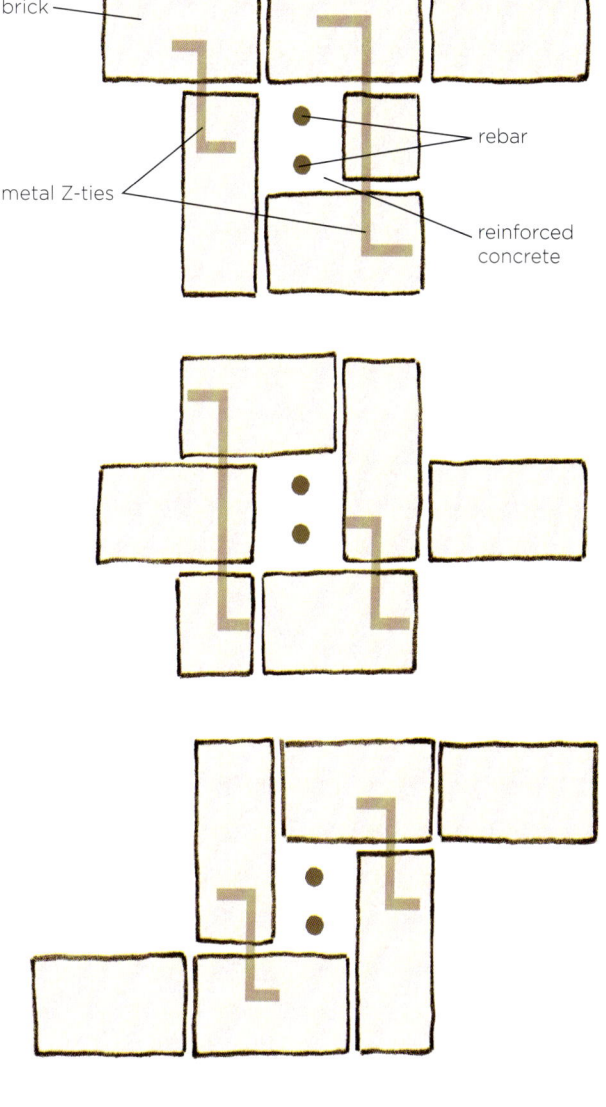

Creative Brickwork

The more you look around for examples of brick walls, the more variety you will find. Though it is a uniform, solid, and typically rectangular material, brick can be used to construct walls that are open, light, rounded, and unobtrusive.

Building with Concrete Block

TAKE SOME CONCRETE, form it into a block, and cure it, and you have created the versatile (if largely uninspiring) construction product known as concrete block. Block is inexpensive, very easy to lay, and durable. Standard block is not pretty, however, and I want to make it clear at the outset that I am not proposing that you give it much, if any, visibility in your own yard. But decorative block can be used to great effect as a fencing material, and standard block can be used as a substrate for colorful, functional, and unique stuccolike finishes.

Manufacturers of concrete block churn out dozens of types, sizes, and even colors. Regular block, like brick, is typically referred to by its nominal size, which includes the actual dimensions of the block plus a ⅜-inch mortar joint. Thus, the standard 8-inch block will measure 8 inches by 16 inches only with the addition of mortar on the bottom and one side, as shown in the illustration. Block is also commonly found in minimal 4-, 6-, 10-, and 12-inch sizes. In each case, the nominal size refers to the width of the block; the actual length (15⅝ inches) and height (7⅝ inches) remain the same on each. Half-length blocks and special shapes and sizes are also available.

Decorative or architectural block can be used to create great-looking freestanding walls. Many of the products on the market are intended for use as retaining walls or veneer, and therefore have only one attractive face. Most fences and walls, however, are expected to look good on both sides, so you will have to shop a bit harder to find some of the products suitable for this purpose. Search online to find a local supplier.

Screen block is not much good for privacy fencing, but it is quite useful for enclosing a swimming pool or adding a degree of intimacy and enclosure around a patio. Ribbed and scored block can add nice depth and unusual patterns to a block wall. Splitface block is particularly nice for many architectural purposes, in my opinion. The rough surface mimics the look and feel of split stone, and I think you will have the best luck locating this product in standard sizes that are suitable for a freestanding wall.

Some of these decorative blocks can be ordered in a range of earth-tone colors. If you decide to use a colored block for your wall, I suggest that you examine samples of the specific color you have in mind before special-ordering a large quantity. And I would strongly encourage you to order more than you need, since you may have trouble locating an exact match should you need to replace a block or two someday.

Tools for building with concrete block are the same as those required for building with brick; see the list on page 154.

Mortared Block Walls

Most decorative block is installed with mortar. There are decorative block products available that can be dry-stacked in such a way as to lock the pieces together, but I do not have any direct experience with them. If you find such a product that interests you, you should rely on the manufacturer to provide thorough instructions for installation.

Building a mortared block wall is almost identical to laying brick, except that it goes much quicker.

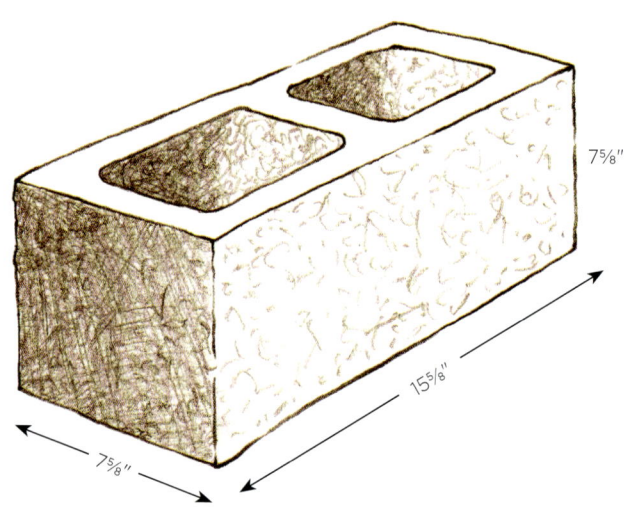

A standard concrete block measures 8 inches by 16 inches only with the addition of a ⅜-inch mortar joint on its top and one side.

The Footing

You will need to prepare a suitable footing that sits below the frost line. If frost is not a problem where you live, I suggest you set the footing deep enough so that the top will be below grade, and thus out of sight when covered. The footing width should be double the width of the block.

Laying Blocks

You will need the same basic tools for block work as described for bricklaying. Bagged mortar mix, made up one full bag at a time, is suitable for constructing non-load-bearing block walls. Mix the mortar with water, as directed on the bag.

FOOTING FOR A CONCRETE BLOCK WALL

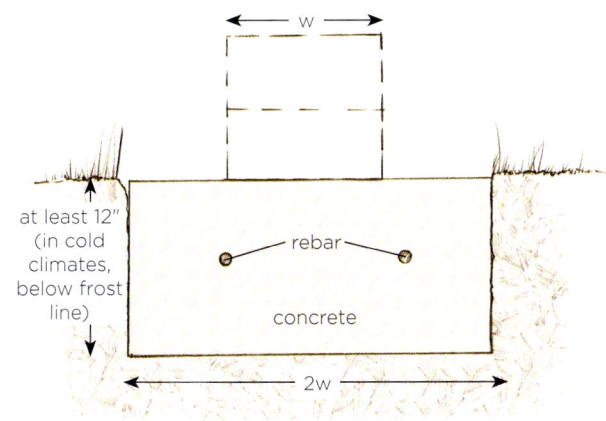

With careful planning, you should be able to avoid having to cut any blocks. Square screen block is normally installed in a stacked pattern, using only full-size blocks, while rectangular block is usually installed in a running bond, requiring half blocks at the corners on every other course. If you do need to cut concrete block, a brick chisel and hammer are suitable for the job, using the same scoring technique as with bricks (see page 141).

Standard blocks these days usually have two cores separated by a web. If you examine the web and edges carefully, you should see that they are wider on one side than the other. The wider side is intended to provide a little more room to spread mortar and should always be installed facing up. The thinner side is then pressed into the mortar.

Plan to use full-size blocks for the first course. Dry-lay the first course from end to end, leaving room between the blocks for mortar (a piece of ⅜-inch wood dowel simplifies this task). Snap chalk lines on the footing as a guide, then move several blocks out of the way and spread a bed of mortar between the chalk lines. For the first course you will want to form a full mortar bed (as wide as one of the concrete blocks).

Snap chalk lines on the footing to guide the first course. Spread a bed of mortar inside the chalk lines, and then form a furrow with your trowel. Press a corner block into the mortar. Keep the blocks level.

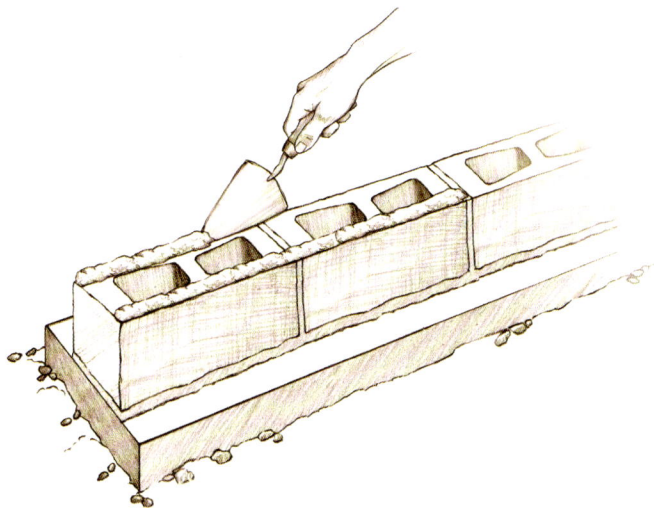

Mortar for the second course, and all subsequent courses, should be applied only along the long edges of the block. Keep the size of the mortar joints consistent, and remove excess mortar from the joints with your trowel.

Use the trowel to furrow the mortar, just as you would with brick, then press the first corner block into place. Use a level to make sure the block is level both lengthwise and widthwise. Use the handle of your trowel, if necessary, to tap the block into a level position.

Set the second block on edge and apply mortar to the edges with your trowel. Then set the block into the mortar bed and press it into the first one. Continue this work sequence for a couple more blocks, then stop and check for level and see that the mortar joints are consistent. Make any necessary adjustments now, while the mortar is still wet. Continue until the first course is complete. As you work, remove excess mortar from the joints with the edge of your trowel. As the mortar begins to stiffen, use a convex jointer to create neat, concave joints, as shown on page 157.

For subsequent courses, mortar should be applied only to the long edges of the block that has already been laid. The easiest way to do this is to imagine that you are putting some jam on the edge of a piece of toast, with the trowel and mortar substituting for a knife and jam. Try to avoid dumping mortar into the cavities, and do not bother mortaring the webs. Set a half-block in the corner of the second row and then use full-size blocks to overlap those in the first course. Finish the course with another half-block.

I suggest that you build up the opposing corners first, to a height of about four or five courses at a time, with each new course stepped back half a block from the previous course, as shown in the illustration on page 158. This way you can stretch a string line from end to end so that the top outside edge of each block can be set against it. This will speed up the work and provide some added insurance that the blocks are being set level and plumb. Even so, check the alignment from time to time with your level.

To lay the last block in each course (the "closure" block), spread mortar on both ends and carefully slip the block into the gap between the adjacent blocks. The top course should be composed of either a special coping block that matches the style of the block in the wall or precast concrete headers.

Reinforcing Screen Block

Note that if you are installing screen block, or any other block, in a stacked pattern (that is, with mortar joints running straight up and down), you should add metal reinforcement to the mortar between each course. Your block supplier should be able to provide you with the proper product, which is needed to add lateral strength to the wall. ■

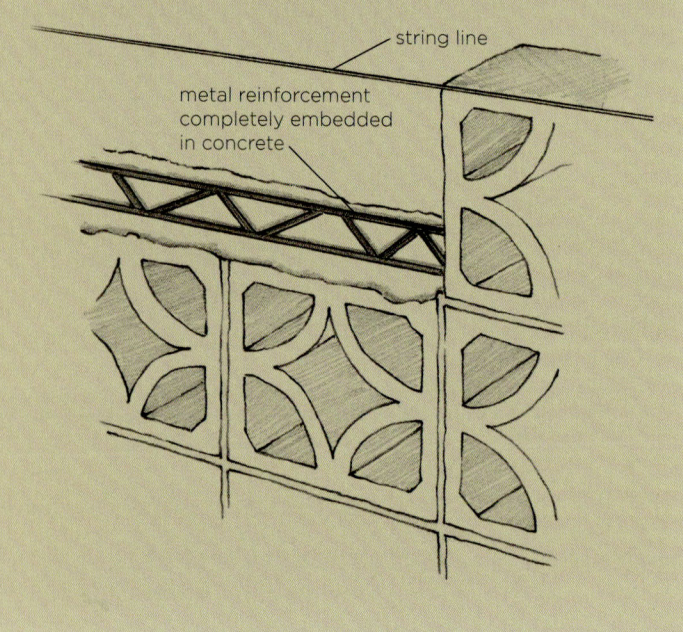

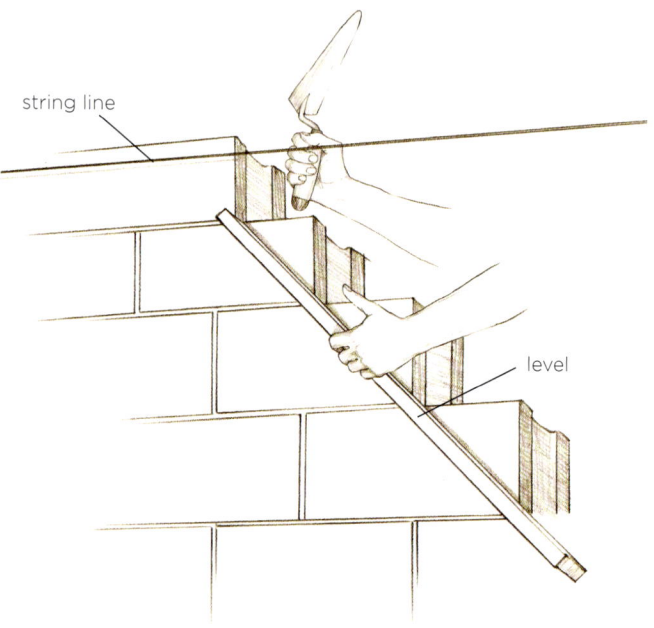

Use a string line and frequent checks with a level to ensure that the wall is going up plumb.

Mortarless Block Walls with Stuccoed Finish

Once you master the basic technique of laying concrete block, you will almost certainly conclude that it is easy work. Now imagine how much simpler still the job would be if you could skip the mortar and just dry-stack the blocks. With a product known as surface-bonding cement, you can do just that. There is a catch, of course. Although the system is frequently referred to as "mortarless block construction," which sounds suspiciously free of labor, it really isn't entirely free of mortar. The first course must be set in a bed of mortar on a concrete footing and, although you do not need to apply mortar to any other joints between blocks, you do need to cover the surface of the wall with a single coat of surface-bonding cement.

Surface-bonding cement is similar to the stucco that covers the walls of houses all over the world, but with a slightly different formulation that includes cement, fiberglass, fine sand, and additives. And unlike traditional stucco, this product requires only a single coat. You can choose from different color blends of the cement mix, or you may want to create a unique tint using separate colorants.

The surface texture can be smooth, rough, or something in between, depending on how you handle the trowel. There is really a lot of room for creativity with this type of wall. I have seen beautiful garden walls in vivid colors with tiles used as a vertical border, as random accent, and as coping (the top layer). If you decide to use tiles, make sure that they are suitable for exterior use, and set them in exterior-grade thinset adhesive fortified with acrylic.

Although the dry-stacked blocks may not seem to offer much strength (which they wouldn't, without the cement coating), the technique has been used successfully to build strong, weatherproof walls for houses and other structures for many years. The footing and first course should be prepared as described for a mortared block wall. After that, blocks are stacked in a running bond pattern without any mortar between courses. This part of the job goes quickly, but it should not be done carelessly. Check the wall for plumb regularly, and I suggest that you use a string line to hold the alignment from one end to the other.

With this type of construction, it is particularly important that the concrete blocks be clean and structurally sound. Wash off any surface dirt or grease, and file down chips or bumps that interfere with a tight fit.

Mix the surface-bonding cement as instructed on the bag. You will need a steel finishing trowel and a groover. A mortar hawk will come in handy, although being the tightwad that I am I use a piece of plywood for this purpose (usually wishing that I had a hawk, mind you). You should also have a garden hose with a nozzle that can apply a light mist to the concrete block. A pair of gloves will come in handy for protecting your hands.

Begin working on the shady side of the wall. Mist one section thoroughly with water to prevent the blocks from absorbing the water out of the cement mix. Working from the bottom of the wall, apply a ⅛-inch-thick layer of cement with your trowel. Hold the trowel at a slight angle, as shown in the illustration, and use upward strokes.

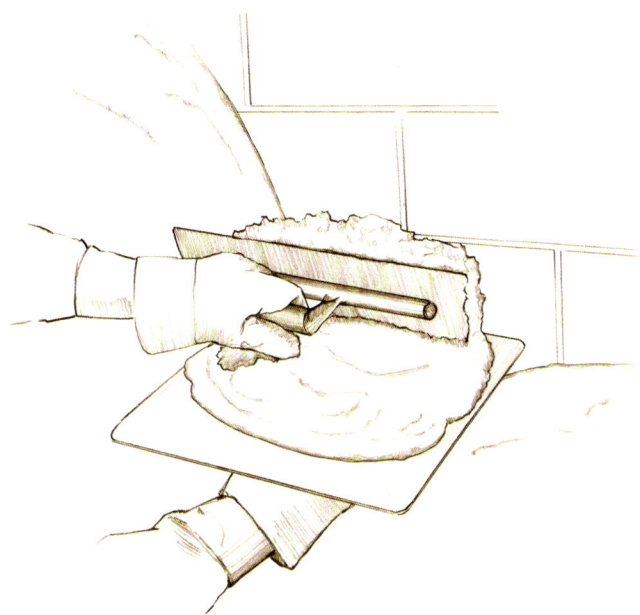

Surface-bonding cement is applied to the wall with a finishing trowel, working from the bottom up. Scoop some cement onto a hawk, then hold the hawk against the wall as you slide the cement off and up with the trowel.

Block Options Concrete block

is available in a range of decorative styles, such as the popular screen shown below. But standard, dull concrete block can be put to exceptionally attractive use, as in the other photographs here, when coated with surface-bonding cement.

From time to time, clean off the trowel and, while it's still wet, smooth the surface into a texture that you like. You should create vertical control joints to handle any cracks that appear in the wall in the future. With control joints, those inevitable cracks will occur in the joint, where it will not be so visible and random. Use a groover to form the joints, moving from top to bottom. A 2-foot-high wall should have control joints every 4 feet, a 4-foot-high wall every 8 feet, and a 6-foot-high wall every 12 feet. Fill the control joints with silicone caulk once the surface has hardened.

When you finish the wall, or wrap up work for the day, cover the wall with plastic. Mist it with water once or twice a day for 3 days.

Control joints keep cracks from appearing on the face of the wall. Use a groover to create straight control joints from top to bottom.

Know Your Cement

For footings, slabs, and similar purposes, concrete is usually mixed with Portland cement, which sets very hard. Masonry cement, on the other hand, is not quite as strong, but it cures to produce a more flexible bond that sticks to surfaces better and is less likely to crack. A Portland cement mix, which is usually sold in bags as "cement mix," is therefore not the best choice with stones, brick, or concrete block. Instead, look for bags of blended "mortar mix" for this type of work. You may also find "mason's mix" that contains no masonry cement at all. Instead, it is composed of Portland cement and lime, which together produce a good combination of strength and flexibility.

Masons often prefer one formula over the other, depending on the specific job at hand, and some prefer to mix their own. I don't think that you need to worry about the specific composition. Buying bags of blended mortar mix, to which you need add only water, is the ideal way to approach stone work for a do-it-yourselfer. For reassurance, read the instructions on the bag thoroughly, and don't hesitate to check the manufacturer's website for descriptions of different mixes. ■

CHAPTER

5

Metal Fences

· ·

METAL IS ADAPTABLE to almost any fencing need that you have, from high-end residential to remote, rural applications. Most types of metal fence can be constructed quickly and require little upkeep and maintenance down the road. This chapter discusses a wide variety of fence styles, which perform an equally wide variety of duties. What unites them is simply that they are all metal.

Metal fences cover the gamut, from high-priced, exquisitely crafted, costly, and durable additions to the front yard, to bargain-basement, utilitarian, and flimsy solutions to an immediate threat to a garden or a need for animal control. Wrought-iron fences grace fine homes and are intended to be seen and admired. Chain-link fences are usually not much to look at, but they provide first-rate security at a reasonable price. And the various types of fencing meant to protect crops and gardens or secure pets and livestock are often more beast than beauty.

Some of the finest fences and gates ever assembled started life in the blacksmith's forge. The exquisite craftsmanship evident in these wrought-iron creations is hard to come by today, and even harder to pay for. Fortunately, more affordable metal fencing products are widely available, and they are much more accommodating to do-it-yourself fence builders.

Ornamental Metal Fences

It is hard to think of any fencing material that offers a greater combination of security, beauty, strength, durability, and see-through visibility than wrought iron. The techniques of metalworking, and society's esteem for the smith, have been with us long enough to have produced the Greek smith-god Hephaistos, his Hindu and Roman equivalents Vishwakarma and Vulcan, the Norse god Thor, and the legendary figure of Anglo Saxon and Old Norse literature, Wayland the Smith.

Early wrought iron was used in weaponry and for digging tools, but it also caught on quickly for decorative purposes, as scepters, hinges, weathercocks, and scores of other uses. I don't know when and where wrought-iron fences and gates first appeared, but examples of thousand-year-old architectural ironwork survive today. Wrought-iron fences adorn and protect private residences and public buildings all over the world. Building these fences, however, required a lot of labor. Each piece of iron had to be heated and shaped or twisted, and then individual pieces had to be joined together. In time, wrought iron was joined (and frequently replaced) by cast iron, for which single molds could be used to create endless quantities of identical pieces.

These days, when looking at ornamental fencing, the first question that ought to come to mind is, Is it real, or is it . . . ? The longstanding appreciation for the craft continues today in the form of widespread attachment to the traditional look of wrought-iron and cast-iron fencing. But most of what passes for wrought iron these days is really a much lighter and less expensive imitation made out of aluminum, steel, or (the new, nonmetallic kid on the block) composite

KIT ASSEMBLY. Ornamental fence kits normally use bolts and screws to attach rails to posts, but the exact mechanism can differ from manufacturer to manufacturer.

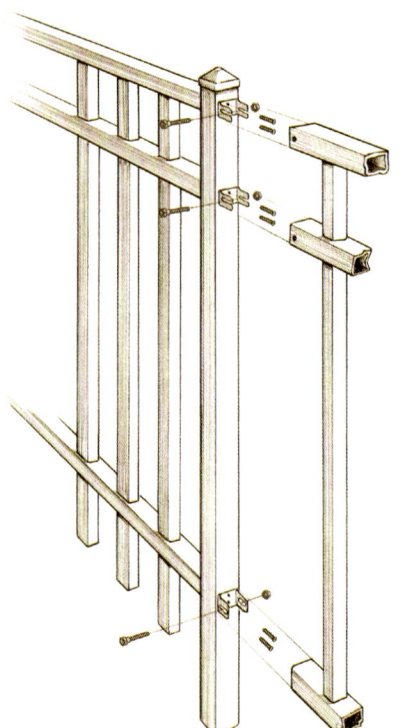

INSET BRACKET ATTACHMENTS

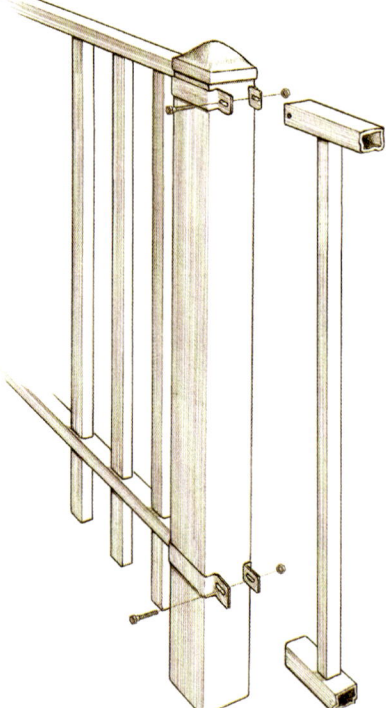

DECORATIVE WRAPAROUND BRACKET ATTACHMENTS

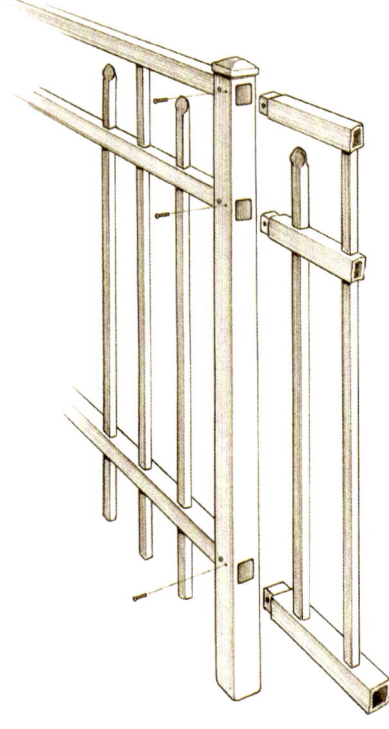

"MORTISED" POSTS WITH "TENON" RAILS

or polymer materials. All of these materials are available for fencing in styles that mimic traditional wrought iron, at least from a distance, as well as in more contemporary styles. They are sold in kits that can be easily transported and assembled by do-it-yourselfers. Not too many years ago, you had to shop around to find suppliers of ornamental fencing, but it is now showing up with increasing frequency in home improvement centers.

Black is obviously the color of choice for such fences, but it is no longer the only one. You can find powder-coated steel and aluminum ornamental fencing in lighter colors that may better complement their setting. These fences seem to be particularly popular as swimming pool enclosures, where brighter colors are more suitable.

Ornamental fencing is sold in kit form, and you should plan to buy all of the materials from the same manufacturer. Each product has its own specific style of assembly, and its own assembly instructions; if you shop around long enough, you might be surprised at the array of styles, sizes, and accessories that are available. Welded steel tends to be somewhat stronger than aluminum, but the latter offers the very great advantage of being rust-resistant and, therefore, requiring less long-term maintenance. Steel fence sections are often welded on site by contractors, or they can be purchased in assembled sections by do-it-yourselfers, while aluminum is more likely to be put together with screws, nuts, and brackets.

The gate assemblies are a particularly nice feature of ornamental fencing. Most manufacturers offer several options for upgrading the strength and security of the fence entryway. Self-closing and self-latching gates, complete with keyed deadbolts in an easily installed package, are a distinct advantage over other gate styles.

In recent years, a new type of ornamental fencing has begun appearing in home improvement centers. It has the look of traditional black wrought iron but contains no metal whatsoever. Instead, it is manufactured from a mixture of polypropylene and fiberglass, which means that it is essentially the same material that is used to make those composite decking boards that have become so popular in recent years. The manufacturer claims that the new composite fencing

Wrought-iron fences are often characterized by the same decorative features that blacksmiths applied to stair railings and other architectural elements.

is fade-resistant, durable, and virtually maintenance-free. It will not rust and never needs to be painted or stained.

The composite fencing system I have looked at is available in 6-foot-long sections in heights of 4 to 5 feet. The sections are joined by concealed fasteners, which is a nice improvement over aluminum fences, and the gates are steel-reinforced, with self-closing hinges and latches. Installation looks pretty easy and is similar to that for other types of ornamental fencing. First, establish the fence line with string stretching from corner to corner. Measure and mark the post-hole locations. Bury posts in holes at least 30 inches deep, centered every 6 feet. After setting the first post, immediately assemble one fence section and slide it into the post. Set the next post in its hole, attach it to the section, and pour concrete in the hole. Regularly check the posts for plumb and the rails for level, then secure the top rail to the posts with screws driven through the inside of the hollow posts. Finally, attach post caps and, if desired, finial caps on the pickets.

I am never eager to recommend construction products that have not proven their worth in real-life settings over time. But composite decking has shown its value over the past decade or so, and the prices and warranties I've seen for composite fencing make it appear to be worth careful consideration. If I were seriously considering a faux-wrought-iron fence for my yard, I would take a close, hard look at the benefits of composite material over steel and aluminum.

An Ornamental Array

Ornamental metal fences are nice to look at in themselves, but as these examples demonstrate, they can also provide a substantial amount of security without obscuring views of what lies beyond.

Chain-Link Fences

First produced in europe in the 1850s, chain link became a fixture of the American landscape through the twentieth century. It offers a great combination of security with visibility, and is strong, long-lasting, and affordable. It stands up well to regular use and is relatively easy to install.

Standard galvanized chain link will not, however, win any beauty contests. There aren't many houses that one would consider good design matches for chain link, by which I mean that it is not a material to be recommended for front-yard fences in most cases. But there are options available that can make the fence more attractive. Color-coated chain link can blend in nicely with many environments. I have seen chain-link fences coated in green, brown, and black that serve their intended function and yet are fairly easy to overlook, which isn't a bad objective.

Colored slats that weave through the mesh are also available. You can buy fencing with slats already installed, or you can add your own later. Slatted chain-link fence is more useful for adding privacy than for boosting visual appeal, in my opinion, but you may well think otherwise. The slats are available in a wide range of colors.

I have also seen one brand of colored chain-link fencing that uses square posts, which look nicer than the standard round posts, if for no other reason than because it is different. And chain link can be improved visually by surrounding the fence line with plants that grow high enough to hide it or vines that climb up and through the mesh.

Chain link can be found in heights ranging from 3 feet, for simple garden barriers, to 12 feet, for tennis courts and public swimming pools. The most common height is 4 feet, which is an easy height for do-it-yourselfers to install. Higher fences can be tricky to install, however, and you might want to leave that work to a fencing contractor.

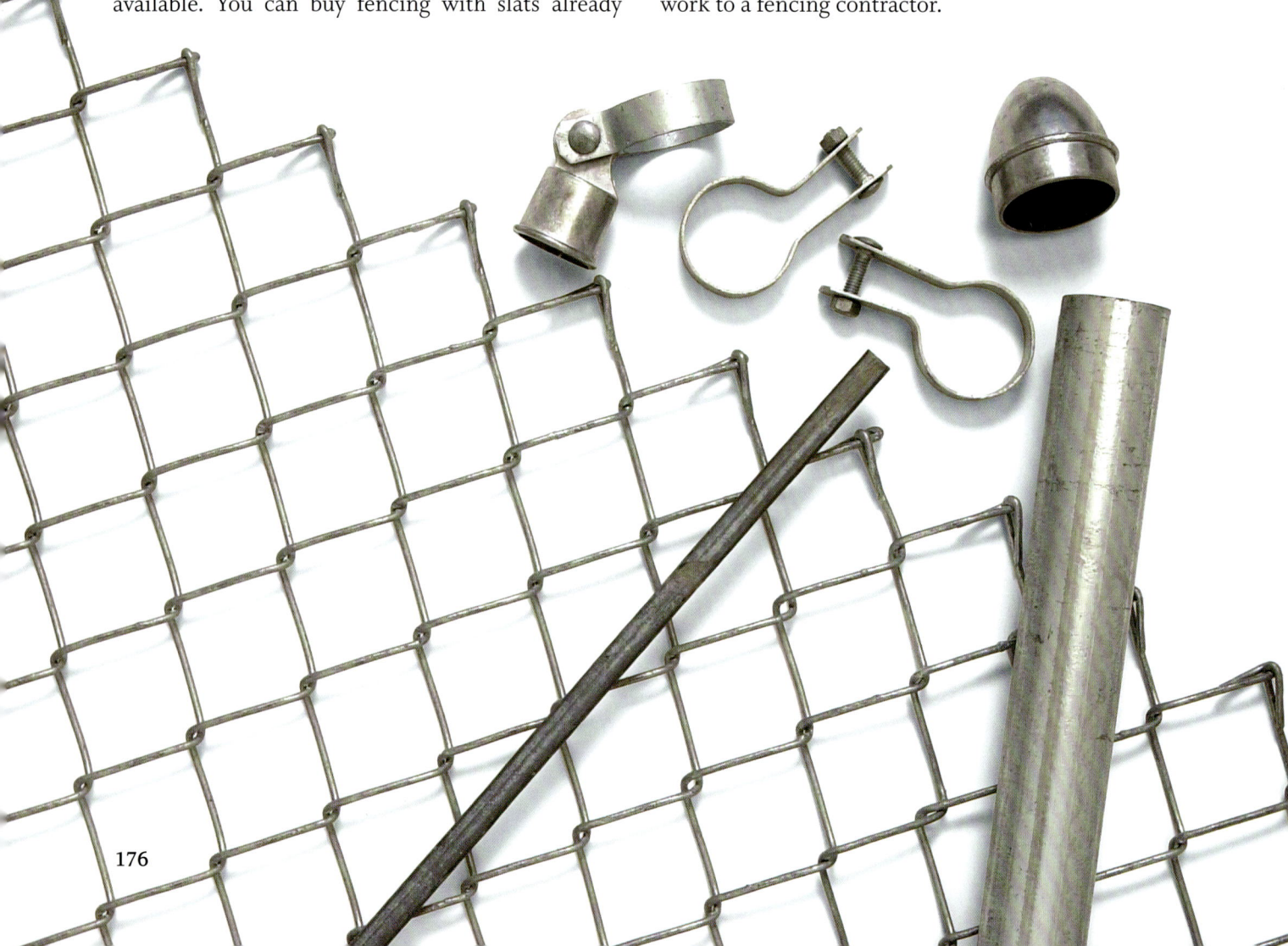

Shopping for Parts

There are many manufacturers of chain link, and although the parts are often interchangeable, I think it makes best sense to buy everything you need from one supplier. There are quality differences, however, so it does pay to shop around and ask some questions.

After deciding on height and surface coating, the key factors to consider are gauge and mesh size. Gauge is a standard measurement of the diameter of wire: The higher the number, the thinner (and weaker) the metal. You can find chain link in gauges ranging from 6 to 14, but for most residential purposes I would suggest something in the range of 9 to 11½. Wire that is thinner than 11½ gauge stands a much better chance of being damaged over time.

The size of the mesh relates to both visibility and strength. Smaller mesh means that more wire must be used in the fence, which adds to its strength (while reducing visibility). Mesh under 1 inch is generally found only in high-security installations. Swimming pools and tennis courts are often enclosed with 1¼- to 1¾-inch mesh, while standard yard fences rely on mesh of 2 inches or more. Obviously, thicker-gauge metal and smaller mesh size will translate into greater

expense, so you will want to choose carefully. One other factor to keep in mind: Kids have a strong preference for bigger mesh, which allows ample room for their toes. Which is to say that the bigger the mesh, the easier the fence is to climb.

The fencing itself (which is actually called "fabric" in the trade) is not the only component that affects strength and durability. The posts and rails that constitute the fence frame are also offered in different strengths and sizes. Manufacturers seem to break down the choices into categories based on the height of the fence and the degree of use it will get (light, medium, or heavy). Terminal posts (those used at ends, corners, and gate openings) are the biggest, followed in size by the intermediate posts and the top rail. Sleeve connectors are used to join top rail sections.

The fabric is joined to the frame with a variety of fittings, including bands, rings, nuts, bolts, and gate hinges and latches. The completed fence is only as strong as these fittings, so you might want to compare the hardware that accompanies different product lines. Make sure that any warranty that comes with the fencing covers all components. And be sure to check with your local building department before making any purchases to see whether you need a

ANATOMY OF A CHAIN-LINK FENCE

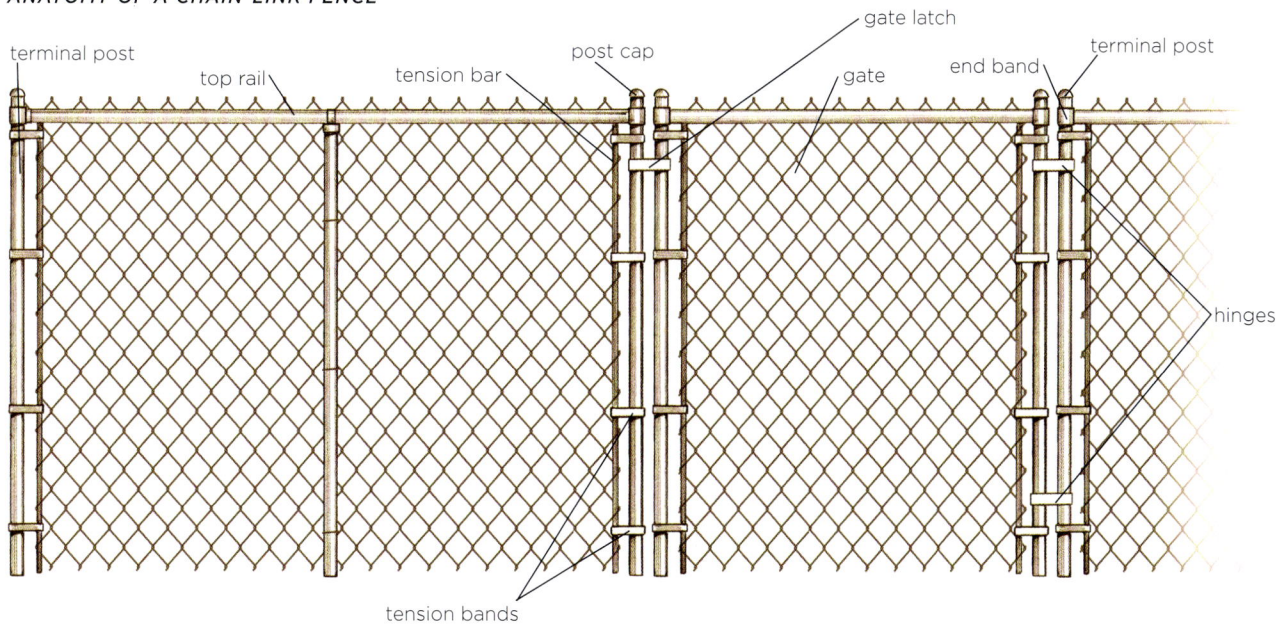

building permit and whether you are limited by other restrictions on height and materials. Once all of the shopping and paperwork are done, the installation is pretty straightforward. Your fencing may or may not come with detailed instructions. The following guidelines should help you through a typical installation but should not be treated as a substitute for any instructions from the manufacturer.

Setting the Posts

The first tasks in installing a chain-link fence are similar to those for a wood fence, and so I suggest that before proceeding you read the sections in Chapter 3 on setting posts, digging holes, and mixing concrete. Set up string lines, as described in Chapter 2, to mark the location of your planned fence, making sure that all corners are square and that you honor any required setback. Drive stakes into the ground to mark locations for all terminal posts and then add stakes for intermediate posts. The fence manufacturer should instruct you on how far apart the posts should be spaced, but remember that you can reduce this maximum spacing a bit to ensure that the posts are evenly spaced. Do not, however, follow this policy with gate posts, which must be spaced exactly as directed.

If you are renting an auger to dig the holes, or hiring someone else to do the job, you will want to dig the holes all at once. But I suggest that you then set all of the terminal posts before setting any intermediate posts. Terminal posts are set in larger-diameter holes, and they need to be a few inches higher than intermediate posts.

Dig 8-inch-diameter holes for the terminal posts and 6-inch-diameter holes for intermediate posts. In cold climates, dig all holes below the frost line. In

LAYING OUT THE FENCE LINE

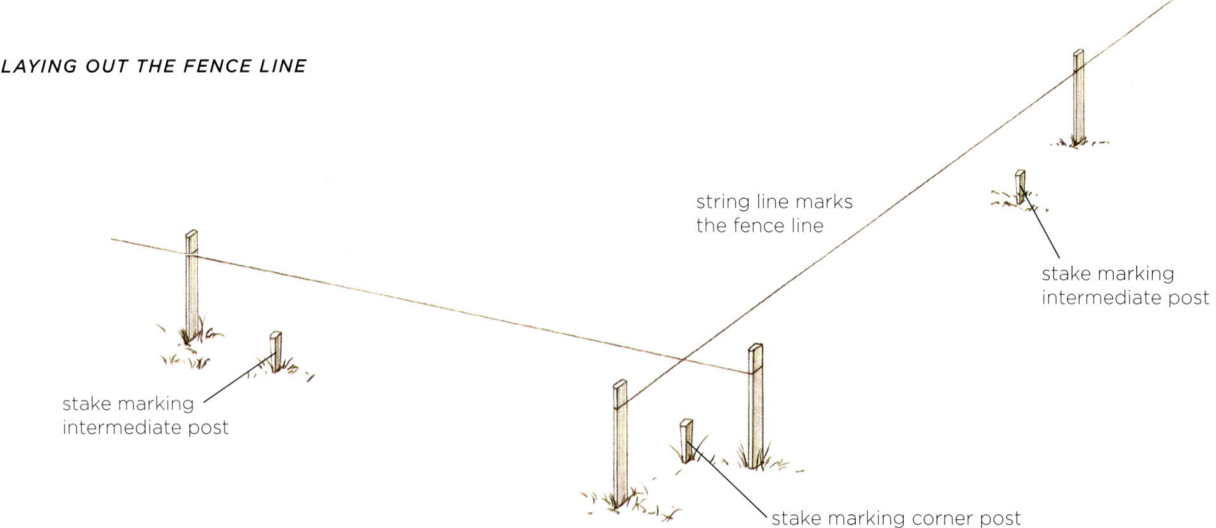

string line marks the fence line

stake marking intermediate post

stake marking intermediate post

stake marking corner post

SETTING THE POSTS

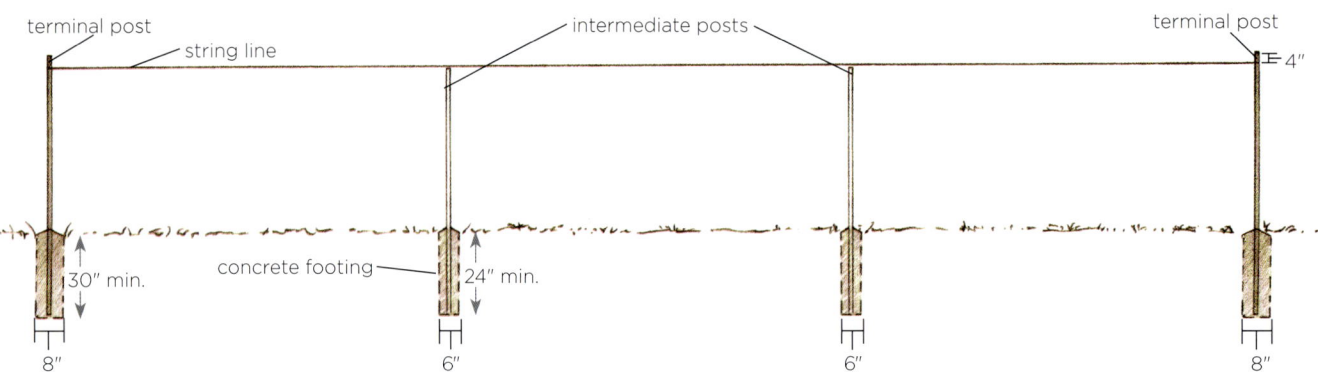

terminal post

string line

intermediate posts

terminal post

4"

concrete footing

30" min.

24" min.

8"

6"

6"

8"

climates where frost is not a concern, terminal posts should be set in 30-inch-deep holes and intermediate posts in 24-inch-deep holes. Brace the terminal posts centered in their holes so that they are plumb, and make sure that the tops of all posts are level. Use some duct tape to hold the posts to the braces if you do not have suitable clamps. The standard instructions are to set terminal posts 2 inches higher than the height of the fencing material, but you should check with your manufacturer for specifics on these dimensions.

If you are installing the fence on uneven terrain, you will have to make some adjustments. On a gradual slope, the posts can all be set the same height aboveground, which allows the fencing to be installed so that it follows the slope perfectly. More pronounced slopes may require that the fencing be installed in a terraced pattern. For more complex sites, I suggest you speak directly with your supplier or the manufacturer.

Mix concrete and pour it carefully into the holes. Use a margin trowel to smooth the concrete and slope it slightly away from the post in each hole. Let the concrete cure for at least a day or two before disturbing the posts. When the concrete has hardened, tie a string line between the posts exactly 4 inches down from the top. The string line must be taut and it should be centered on the posts, not aligned with one side or the other. Set the intermediate posts in their holes in concrete, centered with and at the height of the string line.

CHAIN LINK ON SLOPES. On gradual slopes [TOP], posts can be installed at the same height aboveground and the fencing installed to follow the slope. On steep or irregular slopes [BOTTOM], posts may have to be installed at varying heights and the bottom of the fencing cut to follow the slope.

GRADUAL SLOPE

IRREGULAR SLOPE

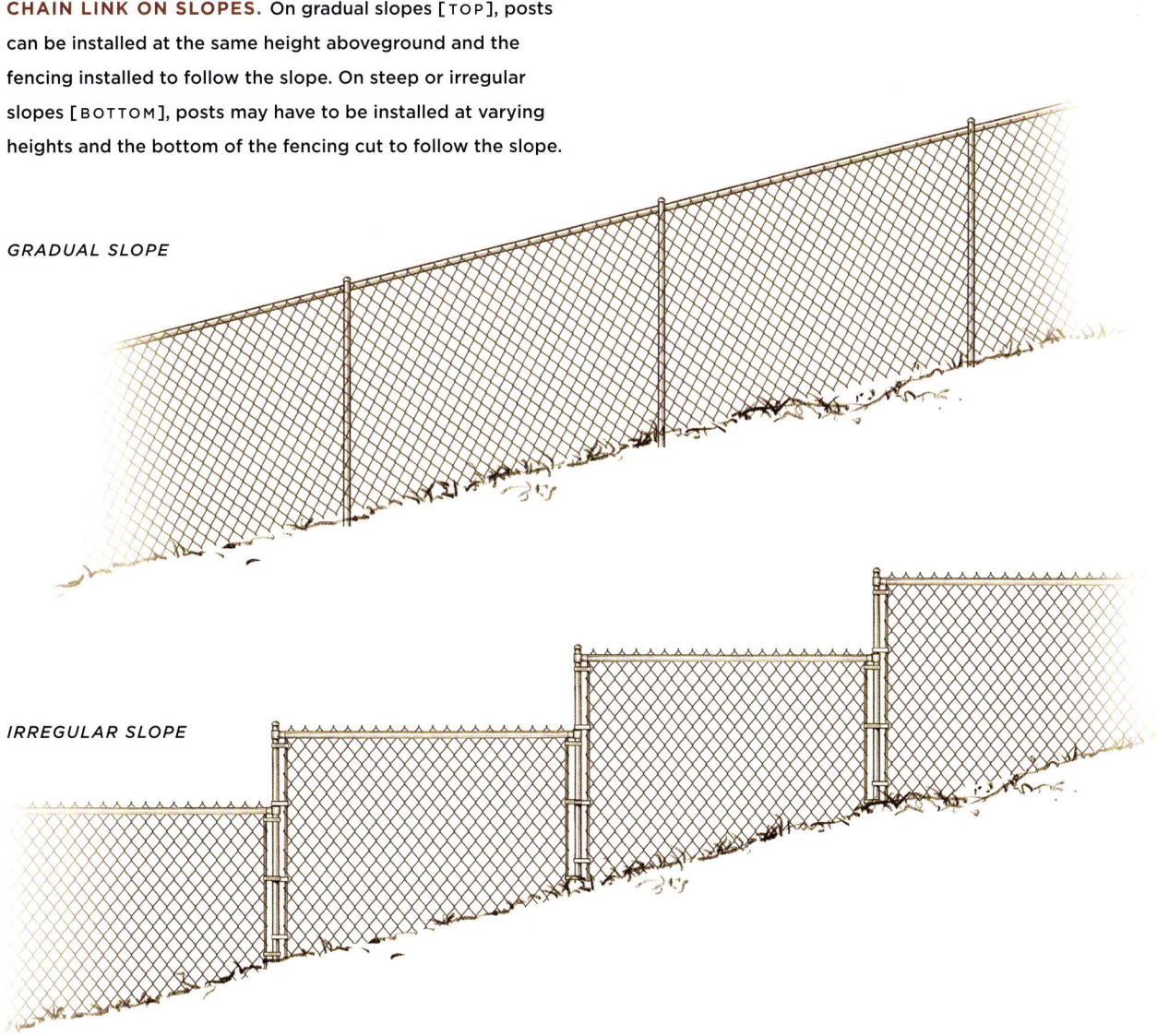

Attaching Fasteners and Rails

Allow the concrete around the intermediate posts to cure for a day or two, then begin attaching the fittings to the terminal posts. Typically, you will need three tension bands on each post up to 5 feet high (adding one for each additional foot of height), along with one rail-end fitting and a post cap. Corner posts require double the number of tension bands and rail-end fittings, since they serve fencing that is heading in two different directions. These bands slip easily over the posts and are secured with bolts and nuts.

TERMINAL POST FITTING

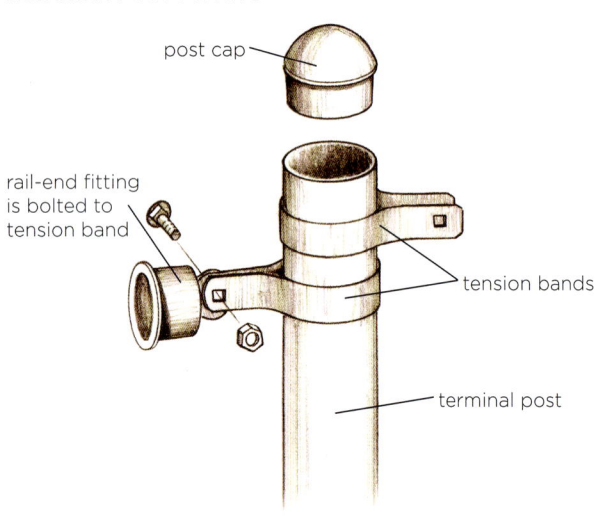

post cap

rail-end fitting is bolted to tension band

tension bands

terminal post

INTERMEDIATE POST FITTING

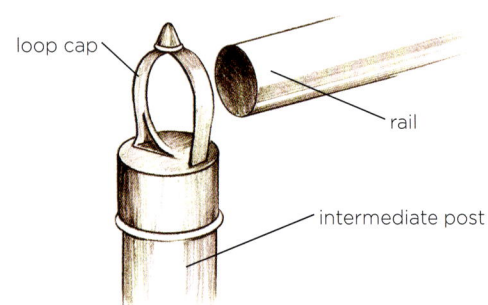

loop cap

rail

intermediate post

SLEEVE CONNECTOR

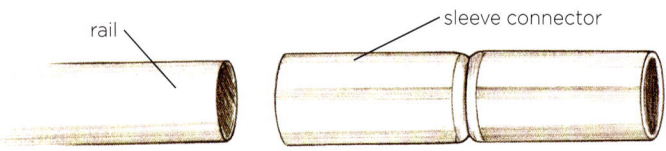

rail

sleeve connector

Set loop caps on all of the intermediate posts and slide the rail through them and into the rail-end fittings. If necessary, cut the rails with a hacksaw and use sleeve connectors to combine two rail sections. (Note that some types of rail are made with smaller diameters on one end, allowing them to slip into another rail, and therefore require no connectors.)

Adding the Fencing

You will probably find this step easier to accomplish if you have a helper to hold up the fencing while you make the connections. Unroll the fencing along the fence line, stretching from one terminal post to the next. Slide a tension bar through the mesh on one end, and then fasten the tension bar to the tension bands with bolts and nuts. Tighten the nuts with a socket wrench while holding the bolt head steady with an adjustable wrench.

The key to a strong, secure chain-link fence is to stretch the fencing tightly to reduce any slack. Your supplier should be able to rent, sell, or loan you a special pulling bar or stretcher to simplify this work. You can do a satisfactory job with the pulling bar alone, but the best results come from using it with a come-along.

Use tension bands to tie the terminal post to the tension bar. Fasten the nuts to the bolts tightly with a socket wrench.

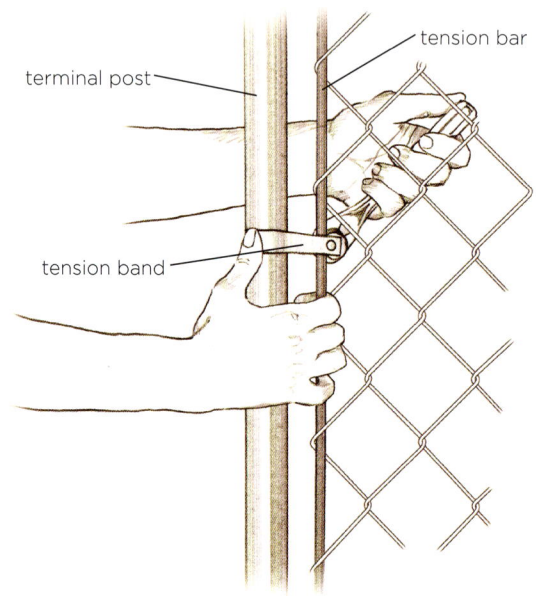

tension bar

terminal post

tension band

Attach the pulling bar to the fence, or to a stretcher or tension bar inserted into the fence temporarily, and then hook one end of the come-along to the pulling bar and the other to the next terminal post. Slowly tighten the fence until it is taut, and then insert a tension bar and attach it to the tension bands on the terminal posts.

If the fencing is too long, use pliers to open a loop at the top and bottom of the fence at the spot you want it to separate. Unweave one of the wires through the links until the fence comes apart. To join two sections of fencing, first remove a strand of wire from the end of one section, and then use that wire to tie the two pieces together.

With the fencing attached to the terminal posts, use tie wire every 12 inches to secure the mesh to the rails and posts. Some suppliers offer special S-shaped ties to simplify this task. Repeat this process for each section of fencing.

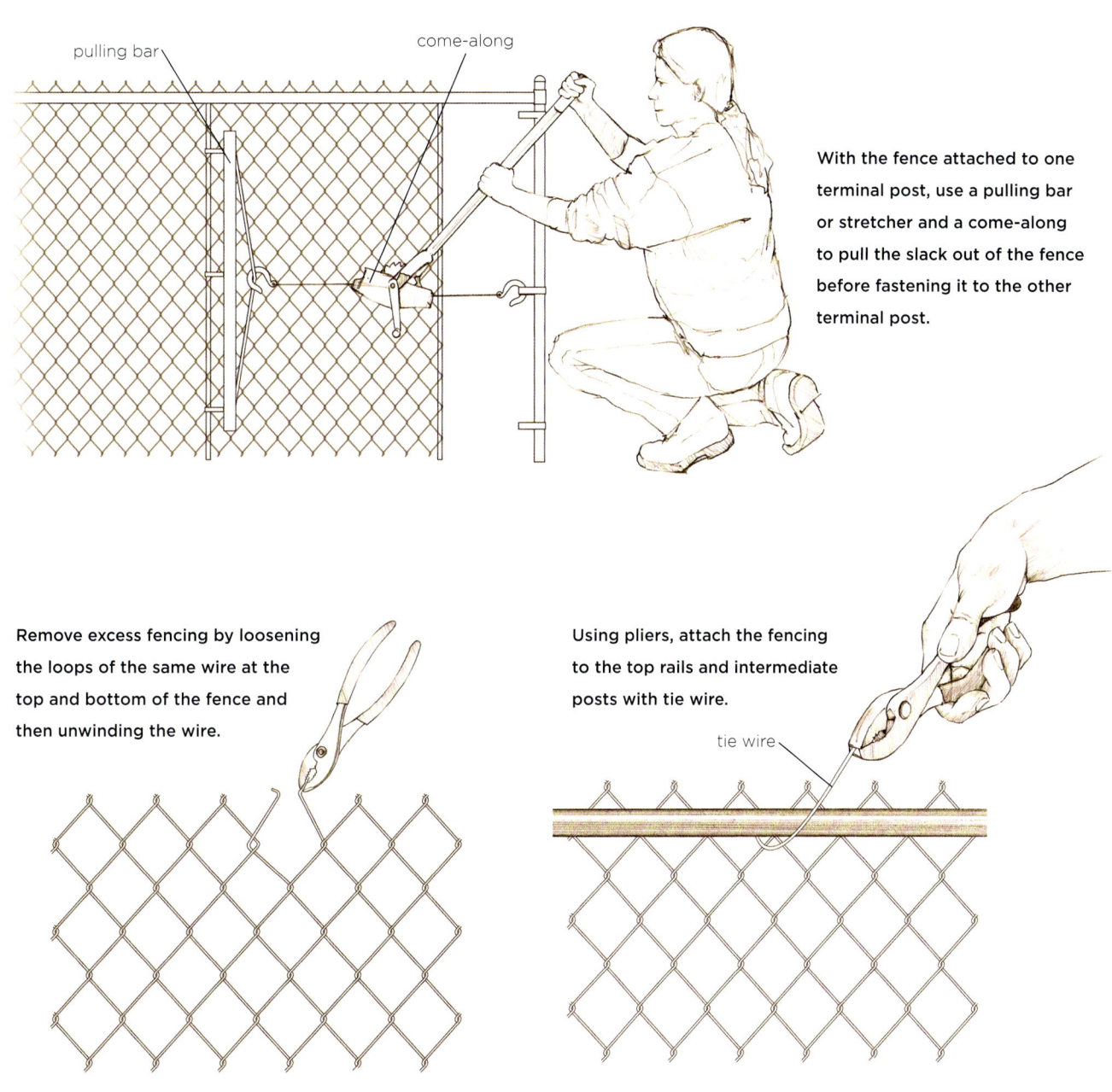

pulling bar

come-along

With the fence attached to one terminal post, use a pulling bar or stretcher and a come-along to pull the slack out of the fence before fastening it to the other terminal post.

Remove excess fencing by loosening the loops of the same wire at the top and bottom of the fence and then unwinding the wire.

Using pliers, attach the fencing to the top rails and intermediate posts with tie wire.

tie wire

Dressing Up Chain Link

A chain-link fence will look nicer and provide substantially more privacy if you add colorful plastic slats to the mesh. The slats can be installed vertically, as shown here, or at a diagonal. You can also buy fence with slats already installed.

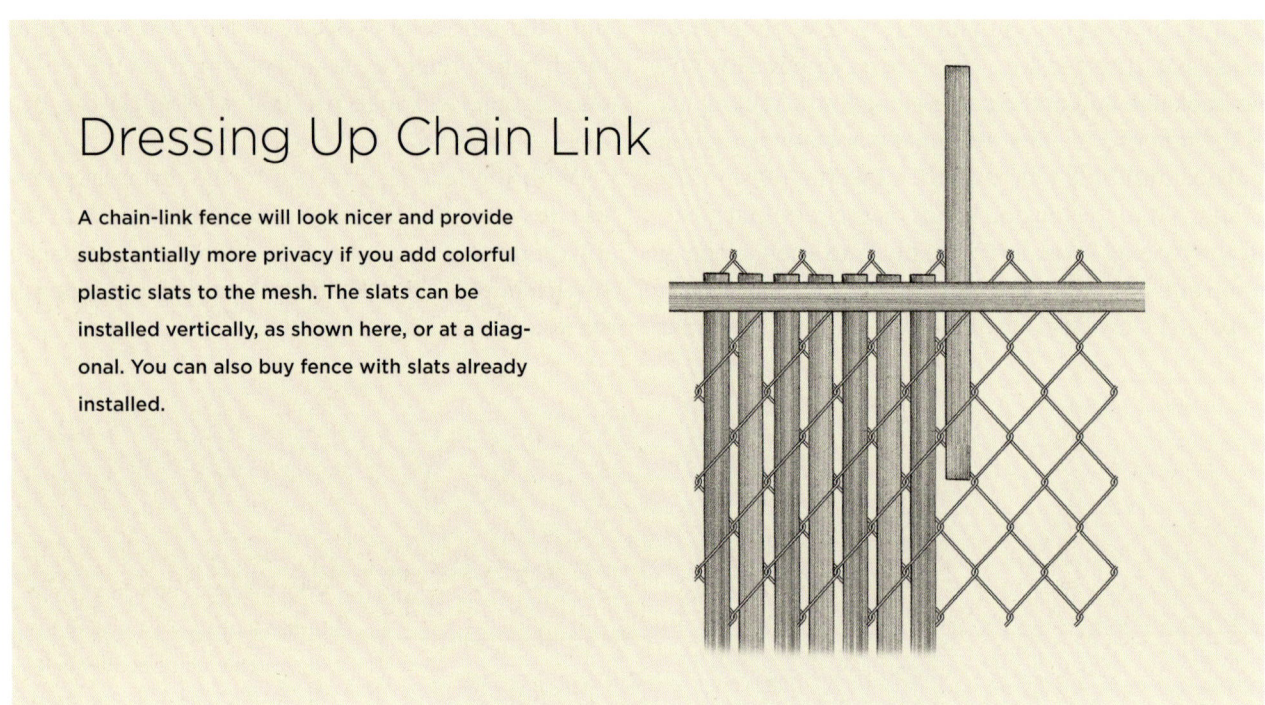

Chain-link fences and warning signs seem to go together. That's because chain link provides good security at a reasonable cost.

Installing the Gate

Virtually everyone who installs a chain-link fence will also install one or more matching gates. Although the gates I have seen look pretty indistinguishable from one product to another, I do suggest that you buy your gate from the same source as the rest of the fence. Choose a gate with the same gauge, mesh size, and frame as the rest of the fence.

Adding a gate is really all about attaching hardware, since you do not need to worry about attaching the fencing material itself to posts and rails. Just be sure that you space the gate posts properly for the gate you choose. As a general rule, if you buy a 42-inch gate, you should leave a 42-inch opening between the posts.

First decide which way you want the gate to swing, then attach the two gate post hinges to the appropriate gate post, about 8 inches from the top and bottom. (Note that a high gate may require additional hinges.) The pins on the hinges should point toward each other, as shown in the illustration. Next, attach the gate frame hinges to the gate frame, and then hang the gate in place (a helper will come in handy here). With the bottom of the gate about 2 inches aboveground, tighten all nuts and bolts on the hinges. Finally, fasten the latch to the other side of the gate so that it is at a comfortable height for users.

Climbing vines can grow up and through chain link, providing an attractive cover for a secure but not so attractive fence.

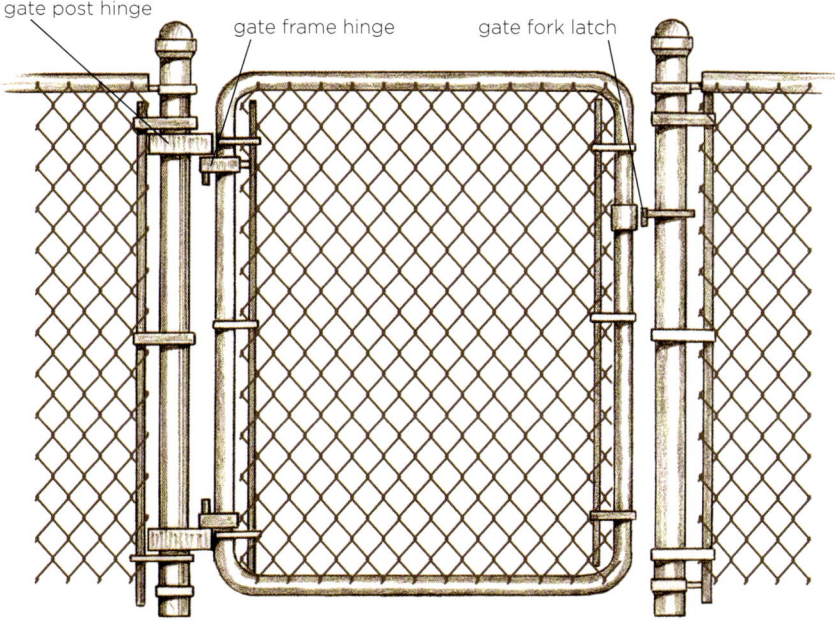

gate post hinge   gate frame hinge   gate fork latch

The first step in properly installing a gate in a chain-link fence is to make sure that the gate posts are spaced properly. The fence manufacturer will supply this dimension. From there, it is a simple matter of attaching the hinges and a latch.

Mesh Fencing

IN TERMS OF SHEER NUMBERS of products—not to mention speed of installation and low cost—nothing beats metal mesh. This is mass-produced fencing boiled down to its most elementary, functional essence. Determine exactly what you want the fencing to keep in or out, then choose the appropriate product. While the on-hand selection may be small at any given store, rest assured that the manufacturer of that limited offering makes many variations on the theme. You should be able to special-order exactly what you need, or you may want to visit a more specialized retailer, such as a garden, landscaping, or agricultural supplier.

Wire mesh has yet to attain any type of architectural chic that I am aware of, although I have long ceased trying to forecast cultural trends. But mesh fencing can serve as a trellis for climbing vines, which can have a particularly wonderful landscaping effect without compromising the basic functionality of the fence.

The only time-consuming part of the installation involves setting posts, but even that chore can be minimized by using metal posts. The great advantage of metal posts is that they are simply driven into the ground—no holes to dig or concrete to prepare. Unfortunately, metal posts aren't nearly as strong as buried wood posts. They are useful for getting a garden enclosure or a temporary fence up quickly, but they should not be used for any fencing that requires much tension or otherwise needs to resist much pressure (such as from heavy animals or adventurous kids).

Choosing the Right Mesh

Mesh fencing can be knotted, welded, or woven, depending on the function of the fence and the gauge of the metal used. Welded and woven fencing is what you are likely to find the most of, but over rolling terrain, knotted fencing can be a great advantage due to its increased flexibility. Some types of knotting create sharp edges, however, and may not be suitable for much contact with people or animals.

The gauge, or thickness, of the wire varies greatly, from the very thin 20- or 24-gauge wire used on netting to the thick 9- or 10-gauge wire used for heavy animal enclosures. The size of the mesh also varies considerably, from ¼-inch squares to 2-inch by 4-inch

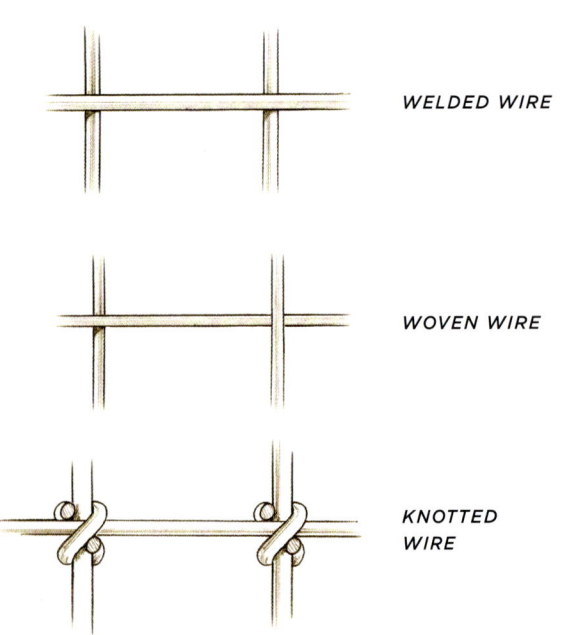

WELDED WIRE

WOVEN WIRE

KNOTTED WIRE

WIRE GAUGES IN END AND LENGTHWISE VIEWS (ACTUAL SIZE)

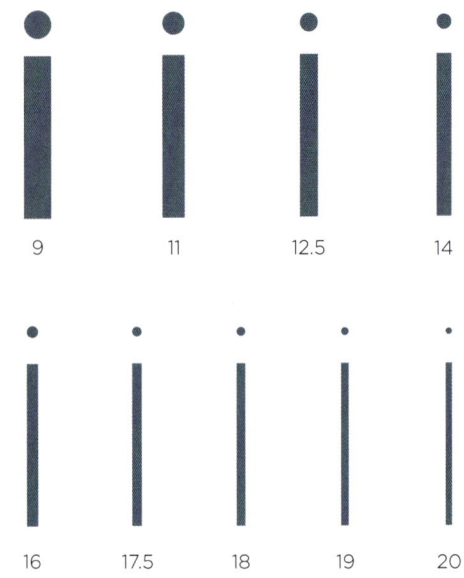

9   11   12.5   14

16   17.5   18   19   20

rectangles or diamonds. Add to these options the various heights of fencing available, ranging from 12 inches to 72 inches, and you can begin to appreciate the scope of product choices. Here I will offer a brief summary of some of the more common classifications of mesh fencing.

Hardware Cloth

This thin (typically 19- to 23-gauge) wire is produced with small mesh sizes (¼ inch or ½ inch square). It is used for light-duty fencing, such as for small rabbit cages and the like.

Poultry Netting

Poultry netting is typically made with 20- or 22-gauge wire, woven into hexagonal shapes. Mesh sizes range from ⅝ inch to 2 inches. The larger mesh is often used to house turkeys, while chicken wire or netting is made with 1-inch mesh. Poultry netting is lightweight and easy to handle, and it is often used for low-cost garden fencing.

Small-Animal Fencing

This material is used to keep rabbits and rodents out of gardens. Its most distinguishing characteristic is that the mesh openings at the bottom of the fence are smaller than those at the top (the assumption being that the critters it is meant to stop are not particularly adept at jumping or climbing). What is commonly called "rabbit netting" is really just a blend of 1-inch-mesh chicken netting on the bottom and 2-inch-mesh turkey netting on top. It is available in heights of about 28 inches. Stronger welded steel fencing uses 16-gauge wire and rectangular mesh, in heights of 28 to 40 inches.

Apron Fencing

This type of mesh fencing has a preformed apron on the bottom that is meant to be laid horizontally a few inches belowground to prevent animals from burrowing under the fence. If you have regular mesh fencing, without an apron, a simple (although somewhat less effective) alternative is to dig a trench several inches deep beneath the fence line and set the bottom of the fencing in the trench.

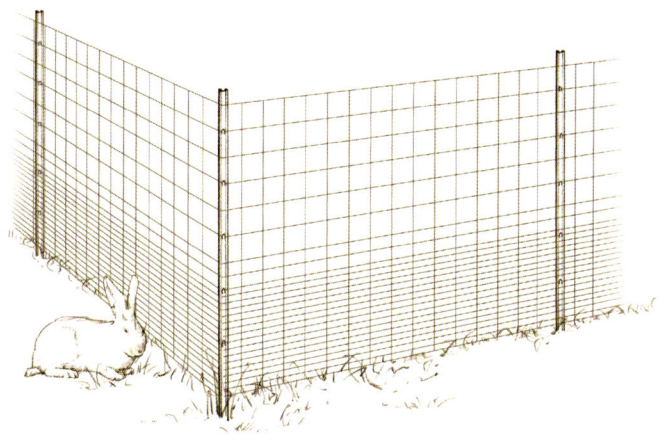

SMALL-ANIMAL FENCING

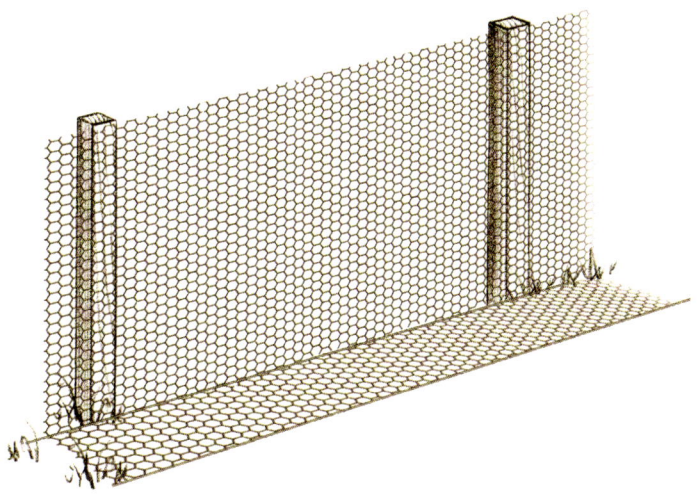

APRON FENCING

General-Purpose Mesh Fencing

Available under a host of descriptive names, such as utility fencing, garden fencing, and kennel fencing, this is typically welded wire fencing with uniform, rectangular mesh that is used for garden and small animal enclosures. You can find mesh sizes ranging from ½ inch × 1 inch to 2 inches × 4 inches, in heights up to 72 inches. Heavy-duty 2-inch mesh makes a less expensive but suitable alternative to chain link. Larger 6-inch-square mesh is commonly used as reinforcement for concrete. It is not galvanized, and so is not suitable for most fencing needs, although I've seen lots of it used to support tomato plants and beans in the garden.

Vinyl-Coated Mesh

Green vinyl-coated 3-inch × 2-inch mesh fencing has a softer and more decorative impact as a garden enclosure than plain galvanized steel.

No-Climb Mesh and Horse Fencing

This heavy 10- or 12-gauge wire in 2-inch × 4-inch mesh is formed in a pattern that prevents horses, cows, sheep, and other hoofed animals from stepping through or using their hoofs to otherwise damage the fence. Standard horse fencing is typically a 2-inch × 4-inch diamond-shaped mesh.

Deer Fencing

This product is, by necessity, made in heights of 7 or 8 feet. It is usually made out of black plastic mesh that has little effect on visibility but is effective in controlling deer pathways. It is often attached to trees, but can also be used with black galvanized steel posts. This fencing will not harm deer, although it is not always successful at stopping a fast running herd.

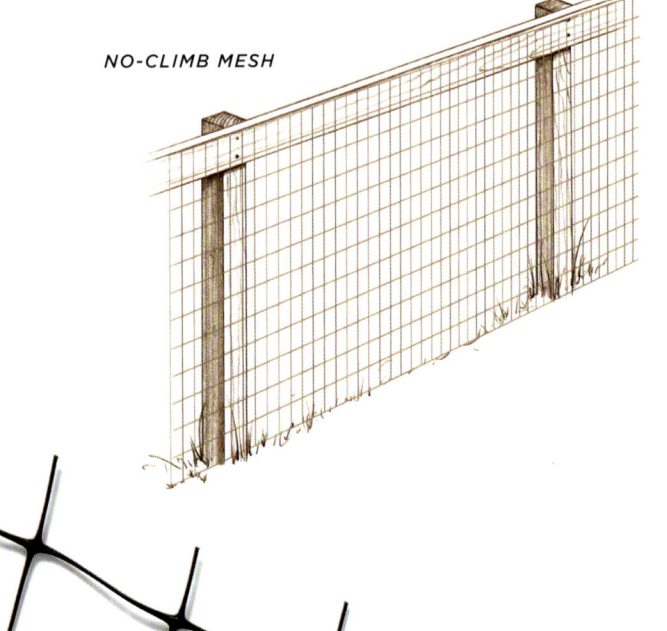

NO-CLIMB MESH

Installing a Mesh Fence

The strongest and most secure construction technique is to attach the mesh fencing to a frame of wood posts and rails. The weakest (and easiest) technique is to attach the fencing to metal posts. A reasonable middle ground for many purposes is to use wood posts for all terminal posts (corners, ends, and gates) and metal for the intermediate or line posts.

First, use a string line to establish a straight fence line, and drive wooden stakes into the ground to mark the location of the posts. There really is no exact science as far as post spacing goes with these types of lightweight fences. Much depends on the size of the posts, the depth at which they are buried, the type of soil they are buried in, the severity of winds, and the specific function of the fence they will be supporting. To ensure a tight fence in residential settings, my recommendation is to space the posts no more than 10 feet apart. For purely agricultural purposes, however, you can usually get away with spacing wood posts up to 16 feet apart and metal posts about 12 feet apart. If you want to research the matter further, talk with a good fence supplier or installer in your area.

Setting Metal Posts

Metal fence posts are often identified by their cross-sectional shape (U-posts, T-posts) and normally have a pointed plate near the bottom that helps hold them in the ground. I suggest that you buy posts long enough so that they can be buried 24 to 30 inches deep (although I acknowledge that this can be difficult to accomplish in some soils). I have seen (well, to be honest, I have purchased) inexpensive metal posts that were so flimsy that you could bend them over your knee. Stay away from these, as they may not even survive the ordeal of being driven into the ground. Posts intended for medium or heavy duty are made with 12- or 13-gauge steel and are worth the extra investment, in my opinion.

Metal fence posts are driven into the ground with a heavy maul or sledgehammer or with a manual or air-powered post driver. Check with your fencing supplier or a local rental yard to see if they have post drivers. You may find it easier to start each post standing on a ladder, and drive the posts so that they remain as close to plumb as possible.

Be sure to install the posts with their hooks or lugs facing the same way, as these are the connectors for the fencing. The general rule of thumb is to install the fencing on the side facing the animals that you wish to keep in or out. Thus, place the fencing on the outside of a garden fence and on the inside of an animal enclosure. Some metal posts require the use of wire clips that are bent into place over the mesh fencing with a pair of pliers. The quicker type of installation comes with posts that have hooks that can be hammered closed after the fencing is slipped into place.

Setting Wood Posts

Information on setting wood posts and installing wood rails can be found in Chapter 3 (pages 76–89 and 90–97, respectively). I prefer to install wood rails on edge rather than flat because they stand much less chance of sagging. Mesh fencing is lightweight, however, so this factor may not be of great concern to you. Still, the temptation to sit on or climb over a fence with flat rails is so much greater than one with rails on edge that you may want to factor in that potential before assembling the fence.

The best way to attach mesh fencing to wood posts and rails is with galvanized U-staples. I suggest using staples no smaller than ¾ inch for this purpose. Roll the fencing from the first post to the second, and then fasten the edge to the first post. Have a helper stretch the fencing as you drive staples every 12 inches or so along the top rail. It is possible to join two sections of fencing together between posts, but I don't recommend doing so. Instead, cut the fencing (use wire cutters) at a post, overlap the next roll of fencing on the same post, and attach the two sections to the post with staples. Once a full roll of fencing has been attached to the top rail, fasten it to the bottom rail or rails.

ATTACHING MESH FENCING TO METAL POST

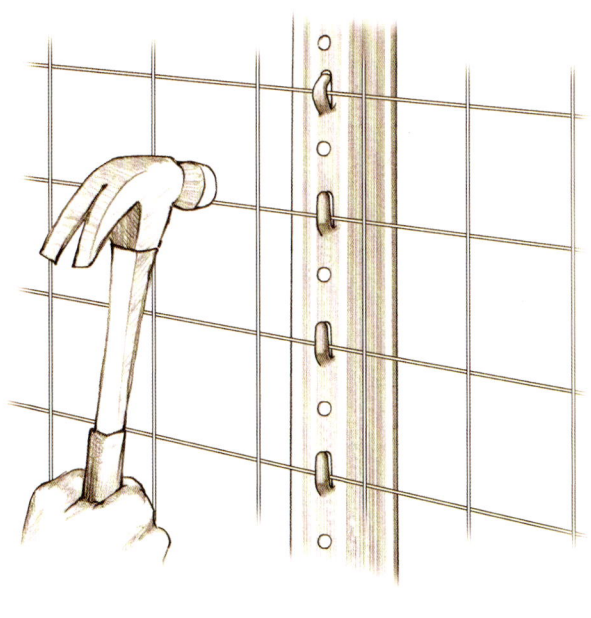

ATTACHING MESH FENCING TO WOOD POST

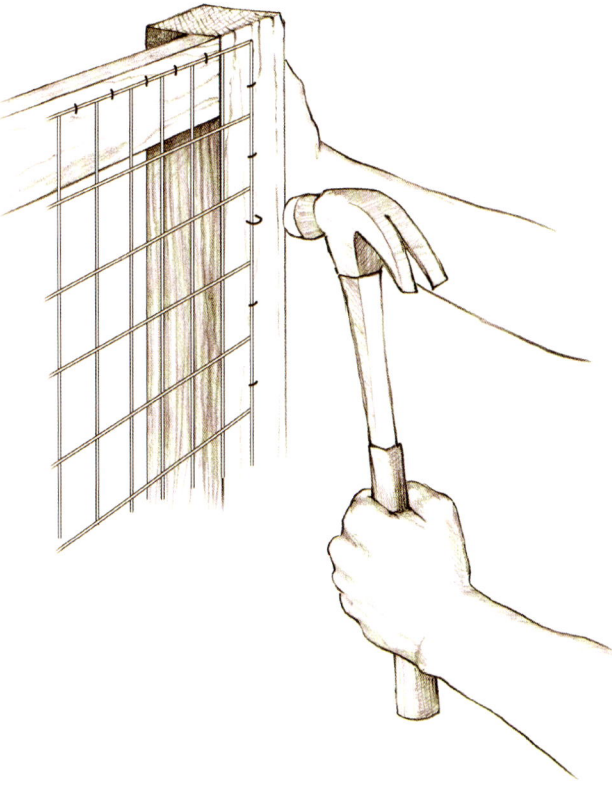

Mesh Plain & Fancy

Mesh fencing is first and foremost a utility-grade product, as shown in the animal enclosure to the right. With a little creativity, however, it can be used for a number of decorative yet functional landscaping and architectural purposes.

Barbed Wire

IT IS SAFE TO SAY that the heyday of barbed wire has passed us by, and I know that I have plenty of company in saying "Good riddance!" Barbed wire played an enormous role in encouraging and supporting the settlement in the nineteenth century of what became the agricultural heartland of North America. It was cheap and easy to install, allowing for the rapid enclosure of thousands of acres of incipient ranch and farmland. The problem with the barbed wire, however, is that it hurts everything and everybody that touches it. Threatening injury is its raison d'être, and I have ruined more than one pair of blue jeans on the stuff.

Barbed wire is still a popular fencing material for cattle, and occasionally for other types of livestock, but it is almost universally banned in residential areas. Although it remains an affordable option, equally effective electric fences (see pages 190–191) can now be installed for even less investment.

Barbed wire is sold in reels of 1,320 feet (or ¼ mile). Manufacturers offer a number of different styles. The wire used ranges from 12½ to 15½ gauge, in either two- or four-point configurations, with the points or barbs spaced 4 or 5 inches apart. As a general rule, thicker wire and a higher number of points make for a more effective and longer-lasting fence. But "high-tensile" barbed wire is now available that is very strong, yet made from thin and lightweight wire. Choose the product based on the size of animals you need to confine. This factor will also determine how many lines of wire you will need on the fence. For mild-mannered cattle, two lines are often enough

Once a ubiquitous and defining feature of the North American countryside, barbed wire continues to keep livestock from straying too far from the farm.

(one at chest height and one at the nose), while bulls may need up to five lines spaced 10 inches apart.

Barbed wire can be attached to either wood or metal posts, in the manner just described for mesh fencing. At each terminal post, wrap the wire tightly around the post a couple of times before fastening it with staples or clips.

BARBED-WIRE STYLES

four-pointed wire

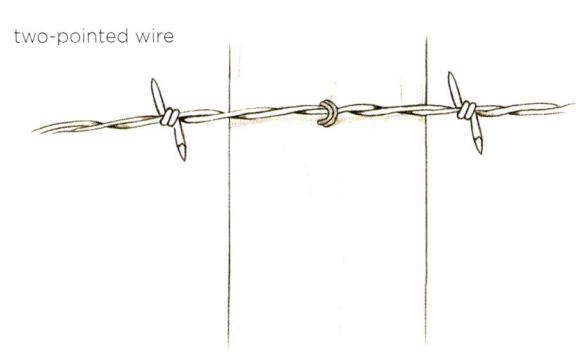

two-pointed wire

Electric Fences

Electric fences are most often used to enclose livestock, but they are also effective at keeping deer and other pests out of the garden and predators like coyotes away from farm animals. If you are unfamiliar with electric fences, you may be of the misconception that they are dangerous. In fact, they are pretty harmless. They give a mild shock, not a life-threatening jolt, to those who come into contact with them. My most intimate experience with an electric fence occurred while I was deer hunting some years ago with a farmer who assured me that the power to the fences on his property was off. I discovered just how mistaken he was when I swung one leg over the fence, permitting the hot wire to hit me in some, well, delicate territory. The shock was real, but I suffered no further injury nor ignominy (aside from an extraordinary chorus of laughter). Needless to say, I now assume that any suspicious wire I encounter near farmland is hot until proven otherwise.

Electric fences are very simple devices consisting of little more than a wire or two attached to insulated posts and an energizer. Installation is a snap. In fact, you can put up a modest-size electric fence quicker than you can set up a volleyball net. But check your local codes first. They may restrict where you can use

electric fences, or even whether you can install one at all, and may also require that you have the fence installed by a professional. Warning signs near the fence are also sometimes required, and they are a good idea even where they aren't legally necessary.

The biggest potential problem with an electric fence is what happens to it when it loses the "electric" component. The single wire (or two) presents absolutely no physical barrier, so as soon as the power is lost the effectiveness of the fence vanishes. The performance of some fences can also be diminished by very dry or snowy conditions.

While electric fences alone are not much to look at, you can combine them with a much more attractive wood fence. In fact, you can make a relatively weak post-and-rail fence into an effective animal enclosure by simply installing a barely noticeable hot wire that will discourage animals from leaning against or trying to pass through the fence.

The Energizer

The energizer (or charger) is the heart and soul of an electric fence. There are many types available, and you will want to choose one that has the range you need and supplies a strong enough jolt to control the size of animal or animals you are concerned with. You should also compare such features as energy consumption, fuse and lightning protection, and warranties. If the fence is likely to be covered with vegetation, which could effectively reduce the jolt that the fence would give to an animal, you should compensate with a more powerful energizer. You don't have to spend a lot of money for a good energizer, but you can make a big mistake by spending too little. Talk to some landowners in your area who have electric fences installed, and plan to do business with an experienced, knowledgeable, and helpful supplier.

There are three basic types of energizer. Standard electric units are plugged into a regular 120-volt outlet. These tend to be the least expensive and the most reliable as long as you buy a UL-approved model with good lightning and surge protection. Regular battery-

A small strip of electrified polytape adds protection to a traditional fence yet remains relatively inconspicuous.

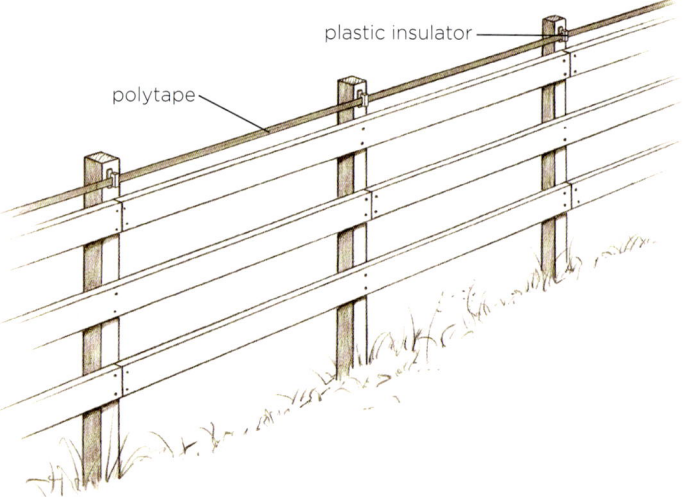

plastic insulator

polytape

operated systems will operate even when household power is off, but the batteries have to be recharged regularly. Solar-powered energizers can be installed far from any electrical power lines. Combination units are available with a battery that will activate when the household current shuts off.

Begin by installing the energizer. Do this yourself only if it is permitted by local code and if you are comfortable working with electrical devices. The energizer should be mounted securely to a post or to the wall of a building near the fence line. Proper grounding of the systems is absolutely critical to safe and efficient operation, and it also happens to be the most common trouble spot. Be sure to follow the manufacturer's instructions carefully and comply with local regulations. Check with the building department to determine whether your installation will need to be inspected.

Conductors and Insulators

A properly sized and installed energizer is just the beginning. Good-quality conductors and insulators are also required to carry the current efficiently through the entire fence. Galvanized steel wire is the traditional choice for a conductor, and 12½ gauge is the most popular size. Polytape, which is essentially electrified plastic ribbon, is another good choice, especially for garden enclosures. Because it is more visible than a strand of wire, it is less likely to be inadvertently touched.

The conductor is supported by posts. Wood and metal posts have long been used to support electric fences. They must be driven into the ground (see pages 76–89 and 186–187) and fitted with insulators to carry the hot wire or wires. Check with the manufacturer for specific instructions on post spacing and insulator attachment.

Fiberglass rods are becoming increasingly popular, and for very good reasons. Fiberglass is self-insulating, and the thin rods can be easily driven into the ground. Some fiberglass rods have a built-in footrest; just set the rod in place and step firmly on the rest to drive it into the ground. The rods are equipped with snaps that can hold steel wire or polytape. This is about as easy as fence building can get.

Electricity for the Garden

The best time to install an electric fence around a garden is before you plant anything. That way you create a barrier before the neighborhood deer, rabbits, and raccoons get into the habit of browsing for lunch in the area. To deter deer from a garden, stretch a strand of polytape around the perimeter about 30 inches aboveground. The deterrent effect will be even better if you tie cloth strips around the polytape every 3 or 4 feet (with the electricity off, of course), and then spray the cloth with deer repellent. To discourage smaller animals, run another strand 6 inches aboveground. ■

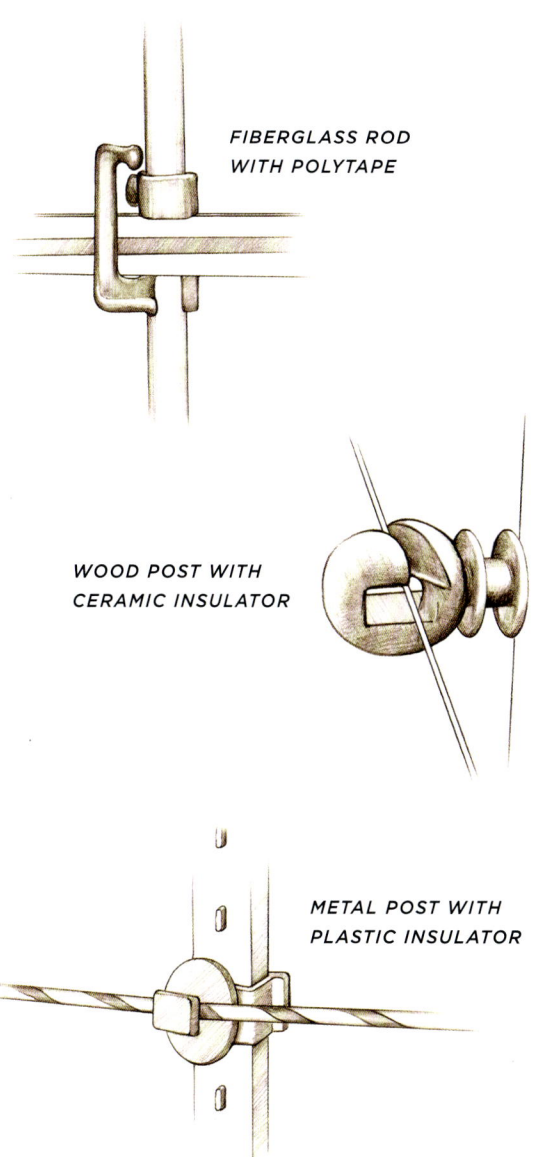

FIBERGLASS ROD
WITH POLYTAPE

WOOD POST WITH
CERAMIC INSULATOR

METAL POST WITH
PLASTIC INSULATOR

CHAPTER 6

Living Fences

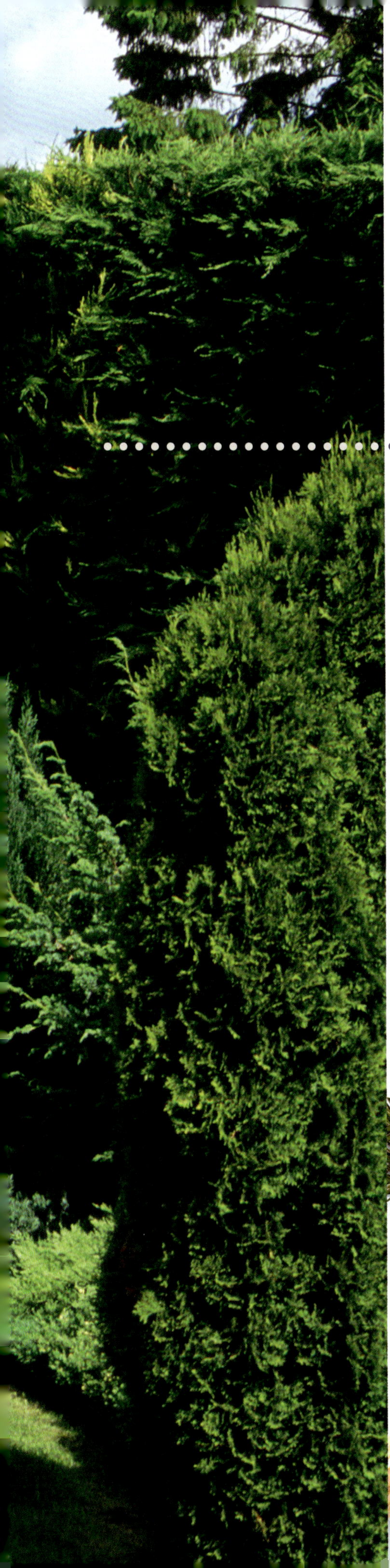

No, this isn't a chapter devoted to horror stories about fences coming to life and terrorizing entire households. The term *living fences* refers to fencelike entities that grow, as opposed to those that are built. Hedges are the most common type of living fence, but they aren't the only possibility. Ornamental grasses and bamboo are worth considering for some purposes.

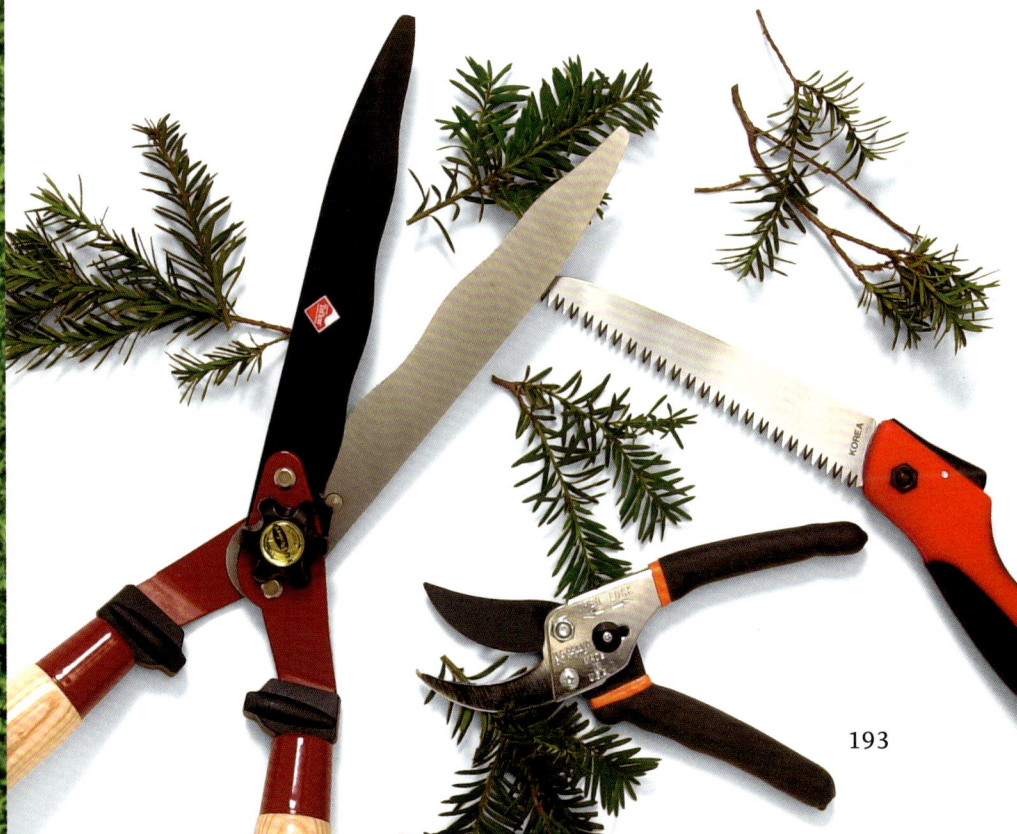

From a purely technical point of view, just about anything that grows high enough to suit your purposes could be used for fencing, but from a more practical perspective the choices can be narrowed down considerably. And, although it is a bit of a stretch to call it "living," a berm can be used to great effect for some fencing purposes, while appearing to be a completely natural part of the landscape.

Whether left to grow tall and untended or carefully trimmed, a living fence can serve as a functional enclosure or as a boundary marker.

If you opt to plant rather than build your fence, do not assume that this choice absolves you from the legal restrictions that may be imposed on, say, a wood fence. You may face similar kinds of code and zoning limitations, especially regarding the height and the necessary setback from your property. See Chapter 2 for more on legal issues and fences. Even if you are allowed to plant right along the property line, think carefully before doing so. Whatever it is that you plant will eventually grow, and any growth that extends into your neighbor's yard may be a technical violation of the boundary. A better option in this situation is to do your planting a good 4 feet back from the property line so that you will be able to work on, and thus control, both sides of the maturing foliage. Another option is to join with your neighbor in planting a joint hedge.

Hedges

"As a boundary fence, especially upon the roadside, there is much to be said in favor of the hedge. Nothing gives a neighborhood such a finished rural aspect as to have the roads bordered by hedges." This sentiment, from the 1892 publication *Fences, Gates, and Bridges* (p. 67), is one that immigrants from England brought to the American colonies. And there are those today who argue that the best way to improve the appearance of a fence is to plant a hedge in front of it. I'm not aligned with that school of thought, but I will confess that, all things considered, living fences are my favorite type of fence. One of the great advantages of a hedge is that if you find that it is too low, you can let it grow higher. Too high? Trim it down. That flexibility is hard to pull off with a built fence.

A hedge (or "green screen") is basically a row of shrubs, and shrubs are essentially trees that have foliage all the way to the ground. A hedge can do just about anything that a regular fence can do, and may actually be able to do a few things that a regular fence cannot do. You may be able legally to grow a hedge higher than you are allowed to build a fence, for example. And well-chosen and properly planted shrubs can create a virtually impenetrable barrier that offers security, privacy, protection from the elements, and beauty all in one package. And, for many people, one of the great advantages of hedges is that creating and caring for one requires that you learn absolutely no carpentry skills (although you may still need to create some type of gate), you don't have to mix any concrete, and you won't face regular paint jobs.

But, you might say, there are so many types of shrubs to choose from. True enough, but it's pretty easy to whittle the choices down once you have a clear idea of what kind of hedge you want. To begin, you'll need to answer some basic questions.

Here, a formal hedge is softened by trimming the individual shrubs into spheres. For this type of yard maintenance, an electric or gas-powered trimmer is almost a necessity.

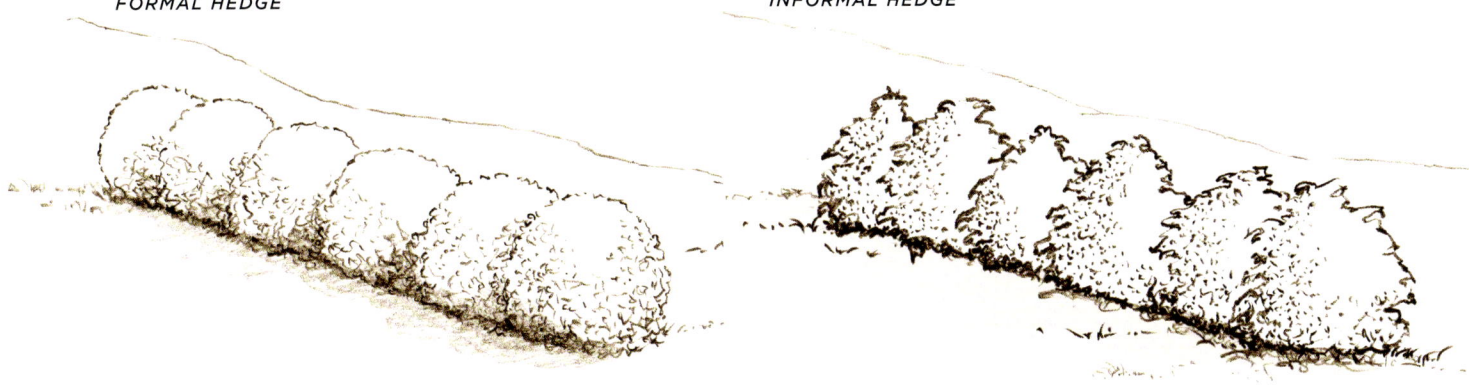

FORMAL HEDGE

INFORMAL HEDGE

How big would you like it to be? This decision involves both the height and the width of your planned hedge. You can find shrubs that grow to just about any height you want. Some grow mostly upright, while others tend to spread out horizontally.

Formal or informal? Formal hedges are created with dense evergreens and are regularly sheared into neat, straight lines. Informal hedges tend to be flowering shrubs that are allowed to grow into their natural shape. The distinction is mostly about the degree of styling they receive rather than about the types of shrubs that are used, but there are some shrubs much better suited to formal hedges. Stop shearing your formal hedge and, voilà, it becomes an informal hedge.

How soon do you need it? Some shrubs can grow to full size in a couple of years, while others will take longer. As a general rule, slower growers are also longer lasters. And, because of their slower growth, they require less frequent pruning and trimming. But don't be afraid to look at fast-growing shrubs if you need a living fence in a hurry. They tend to cost less than slow-growing shrubs and, as long as you can handle the extra long-term maintenance commitment, you can have your fence in place much sooner.

Forever green, or seasonal color? Hedges can be created with evergreen or deciduous shrubs. The former keep their foliage year-round, while the latter will provide a seasonal change of scenery. Formal hedges, and those intended for maximum privacy and security, are generally made with evergreens.

Got salt? Hedges are often planted near the street, and snow-covered streets are frequently salted to prevent dangerous icy conditions. Some common shrubs, however, are easily damaged by salt and should be avoided for street-side hedges.

Fruiting or not? Shrubs that produce fruit tend to be some of the most colorful, not to mention tasty. But keep in mind that birds may be attracted to the fruit, perhaps in droves. That can be either a lovely benefit or a royal pain, depending on your situation.

Which hardiness zone do you live in? Plants perform differently from one climate to another. Most nurseries and garden centers can tell you which USDA hardiness zones each of the plants they sell are suitable for. To find out which USDA hardiness zone you live in, log on to www.usna.usda.gov and click on the "USDA Plant Hardiness Zone Map." Stick with plants that have demonstrated good performance in your zone. Adventurous gardeners may like to stretch the limits of suggested hardiness zones. But it's one thing to experiment with a few dollars' worth of flower or vegetable seeds or seedlings, and quite another when you are spending serious money for a large number of shrubs.

What type of soil do you have? Growing conditions also take into account the type of soil (well drained versus frequently wet; acidic or alkaline). It is possible to adjust soil conditions, but it usually makes more sense to match the plants to the existing soil. If you do not know what type of soil you have, you

STRAIGHT-TRENCH HEDGE
FOR A CONTINUOUS SCREEN

STAGGERED HEDGE FOR A WIND-
BREAK OR TO BLOCK A VIEW

can determine the pH (acidity or alkalinity) with a test kit from the local nursery or landscaping store. In some states, the Cooperative Extension Service may provide more extensive soil testing that can determine both pH and any nutrient deficiencies.

How much sun will the hedge receive? Will the hedge be located in full sun, frequent shade, or a bit of both? Different plants have different needs as far as sun and shade are concerned.

Create a Landscaping Plan

When planting individual trees or shrubs, careful thought to location is not always important. With a hedge, however, careful planting is mandatory. And the more formal you intend to make the hedge, the more care you should take in planting.

There are two approaches that can be taken with hedges. If you want to maximize privacy and security in a formal hedge, plant the shrubs in a trench and space them so that they grow into a full, dense, and seemingly continuous screen. If your primary objective is to block a view or block some wind, opt instead for an informal, staggered hedge, in which plants are placed in individual holes. The staggered approach requires a bigger footprint, however, so do give careful thought to how much yard space you are willing to give over to your hedge before going that route.

A straight-trench hedge of tall-growing shrubs can provide an attractive background to a flower garden while simultaneously hiding less-welcome views on the other side.

Choosing Shrubs

It is a good idea to read up on the many choices in shrubs that are available, but there is simply no substitute for talking with experienced landscapers or a knowledgeable person at your local nursery or garden supplier. Where you buy can often be just as important as what you buy. I tend to place a premium on plants that are native to my region and have been grown nearby rather than those that were trucked in from some distant shrub farm. The closer to home that your plants began life, the more likely they will be to adapt to your site.

Although plant types are usually identified as either evergreen (keep their foliage year-round) or deciduous (lose their leaves in fall), these distinctions can be a bit misleading. For example, botanists today classify both the primarily evergreen rhododendron and the primarily deciduous azalea under the single genus *Rhododendron*. This genus contains a staggering variety of plants that range in mature height from a few inches to over 50 feet. Rhododendrons are often the least expensive and most widely available evergreen shrubs that you will find. When properly matched to the site, they can grow quickly and produce dense foliage. When planted in the wrong site, however, they can be more trouble than they are worth. They like cool, acid soil that is full of organic matter and drains well. Most prefer partial shade to full sun.

Evergreen Shrubs

The following evergreen shrubs are good choices for hedges. Some may be more popular in your area or more suitable for your climate than others.

ARBORVITAE (*Thuja* spp.). The shrubs are neat and symmetrical and grow particularly well in well-drained soil. Eastern arborvitae (*T. occidentalis*) may be the best all-around choice for high, thin hedges, except in very hot or dry conditions. It is suitable for both formal and informal hedges. The varieties 'Emerald' and 'Hetz Wintergreen' grow particularly upright and thin, although the latter will grow higher and wider if left untrimmed. The 'Globosa' variety makes a great informal hedge that, if planted close together, may not need to be pruned at all.

YEW (*Taxus* spp.). A good choice for partially shaded and moist locations. Yews tend to be very tolerant of regular shearing and pruning. These dark green shrubs can also handle a wide variety of soil types. They live long lives but do take some time to reach maturity. The hybrids in the *T. media* family are the most common. One of them, Hick's Yew ('Hicksii'), can create a narrow, upright living fence that can reach a height of 10 to 12 feet.

BOXWOOD (*Buxus* spp.). The small-leaved varieties are easy to grow and create aromatic foliage in warm weather. They will grow 3 to 6 feet high unless trimmed and are useful in both formal and informal hedges. Upright varieties can reach 10 feet in height. Boxwood has long been used as a border planting for formal gardens and estates, and it is also a popular choice for topiary.

OREGON HOLLY-GRAPE (*Mahonia aquifolium*). Depending on the specific variety, this plant can grow from 2 to 6 feet in height. It produces yellow flowers in the spring, which are followed by blue-black berries and red foliage in the fall. It prunes well and makes a nice, fairly low barrier that tolerates full sun and requires little water.

HOLLY (*Ilex* spp.). Holly can be found in sizes and varieties that can suit almost any situation. Dwarfs may grow to only a foot in height, while tree varieties can exceed 50 feet. Hollies are dioecious, meaning that plants are either male or female. You must plant both sexes near each other for the females to produce fruit. Most hollies like full sun for good berry production and compact growth. They prefer slightly acidic soil with good drainage. *I. aquipernyi* is a good choice for a large hedge. It grows fast and creates thick, hard-to-penetrate foliage. With occasional pruning, it is suitable for a semiformal hedge. The shorter inkberry (*I. glabra*) is suited for a more informal look. It has spineless leaves and small black berries. Dwarf varieties can be sheared to a hedge height of 2 feet. The common Japanese holly (*I. crenata*) is a good choice for a formal hedge in cold climates. Its very small leaves are easy to trim and shape. This variety also grows fast, dense, and narrow.

Deciduous Shrubs

Some good choices for deciduous shrubs that can be used for hedges are as follows.

BEECH (*Fagus* spp.). Most often known, and grown, as landscape trees, some beeches make ideal hedges. European beech (*F. sylvatica*) has colorful gold-bronze fall leaves. To create a dense, hard-to-penetrate hedge, plant beeches close together and keep them trimmed as low as 4 feet.

WINGED EUONYMUS (*Euonymus alatus*). Also known as burning bush, this very adaptable plant produces hot pink leaves in the fall if it is given full sun. The dwarf version's leaves are red. Either type works well as an informal, unclipped hedge that will be somewhat rounded at the top and open at the bottom when planted 5 feet apart. With annual or biannual clipping, this plant will also make a fine formal or semiformal hedge.

EUROPEAN CRANBERRY BUSH (*Viburnum opulus*). This fairly flat-topped plant produces clusters of white flowers in late spring, with some colorful fruit and foliage in the fall. The 'Nanum' variety is a great choice for a low, wide, untrimmed hedge. It grows to only about 2 feet but does not produce flowers or fruit.

DWARF PURPLE OSIER (*Salix purpurea* 'Gracilis'). Also known as Alaska blue willow, this shrub has purplish branches and produces yellow fall colors. It is a popular choice for a low, clipped hedge.

YEW

HOLLY

BOXWOOD

BEECH

EUROPEAN CRANBERRY

ARBORVITAE

WINGED EUONYMUS

Planting and Caring for Shrubs

Hedge plants are sold in one of three ways: They are grown in a field and sold bare-root, grown in a field and sold balled-and-burlapped (B&B), or grown and sold in containers. The differences usually have more to do with locale and plant type than anything else. In my neck of the woods, deciduous shrubs are usually raised and sold in plastic or peat containers. Container plants are available throughout the planting season, in various sizes. Bigger containers generally mean more mature plants, at a bigger price. But paying that extra amount sometimes allows you to buy plants with their leaf colors, flowers, and fruit already evident, so you will know what they are going to look like in years to come. They are easy to move around, and they do not have to be planted right away. But try to stay away from rootbound plants, which you can recognize by visible roots growing through the holes in the bottom of the container or above the soil. Bare-root plants are usually available in late winter or early spring. They tend to cost less than the other types, and they also tend to become established faster and grow better. But they should be planted as soon as possible after they have been purchased. Balled-and-burlapped shrubs are also best bought and planted in late winter or early spring. Although planting each type is largely the same, there are a few differences that need to be noted.

Digging

For a staggered hedge, carefully measure and mark the hole locations with wood stakes. Just because it's staggered doesn't mean it should be haphazard. If you are planting a long hedge, you might want to set up a string line down the center of the hedge, and then stagger the holes using the string as a consistent reference. Dig holes as deep as the containers or root-balls and about twice as wide.

The first step in planting a trenched hedge is to remove the sod. If you are careful, you can lift up strips of sod and reuse them elsewhere in your yard. Then dig a trench the full length of the hedge. Dig about 8 to 10 inches deep, or approximately to the depth of the root-ball or container, and about 2 feet wide. Sure, you could dig individual holes for each plant, but digging the trench will probably be quicker, and it allows you to very carefully line up the plants before

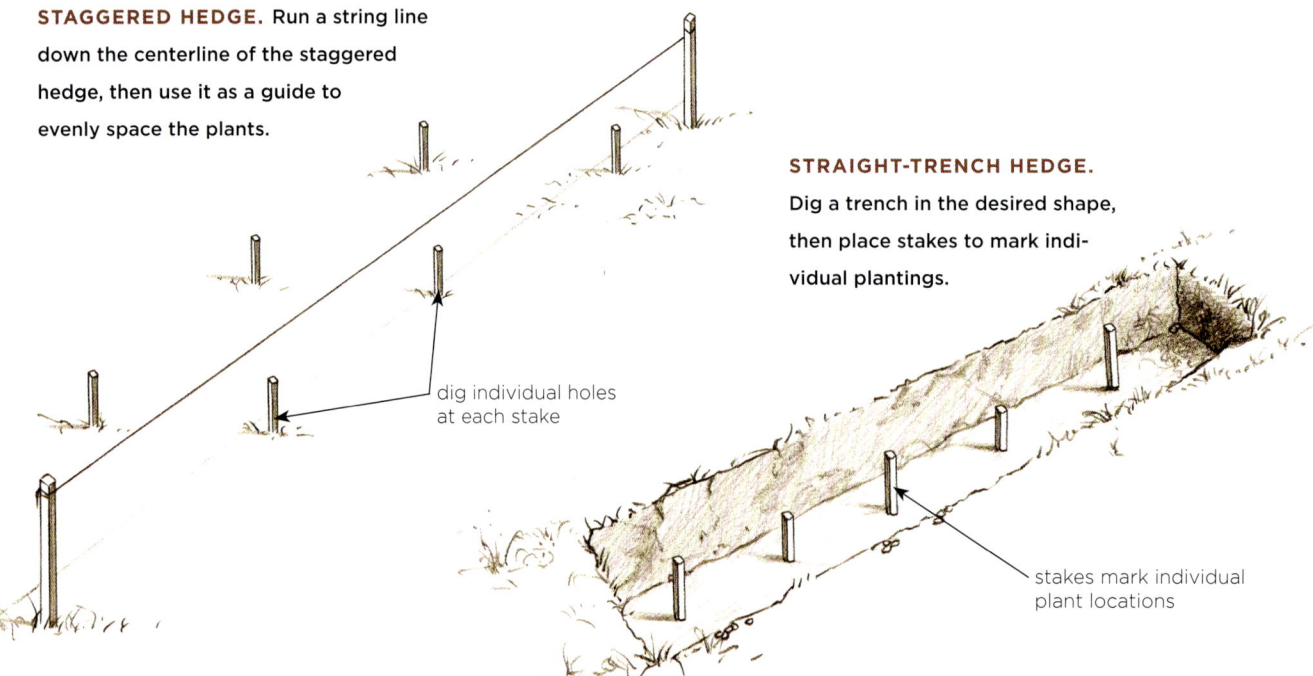

STAGGERED HEDGE. Run a string line down the centerline of the staggered hedge, then use it as a guide to evenly space the plants.

dig individual holes at each stake

STRAIGHT-TRENCH HEDGE. Dig a trench in the desired shape, then place stakes to mark individual plantings.

stakes mark individual plant locations

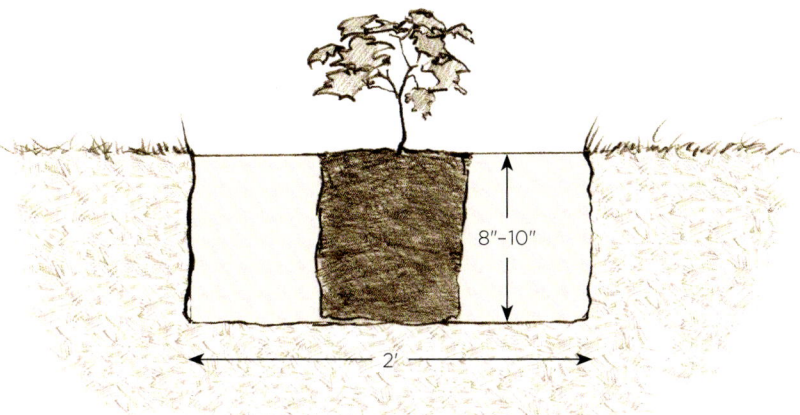

The trench for a hedge should be as deep as the height of the root-ball or container, and at least twice as wide. The dimensions given here are typical.

burying their roots. Place small stakes or sticks in the trench to pinpoint the location of each shrub. Proper spacing is not really a science. It depends on the type of plant as well as the type of hedge you want. As a general rule, shrubs for a 3- to 4-foot-high hedge should be spaced about 18 inches apart, while tall hedges can be spaced about twice as far apart. Talk over your plans with your supplier before making a final decision. Keep the spacing consistent.

Planting

For container shrubs, always move or carry the plant by the container itself, not by the plant. It's usually better to think in terms of removing the container from the plant, rather than removing the plant from the container. Most containers can be cut or torn away. Tease out the roots from the root-ball (if they've become entangled with each other, you can cut them free with a sharp knife). Set the root-ball in place in the trench or hole and pack some soil around the roots to hold it upright.

Handle a balled-and-burlapped plant carefully. Carry the plant from the bottom of the root-ball (it can be quite heavy!), and avoid dropping it. If it has a wire basket around the root-ball, cut the wire with clippers and remove it. Loosen the soil on the sides and firm up the soil on the bottom. Remove any ties and synthetic coverings. Natural burlap does not need to be removed, but it should be folded back to reveal the top third of the ball. Set the ball in place in the trench or hole.

For container-grown plants, remove the plant from the container and tease the roots before setting the root-ball in place.

Remove the wire basket from a balled-and-burlapped plant, if necessary, and remove any synthetic coverings or ties. Burlap does not need to be removed but should be folded back. Carefully set the plant in place.

Mix It Up

Unlike a built fence, a good hedge can seem like a natural part of the environment (unless, of course, you are a landscaper, in which case you are likely to see a human hand behind every blade of grass). To add to the sense of "naturalness," consider turning your basic hedge into a mixed, layered border. Using a variety of plants can expand the function and feeling that a hedge imparts. Mix fast-growing with slow-growing plants, and you can have a "hedge in a hurry" with the former while the latter grow to maturity. Combine evergreen, deciduous, and flowering plants to create colorful seasonal change. Set some midsize shrubs in front of larger shrubs or trees to add depth and variety in height to your hedge. Or line the hedge with a flower or herb garden or some ornamental grasses. A truly natural environment is full of contrast, variety, and seasonal change. ■

With bare-root plants, you do not want to let the roots dry out. It is a good idea to soak the roots overnight in a bucket of water before planting. Form a small mound in the trench or hole, and then spread out the roots over the mound. Set the plant at the same height or just a bit higher than it was in the field. Cut any roots that are broken or longer than the others. Hold the plant steady as you fill and tamp the soil around it.

The roots of a bare-root plant should be spread over a small mound created in the hole or trench. Hold the plant in place while you cover the roots with soil.

If you're setting each plant in an individual hole, fill the hole three-quarters full of soil before moving on to the next plant. If you're setting the plants together in a trench, pack enough soil around each plant's roots to hold it upright before moving on to the next one. When all plants have been set in the trench, check the spacing and alignment and then begin filling it with soil, packing gradually with your hands or feet.

Give the plants a good, slow watering, then add more soil to top off the hole or trench, tamp gently, and water again. Surround the shrubs with a 4-inch-thick layer of mulch. Plan to water the shrubs weekly (or less if it rains) for the first month, then gradually reduce the watering until the ground freezes. Most shrubs will not need to be watered once they've become established.

Trimming and Shearing

One of the biggest mistakes aspiring hedge-makers commit is to plant, water, mulch, and then walk away, thinking that those nice, young shrubs need some time to grow before requiring any further care. In fact, with most shrubs, you should begin training and pruning the first year. Pruning is vital to shaping your hedge by controlling growth. When you prune, you stop growth in one branch and, thus, encourage it in another. Removing dead and diseased branches aids the health of your hedge. If you let the hedge become too dense and the branches too thick and snarled, sun and air can no longer reach the interior growth.

Conifers generally do not require much early pruning, but it is a good idea to cut back the top one-third of each plant right away. Then let the plants grow undisturbed for the first year. With narrowleaf (needled) evergreens, plan to remove (by pinching) about half of any new stem growth in the spring.

Trim away the top third of a newly planted conifer, and then refrain from additional pruning or shearing for a full year.

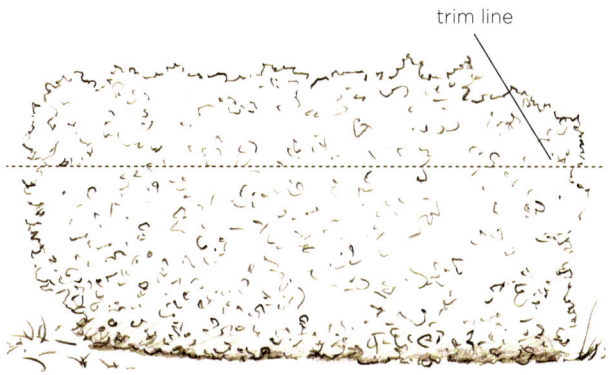

With needled evergreens, pinch off about half of new stem growth in the spring to encourage bushy growth.

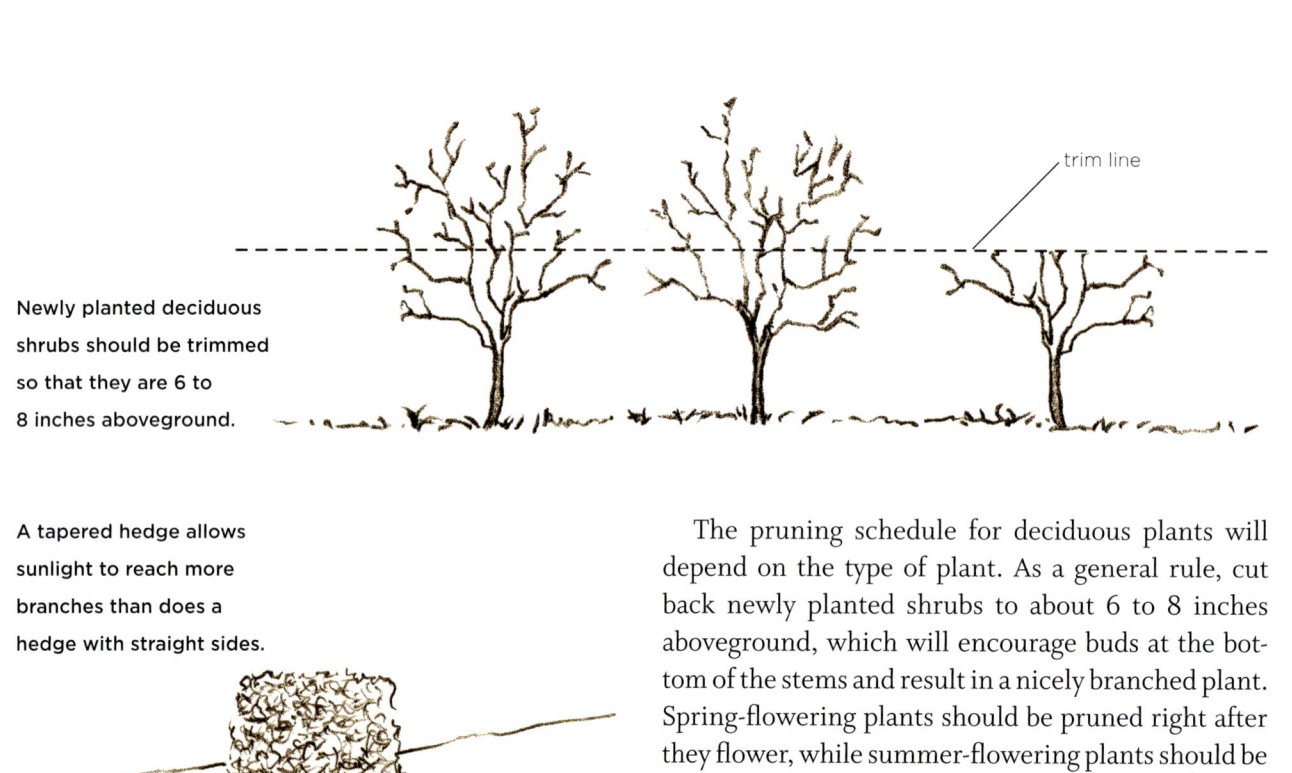

Newly planted deciduous shrubs should be trimmed so that they are 6 to 8 inches aboveground.

trim line

A tapered hedge allows sunlight to reach more branches than does a hedge with straight sides.

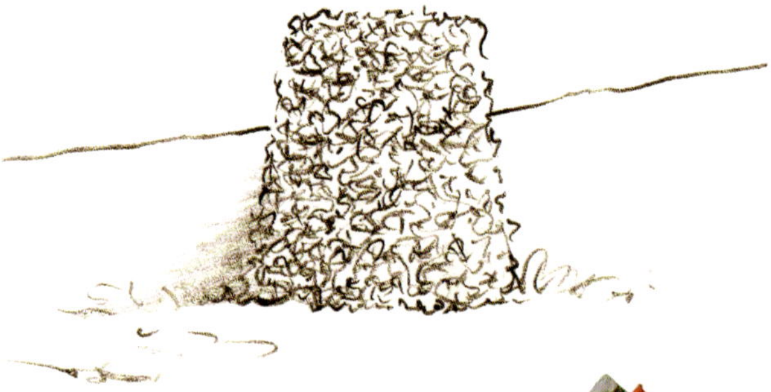

The pruning schedule for deciduous plants will depend on the type of plant. As a general rule, cut back newly planted shrubs to about 6 to 8 inches aboveground, which will encourage buds at the bottom of the stems and result in a nicely branched plant. Spring-flowering plants should be pruned right after they flower, while summer-flowering plants should be pruned in late winter. For long-term pruning needs, I suggest that you talk with a knowledgeable supplier or consult a good book on landscaping.

Most shrubs can be sheared with electric or manual clippers, and regular shearing is absolutely essential for maintaining a formal hedge. The thick leaves on evergreen shrubs such as hollies can be damaged by electric shears, however, and so should be trimmed by hand.

For healthy, long-lived hedge plants, forget those images you may have in your mind of high, formal hedges with sides sheared to a perfectly vertical plane. The best objective is to shape plants so that they are wider at the bottom than at the top. This shaping allows sunlight to reach the greatest quantity of branches. A wide base on the plants also adds to the privacy they afford and better controls weed growth.

Green with Envy

Careful selection of trees and shrubs combined with skillful design and routine maintenance can create boundaries, entryways, and landscaping accents that function as fences and gates without disrupting the natural surroundings.

Grasses

WE ARE ALL ACCUSTOMED to cutting grass, not growing it for decorative purposes. Perhaps that association helps explain why ornamental grasses are not used more often in residential landscapes. If you are of that mind, you might be surprised at what you are missing. No, you cannot achieve a suitable effect by simply letting your lawn grass grow. Scan some gardening and landscaping catalogs, books, and magazines to get an idea of the available colors and textures.

A well-stocked nursery may have dozens of varieties of ornamental grasses for you to choose from, and mail-order sources offer even more. You can find solid colors and stripes, bright blues and blood reds, and silvers as well as greens. Some ornamental grasses grow as high as 10 or 12 feet. Some keep their foliage and remain upright all year long. And most grasses mix well, so you can create a layered effect with different heights and colors.

You are not going to keep trespassers (two- or four-legged) out of your yard with ornamental grass, but you can redirect them with an awfully nice-looking border. Most grasses are not too picky about soil type, although some are more tolerant of moist and shady settings than others. And some grasses are invasive, although you can keep them from spreading by removing unwanted seedlings as they appear.

It is best to buy potted clumps of grass and transplant them, although it is possible to grow ornamental grass from seed. Most grasses do best when they are transplanted in the spring, although cool-season varieties can be started in late summer or early fall. Check with your supplier for specific instructions for the grass or grasses you are planting. As a general rule, you should plan to work some compost or other organic matter into the soil and add fertilizer before planting. Space the plants as far apart as they will grow tall; thus, grasses that will grow 4 feet high should be planted 4 feet apart. Set the clumps in holes, but do not bury the crowns. A 4- to 6-inch layer of mulch will help hold down weeds and conserve water, but be sure that you don't cover the crowns of the plants.

Maintenance could not be easier. Warm-season grasses, which go dormant in the fall, should be cut back just as soon as you see new growth appearing at the base. Cool-season evergreen grasses do not have to be cut back every year, but do plan to remove dead foliage. Watering needs vary considerably, so check with your supplier for recommendations.

Set potted clumps of ornamental grasses in place. Once you are satisfied with the spacing and arrangement of the different varieties, dig all the necessary holes. Remove the containers and set the plants in the holes.

Grass without the Mower

Ornamental grasses can be used to define boundaries, screen views, and add color and texture to just about any landscape. Though the finished product may well draw compliments for its "natural" appearance, a functional grass barrier requires careful planning and care.

Bamboo

Bamboo is a giant grass that is both beautiful and easy to grow. "Running" varieties are best for living fences. Their underground stems (rhizomes) stray from the parent plant and send up vertical shoots. They are hardy, tolerate a wide variety of soil types, and can grow into a sizeable fence in about three years. But you do need to take steps to prevent running bamboo from running rampant through your yard. One strategy is to dig a 2- to 3-foot trench around the planting area and line it with 30-mil plastic, or even fill it with concrete. Dig the trench walls at a slight angle, as shown below. Alternatively, dig a 1- to 2-foot trench around the plant and cut any stems that grow into it. Fill the trench with mulch, and periodically fish around with your hands to find and cut new stems. Don't wait until the stems have started spreading before digging the trench, though, as they can quickly become very hard to cut.

Unlike the case with most other types of plants, when you're buying containers of bamboo, rootbound plants are the better choice, as they indicate a fast grower. Bamboo should be watered once a week. Add a high-nitrogen fertilizer in early spring and a balanced lawn fertilizer in the fall.

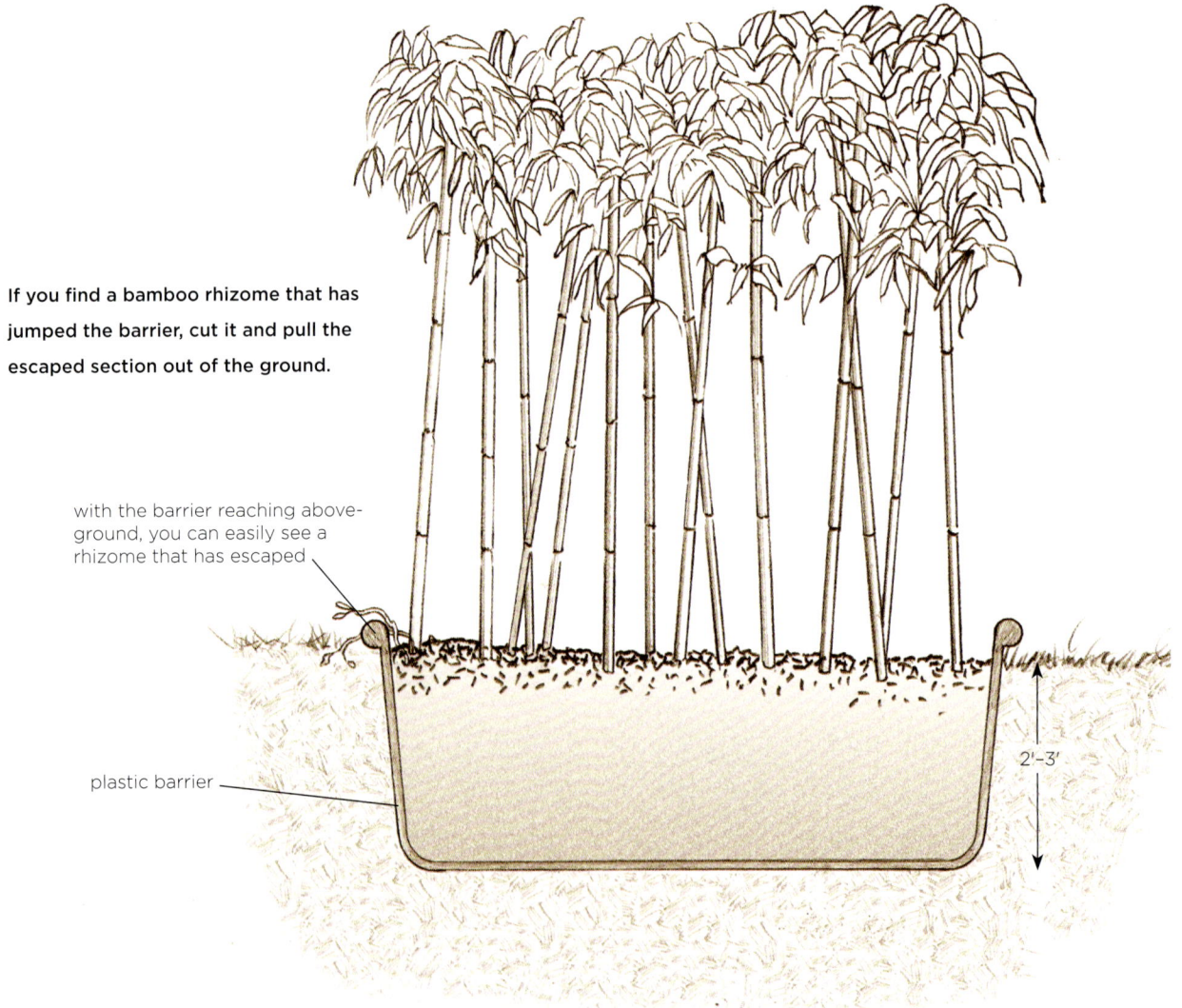

If you find a bamboo rhizome that has jumped the barrier, cut it and pull the escaped section out of the ground.

with the barrier reaching above-ground, you can easily see a rhizome that has escaped

plastic barrier

2'–3'

Fencing with Berms

A BERM JUST MIGHT BE the most subtle type of barrier. A berm is nothing more than an intentionally designed embankment or ridge composed of fill covered with loamy soil and plantings. Berms can be as short as a few feet high or can reach 10 feet or more in height. With thoughtful plantings, a berm can serve as a privacy screen between your yard and the street, the neighbors, or an unattractive view. You can add even greater height to a berm by topping it with a hedge, tall plants, or even some type of fence. And don't just think in terms of a single berm; you can achieve even greater impact with a series of berms around the property, with different heights, lengths, shapes, and coverings.

Compared with a wood fence, a berm can be nearly maintenance-free, and even a fairly high berm can look much less intrusive and artificial than a vertical wall. Small berms are fairly easy do-it-yourself projects, provided you can have some fill delivered right to the site. Plan to pile dirt in a wavy pattern rather than a straight line to make it appear more natural. Add 12 to 18 inches of dirt, then compact it with your feet. Take extra care in compacting the top layer so

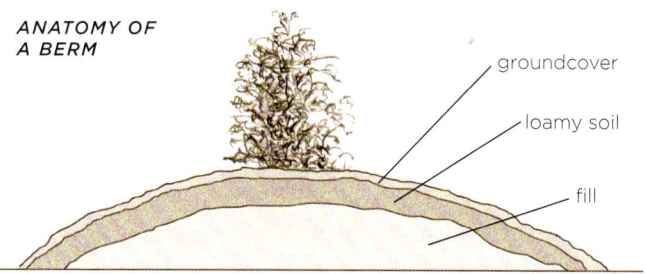

ANATOMY OF A BERM

groundcover

loamy soil

fill

that it doesn't wash away with the next rain. Seed with grass or add mulch or plantings as soon as possible to prevent erosion.

Larger berms really ought to be graded by a contractor. If the idea of a berm appeals to you but the work itself seems daunting, talk to a landscaper or landscape architect.

Obviously, a berm might look downright odd in a small yard and surrounded by nothing but flat land. But if your local landscape has some rolling curvature to it, study the undulations for inspiration on what might be possible in your own property.

A modest berm topped with a fence creates a large visual barrier to the street traffic on the other side.

Gates

FENCES THAT ARE INTENDED TO mark a boundary, define a space, provide some pizzazz to the landscape, or simply direct foot traffic usually do not need a gate. For fences intended to provide security and foolproof enclosure, however, gates are not optional. Instead, they are the "business end" of the fence, the doorway that makes the fence a functional item.

Fences are pretty simple structures. Gates, however, can be fairly complex. While a fence gains most of its stability by its direct connection with the earth, a gate's critical connection is the hardware that ties it to the fence. Fences sit, while gates swing. Weak gates sag, refuse to close properly, and distract from the appearance and appeal of the entire fence. All of which is mentioned to emphasize the need to plan the gate style and construction differently from the rest of the fence.

Gates can be designed to look different from the fence [LEFT] or to exactly mimic the style of the fence [RIGHT]. Either way, they need to be built to take heavy use, day after day, year after year.

Design Decisions

GATES ARE USUALLY MADE from metal or wood, but when making your own, you are pretty much restricted to the latter. With ornamental metal fences, gates can be purchased along with the fence, and they should be assembled and installed as directed by the manufacturer. If you are building a wire-mesh fence, you can use the same material to cover the gate, but it will still require that you build a solid wood frame and attach it to wood posts.

A cheerful gate on a well-defined path beckons visitors forward, inviting them into the unseen yet welcoming landscape beyond.

Wrought iron is a pricey material for a fence, but a wrought-iron gate can be a terrific and reasonably affordable addition to a stone or brick fence, as long as you can find a source for one. If you choose to use a wrought-iron gate, however, make sure to consider carefully—and in advance—how it will be attached to the fence. Fastening dissimilar materials is discussed below, but if you plan to use a custom-designed metal gate, you should discuss all fastening matters directly with the designer or fabricator.

So, this chapter will be devoted to building wood gates, which can be used with just about any type of fence. With the choice of material settled, the big question that remains is what style of gate to build. With front-yard wood fences, I tend to favor wood gates that closely resemble the fence but have a feature or two that make them stand out. This can be done by using the same wood, finishing it the same way, adapting the same general design, and then making a subtle change or two. The change can be as simple as making the gate a bit higher or lower than the fence, or placing the gate between two particularly prominent posts. This approach generally results in a gate that looks like it belongs with the fence stylistically, but allows it to stand apart just a bit. The easier it is to identify the gate, the more welcoming it is likely to feel for visitors.

On the other hand, if a welcoming entry is what you want to avoid (for privacy reasons, or due to some unspecified antisocial purpose), then your goal should be to match the gate exactly to the fence, so that they appear to constitute a seamless barrier. A seamless look can also complement certain house styles.

A gate that serves a living fence is an independent structure, with no direct tie to the adjacent hedge. The gate is simply attached to two gateposts. This approach is worth considering for any type of fence, in my opinion. Attaching a gate to buried wood posts is the strongest option available, and the easiest to maintain and repair. All of the hardware and fasteners are within easy reach and can be repaired or replaced as needed quite easily.

Being made of the same materials allows gates to match their fences. But as the examples on this page illustrate, the results can take very different forms, from an open arch, to a nicely conceived rustic addition, to a perfectly mated picket design that, from a distance, is hard to differentiate from the fence.

How Wide?

You can build a gate about as wide or as narrow as you want. But, being of a practical mind, I suggest that you plan to make it 4 feet wide, and then see if there is any reason to adjust that plan. There well may be good reasons, but in most cases there won't be.

Begin by considering what will need to get through the gate. More than one fence builder has constructed a gate that was too narrow for at least some important object to pass through. Garden carts and lawn tractors may need to get in and out on a regular basis. The deck on my own garden tractor, for example, can just barely squeeze through a 5-foot-wide opening. If you move or decide at some later date to buy that grand piano you've always wanted, a narrow gate opening could prove troublesome. At 4 feet, the gate will most likely be wider than the exterior doors on your house, which should mean that they could handle anything currently in your house. And most people (the author excepted) are smart enough to buy houses with yards that can be kept groomed with mowers that will pass comfortably through a 4-foot opening.

An advantage of keeping the opening at or less than 4 feet is that it can be handled by a single gate. Once you start increasing the size, you need to start thinking about installing a double gate. Obviously, if the gate will be spanning a driveway, it will have to be considerably wider than 4 feet. For particularly large, heavy gates, I think it is smart to talk with a professional contractor rather than tackle the project yourself.

Locating the Gate

Deciding where to locate the gate is usually a no-brainer, but not always. An existing sidewalk, driveway, or pathway of any kind will normally dictate where the gate ought to go. If, however, you are in the infant stages of landscaping your property, you may need to give the matter some serious thought. If you have a straight sidewalk leading directly to the front entrance of your house, the gate would align with those existing elements. But if the entry sidewalk has yet to be built, or if you've thought about changing it, you may want to consider offsetting the gate from the house entry, as shown in the illustration. An offset gate allows for the possibility of some additional privacy if you plant some shrubs along the sidewalk and fence line. On the other hand, obscuring the view of the gate from the house may not be what you want to accomplish at all.

Another nice design trick is to set the gate in from the fence a few feet. This not only calls additional attention to the entry, even if the gate is designed to look like the rest of the fence, but it also sets the entry out of the flow of sidewalk traffic, which can be a real blessing if you live on a well-traveled road.

Planning the Swing

I know from experience that, when it comes to hanging doors and installing gates, do-it-yourselfers are prone to overlooking the issue of swing. I think it may

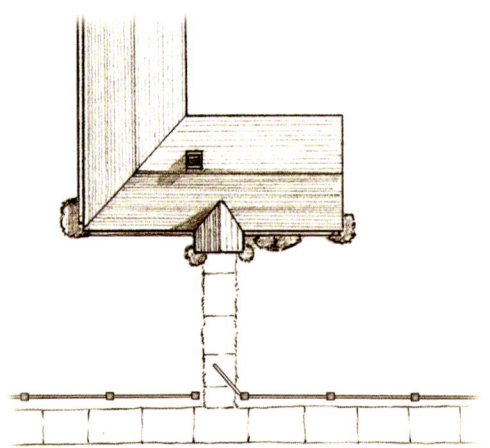

GATE ALIGNED WITH THE HOUSE ENTRANCE

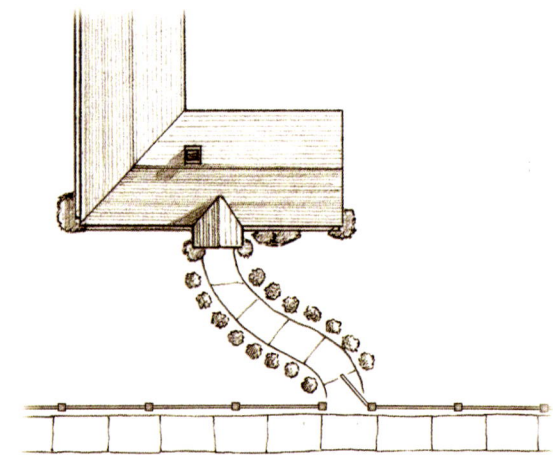

GATE OFFSET FROM THE HOUSE ENTRANCE

be one of those subjects that is so obviously important that it sometimes gets lost amid more troublesome matters. Talk to a professional contractor about such subjects, however, and you are likely to confront the following questions almost immediately:

- Do you want the gate to swing out, or in, or both ways?
- Do you want the gate to be hinged on the right side, or the left side?

In your house, the doors most likely swing into rooms. If you have the misfortune of having doors that instead swing out into hallways, you well know what a pain they can be in disrupting traffic and every so often taking a nasty whack on your nose. With gates, the most popular swing direction is in toward the house, rather than out (into sidewalk traffic, for example). But, where the potential for traffic disruption is nonexistent, you really do not need to feel bound by the tradition. Other factors can be more important. If your fence faces a modest slope, for example, the gate may only be able to swing open in the downhill direction. Some other physical obstruction may force a similar decision. In other words, once you decide which way you want the gate to swing, make sure that it can swing that way.

It may take a little more work and special hardware, but a gate can also be installed to swing both in and out. Heavily used chain-link gates are often designed this way, with a butterfly latch that stops the gate from swinging back and forth at will. If you expect to routinely approach the gate with a baby in one arm and a bag of baby accessories in the other, you might really appreciate a gate latch that can be operated with an elbow and a gate that can swing in either direction. In most cases, however, it is best to restrict the swing to one side of the fence.

Another consideration regarding gate swing is aesthetic. With most gate hardware, the hinges are mounted on the side in which the gate swings. So, if you plan for the gate to swing toward the house, the hardware will not be visible from the other (usually more public) side. This is normally a good thing, unless, that is, you decide to invest in some particularly expensive hardware that you actually want to show off.

Once the first question is settled, you then need to address the second one. Usually, deciding whether to place the hinges on the right or left side of the gate is a matter that can be settled with a coin flip. Sometimes, though, convenience or necessity can take the guesswork out of the choice. Be sure to take a close look at the full swing path you are contemplating to see if there are any obstructions that might impede travel in one or the other direction.

Gateposts

Gateposts were discussed at some length in Chapter 3, and I will not here go through that information again. But, given the importance of the subject, I do think a little repetition might be in order.

Gateposts are different from all other fence posts. Sure, with most fences, they help support the fence. But they also support the gate, which is routinely swinging back and forth, and is frequently knocked around by people and objects passing through. The post holding the gate hinges is under the greatest strain, but both gateposts need to be treated with more care than other posts. That means that gateposts should meet one or, preferably, more of the following criteria relative to other posts. They should be

- thicker;
- buried deeper;
- backfilled more securely; and
- perfectly plumb.

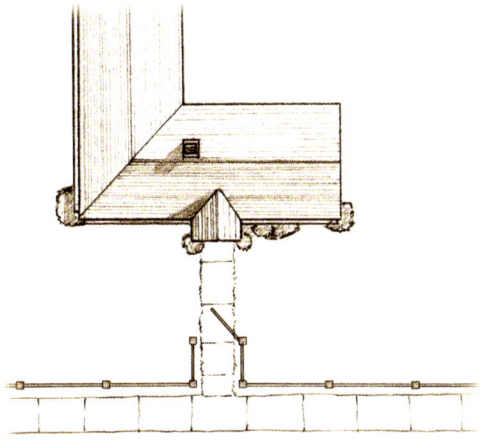

GATE SET IN FROM FENCE LINE

Regardless of what style of backfill you use on other posts, or the depth at which you bury them, I suggest that gateposts should be placed 6 to 12 inches below the frost line in your area, and that they be set in concrete. If frost is not an issue where you live, place the posts at least 30 inches deep.

In terms of size, plan to use 4×4 posts for gates up to 3 feet in width, 6×6s for gates 3 to 5 feet wide, and 8×8s for anything larger than that. These dimensions can be fudged a bit, depending on the weight of the gate and the amount of use it will receive, but you will almost certainly not regret using bigger posts at the gate. For fence styles with visible posts, I think thicker (and higher) gateposts almost always look good.

Gateposts are also ideal candidates for boxing in, as discussed on pages 102–103. This technique allows you to make the post appear to be substantially larger than the structural post buried in the ground. Gateposts also lend themselves to being dressed up with post caps and other decorative touches.

Gate Frame Options

There are two basic styles of wood gate frames. A full perimeter frame is usually the strongest, and it can be adapted to just about any type of fence style. A Z-frame, though generally not quite as strong, does offer the advantages of simplicity and a light appearance. I would not recommend using a Z-frame on any gate wider than 3 feet, but for small, informal gates it is just fine. If you would like to combine a thin profile with maximum strength, build a perimeter frame with the 2×4s set flat rather than on edge. An important component of each of these frames is the diagonal brace, the bottom of which should always rest on the hinge side of the gate.

One potential problem with a thick gate frame is that it requires more clearance on the hinge side between the gate and the post than a thinner gate. Otherwise, the gate may not be able to clear the post as it opens.

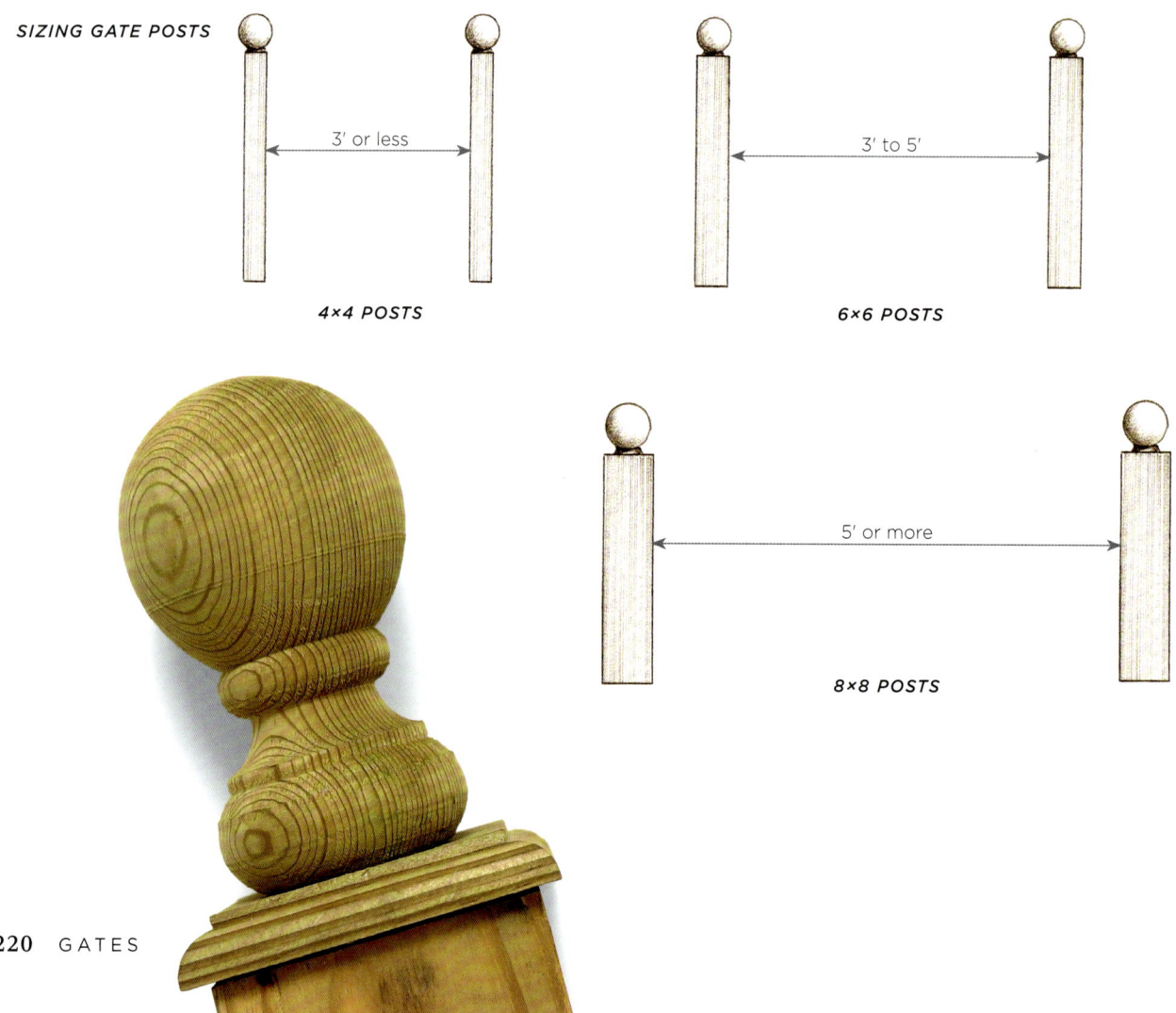

SIZING GATE POSTS

3' or less

4×4 POSTS

3' to 5'

6×6 POSTS

5' or more

8×8 POSTS

With a perimeter frame of 2×4s on edge, you will need to create half-lap joints at the corners. There are many ways to cut half-lap joints, depending on the tools you have available at the time. Probably the quickest method is to use a bandsaw or reciprocating saw to make the two necessary cuts on each board end. A router with a suitable flat-bottom bit or a table saw equipped with a dado blade will produce the cleanest results. If your power tool collection is more modest, though, you can do a perfectly acceptable job with a circular saw, as demonstrated on page 222.

Gate frames are sometimes built with 1×4s or 1×6s, but I would use such thin boards only if the gate was going to be very light and very small, and if the thin boards were vital to the appearance of the finished structure. A compromise between 1× and 2× lumber would be 5⁄4 (called "five-quarter") wood, which measures about 1 inch thick.

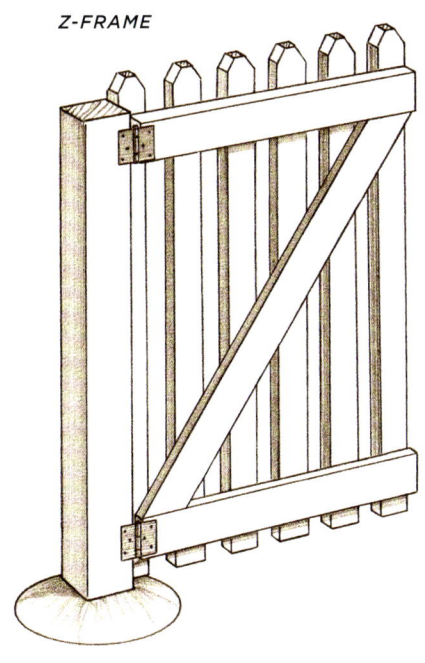

Z-FRAME

PERIMETER FRAME

FLAT PERIMETER FRAME

half-lap joint

Making Half-Lap Joints

If you have a table saw but no dado blade, you can use this same approach. When using a circular saw, I strongly suggest that you clamp the 2×4, flat side down, to a sawhorse or some other sturdy work surface. That way you can keep both hands on the saw, which will better your chances of making clean, consistent cuts.

1 From the bottom of the board, measure up 3½ inches and mark the board.

2 Assuming that your 2×4 is exactly 1½ inches thick (which it should be, but measure first), set the saw blade to cut exactly ¾ inch deep. Beginning at the end of the post, make a series of closely spaced cuts until you reach the mark.

3 With a hammer, knock out the thin pieces of wood. Then grab a chisel and clean out the notched area. Repeat this process on each end of each board. Attach the boards at the laps with glue and 1¼-inch screws. ■

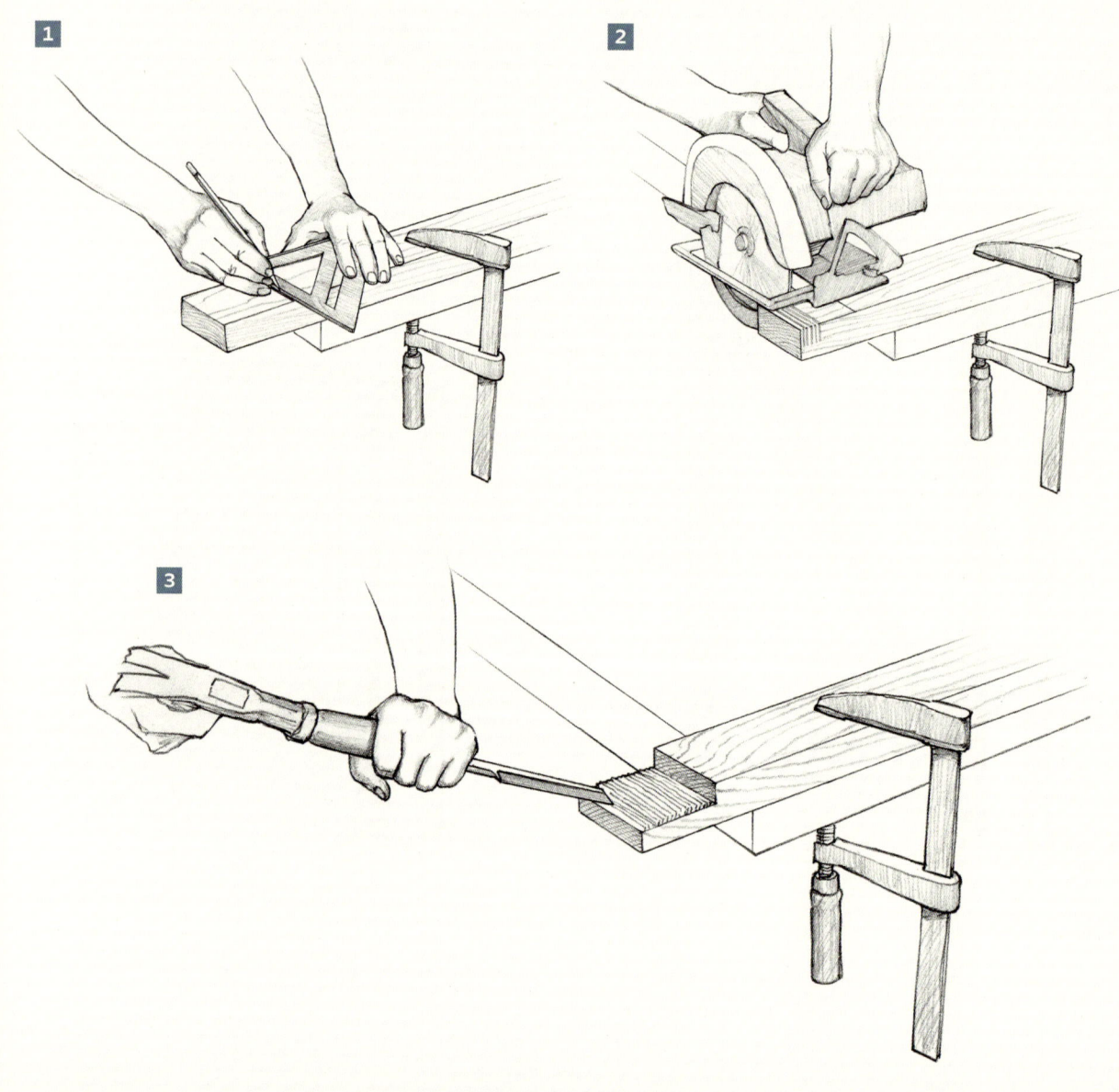

Diagonal bracing is a critical element of the frame. Many gate builders, however, do not seem to understand the specific function of the bracing, given the many improperly braced gates that I see. The hinge side is tied directly to a post, but the latch side spends much of its time swinging freely, with no direct support on the bottom. Loads (which include gravity as well as physical objects you might place on the gate) always try to find their way to the earth. With a properly positioned brace, loads on the top edge of the latch side will be directed down the brace and toward the bottom hinge, from the weakest corner to the strongest corner of the gate. When, however, a wood brace is angled in the reverse direction, from the top of the hinge side to the bottom of the latch side, it actually encourages the gate to sag. Note that for the load to be directed efficiently, the bracing must be installed tight against the horizontal boards.

Note, also, that this advice applies to wood bracing, but it does not apply to cable bracing, such as is found in gate repair kits that contain some cable, a turnbuckle, and fasteners. The two types of bracing, while performing a similar function, work under two very different principles. The wood bracing relies on compression, which is an inward, crushing force. The adjustable cable bracing works under tension, which is a force involving pulling or stretching. So the cable brace must be installed in just the opposite direction as a wood brace. Attach the top of the cable alongside the top hinge.

DIAGONAL BRACING WITH WOOD

DIAGONAL BRACING WITH CABLE

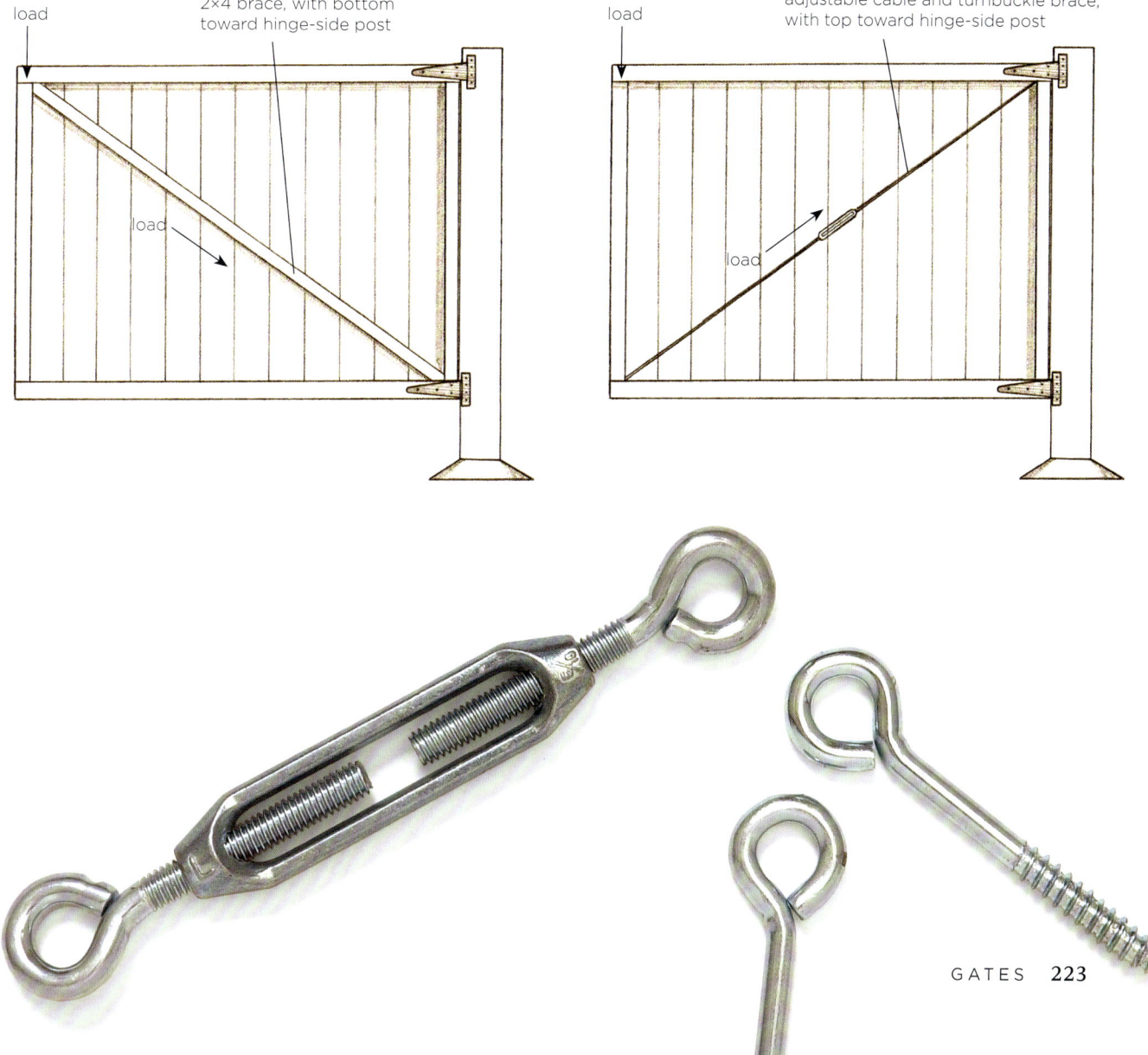

Gate Stops

When I see an ill-fitting gate, one of the first things I look for is whether it is equipped with a stop. Gate stops are another of those simple devices that too often get overlooked in the rush to finish the job. Standard gate hinges are not designed to stop the gate from swinging beyond its intended arc. You need to make plans for stopping the gate at the latch with some type of physical barrier. This need can be met if you use an automatic gate latch, which stops and secures the gate without any special effort on the part of the user. With other types of latches, however, you need to add a special piece of wood to stop the swing or else design the gate or fence with an integral stop, as shown in the illustrations.

STOP ATTACHED TO FACE OF POST

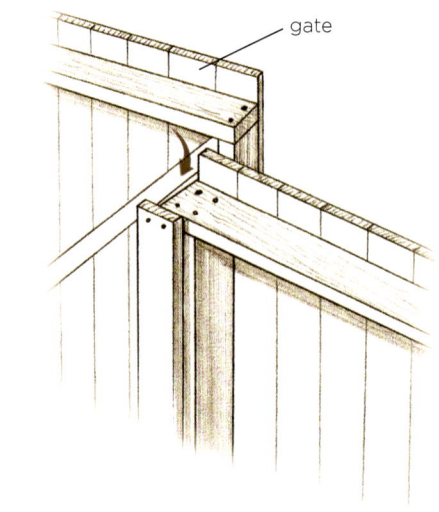

SIDING ON GATE ACTS AS STOP

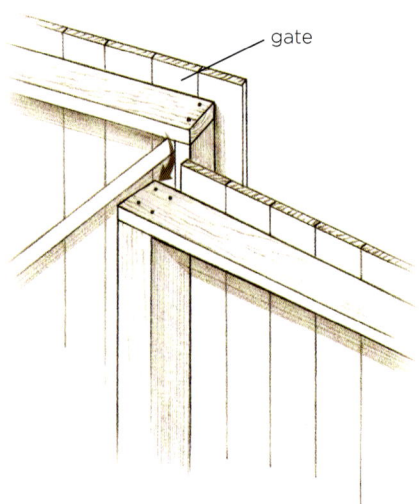

SIDING ON FENCE ACTS AS STOP

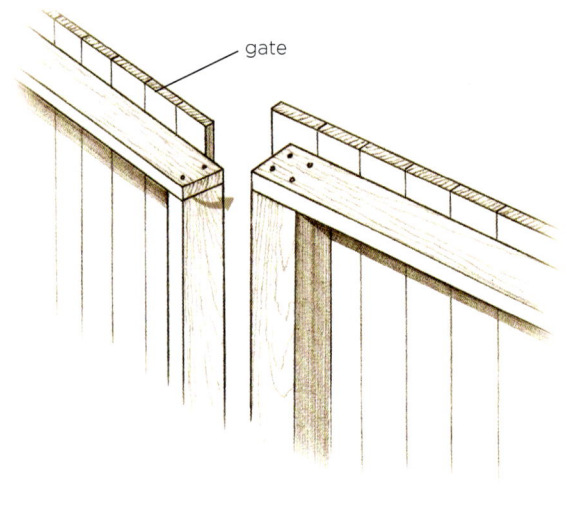

STOP ATTACHED TO GATEPOST

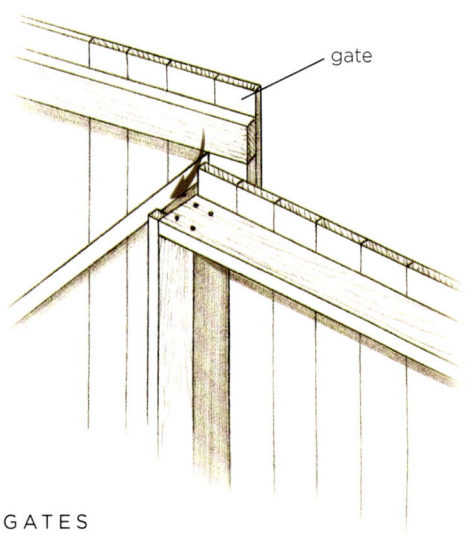

AUTOMATIC LATCH FUNCTIONS AS STOP

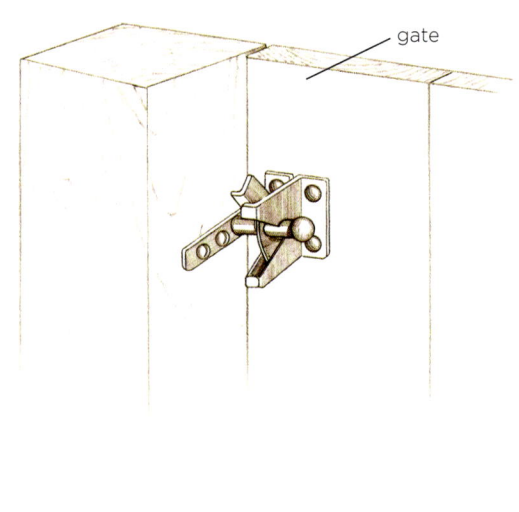

Hardware

THERE ARE MANY TYPES of gate hardware available, with varying requirements for clearances and differing installation needs. Select your hardware before you build your fence. Learn how it needs to be installed, and how much space on the side of the gate it needs to function properly, and then finalize the dimensions and style of your gate.

You can usually find a pretty good selection of hardware at one of the large home improvement stores, while smaller lumberyards and hardware stores often have a limited selection but are able to order through catalogs. The two manufacturers of gate hardware that I encounter the most are Stanley and National. Both have websites that display their full product lines, and you can buy the hardware online or through mail-order suppliers if you can't find what you want locally.

Most of the readily available gate hardware is intended for attaching wood to wood. You can use this hardware to fit a wood gate to a brick or stone fence by first attaching a solid board (such as a 2×4 or 2×6) to the fence with suitable masonry anchors, then adding the mounting hardware. Gates can also be attached to stone or brick fences using hinges that are set in mortar joints (while the mortar is still wet, of course, which requires that you do the work while building the fence itself), or by mounting hinges with special masonry screws.

Gate hardware tends to be either ornamental or utilitarian. Ornamental hardware is usually black and formed into attractive shapes. Utilitarian hardware has a dull galvanized or shiny zinc-plated finish and is not really intended to be admired for its looks. These distinctions are not absolute, however, as you can find some styles of hardware offered in all of these colors and finishes. Personally, I can't think of any situation in which black-coated hardware would not look better than a galvanized finish, and the price differences are usually pretty minimal.

Hinges

T-hinges are the most basic style of gate hinge. They usually have a nonremovable pin with a vertically oriented "mounting leaf" (the side that attaches to the post) and a horizontally oriented "door leaf" (the side that attaches to the gate).

Strap hinges have long, identical leaves on each side of the pivoting pin. They are most common on large gates but can be used on any size of gate. They often have removable pins.

Spring hinges are useful if you want your gate to be self-closing. Most spring hinges made for gates are adjustable, and I suggest that you stay away from any product that isn't. A spring with too much tension can be a nuisance at best, slamming shut with enough force to awaken sleeping babies and prematurely loosen hinges, and downright dangerous for kids running in and out. Even some of the low-cost spring-loaded gate hinges that I have seen over the years operate within an adjustable spectrum that ranges from a hard slam to a moderately hard slam. One product that avoids this limitation is the Kant-Slam hydraulic gate closer. It costs significantly more than other styles, but you only need one for each gate, along with one or two regular hinges. The Kant-Slam looks

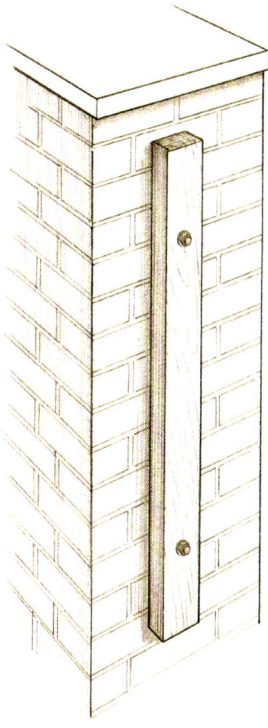

To install hardware on a masonry gatepost, a board can be fastened to the post, to which hardware can be attached.

a bit intimidating, but it is quite easy to install. It has a couple of springs on either side of an oil-filled piston regulator. The closing speed can be easily and accurately adjusted by turning a screw on top of the piston.

Screw and bolt hinges have both practical and aesthetic advantages. One style has a standard strap hinge that attaches to the gate, but the mounting leaf is replaced with either a screw hook or a bolt hook that needs to be inserted in a hole in the post. The screw hook operates like a lag screw, with the end of the threads remaining embedded in the post, while the bolt hook passes through the post and is secured with a washer and nut on the other side. Both styles can handle heavy gates and can be mounted to allow the gate to swing in both directions. Gates can usually be removed and replaced quite easily with this type of hardware, which may or may not be an advantage to you. Screw hooks are a great convenience if you use round gateposts, but I tend to favor bolt hooks for dependable, long-term service. If a bolt hook loosens, you can usually fix it by tightening the nut. If a screw hook loosens, it means that the threads are no longer tight with the wood inside the post, which is more difficult to rectify. A variation on the screw hook is a surface-mounted pintle, which is attached to the post with screws and is mated with a matching eye hook on the gate. The lack of visible hardware on the surface of the gate or post is one of the benefits of screw and bolt hinges for many people. You can also find hinges with a screw hook and an eye hinge. The former is screwed into the post, and the latter into the side of the gate, resulting in no visible surface-mounted hardware at all.

Deciding which size of hinge to use is a bit of edu-

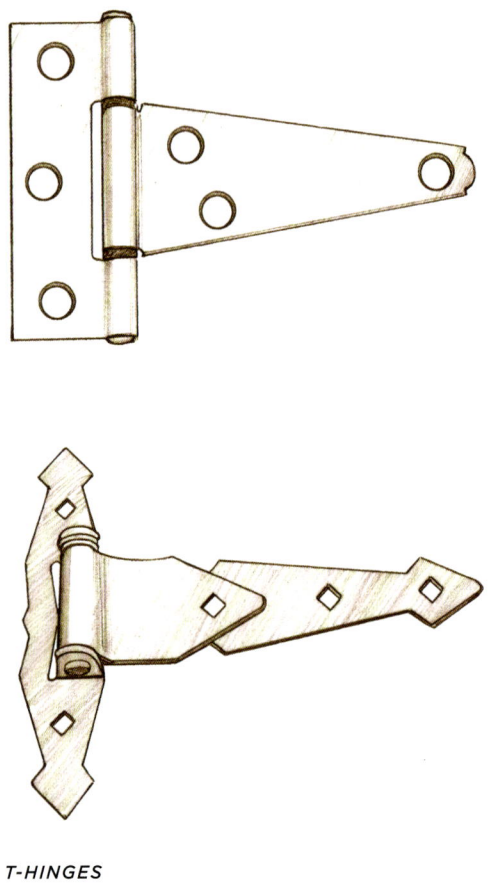

T-HINGES

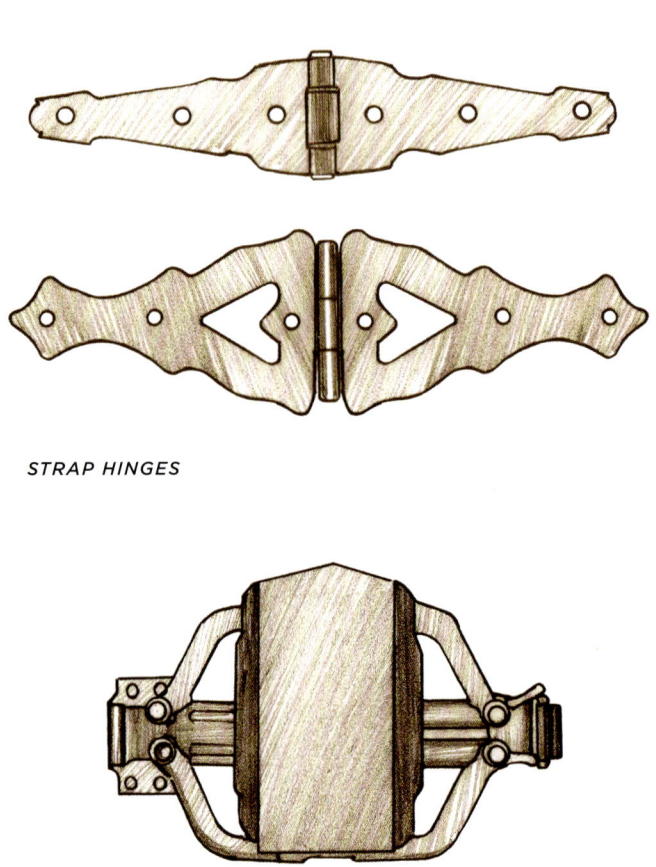

STRAP HINGES

KANT-SLAM

cated guesswork, as far as I am concerned. As a general rule, larger and thicker hinges equate to stronger hinges. The screws intended to be used with a hinge are usually included with it; hinges sold with a small number of short, skinny screws tend to be weaker than those sold with more, and larger, screws. Good-quality hinges are made from steel plate that is about ⅛ inch thick. The difference in price between a pair of T-hinges with 4-inch door leaves and a pair with 6-inch door leaves is likely to be not much more than one dollar. Don't be cheap with your hardware. Use the biggest hinges that will both fit and look good on your gate.

A typical T-hinge will have three holes in each leaf, and will come with six screws to fill each of those holes. Some people seem to think that all of those holes are there to give you a choice on where to locate screws. Not so! Those holes are there so that you can drive screws through each and every one of them. Using fewer than the intended number of screws serves to do one thing: It weakens the connection. So, use the full quantity of screws included with the hinge, but you may want to find screws other than those included in the package. I have a small drawer full of small screws in plastic bags that accompanied all kinds of purchases I've made over the years. These screws are invariably too short, too thin, or too cheap (quality-wise, that is) for my tastes. If you buy a hinge that is packaged with 1-inch-long screws that you plan to install on a 4×4 post, throw the screws out (or, if you are like me, throw them into that drawer), and go buy longer, thicker, stronger screws or bolts made for exterior use.

STRAP HINGE WITH SCREW HOOK

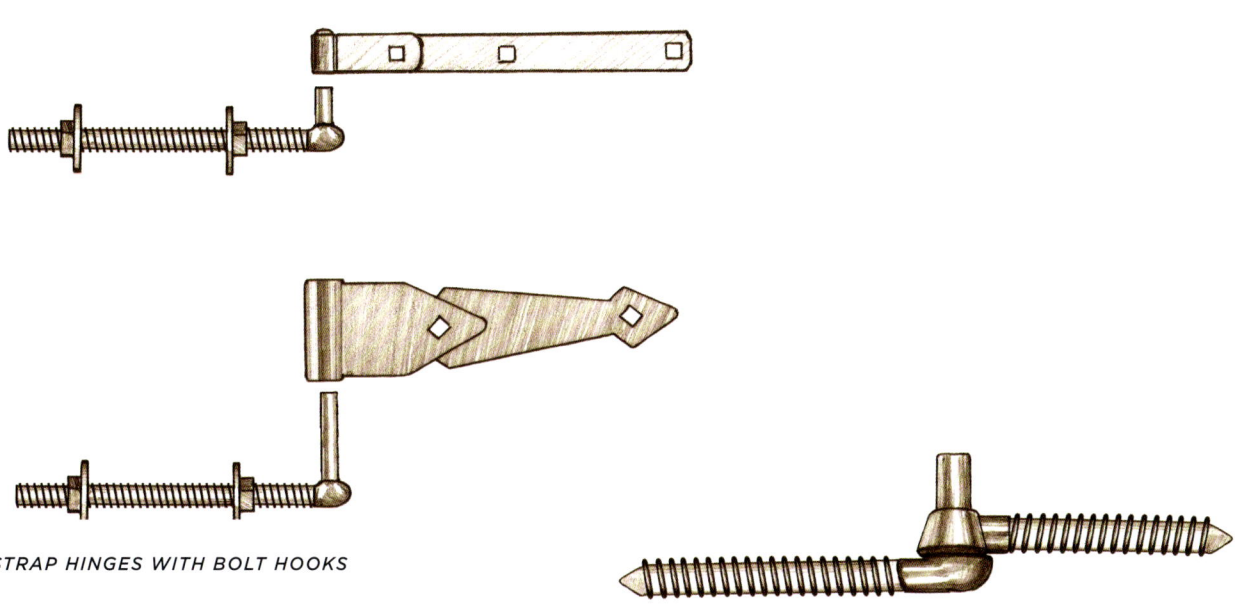

STRAP HINGES WITH BOLT HOOKS

"INVISIBLE" PINTLE HINGE

HOOK AND EYE

SLIDING LATCH

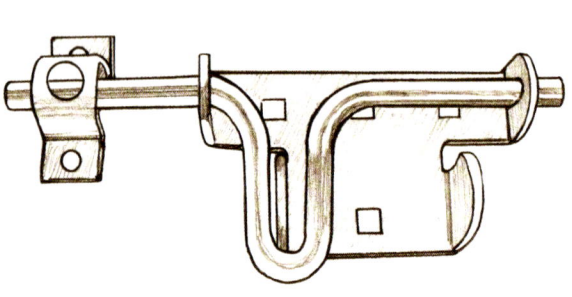

SLIDE-ACTION BOLT

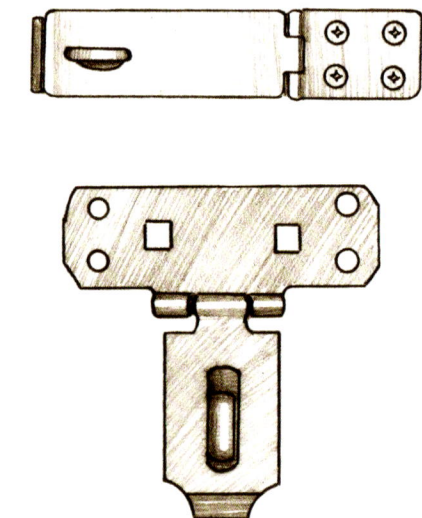

HASP-AND-STAPLE LATCHES

Latches

The most important things to think about when choosing a gate latch are security, ease of use, and appearance. Beyond that, ease of installation may also be a factor, as some styles of latches require more work than others, and the size or style of your gate may limit the choices you can make. With short gates, you can mount a latch on the inside that can be reached easily enough from the outside. But for higher gates, you will need a latch that can be operated from both sides of the gate.

Where security is of little concern, simple is often better. A hook and eye is about as basic and inexpensive as you can get. A piece of rope or a bungee cord can be effective means of keeping a gate closed. For a rustic look on round posts, a loop latch is a nice touch and a snap to install.

Sliding latches constitute a modest upgrade in strength and security. The simplest sliding latches have a round or rectangular bar that slides straight into a catch or strike plate. Slide-action bolt latches are a little more substantial. You lift and slide the bolt either into a strike plate or through a handle that is also used for opening and closing the gate. This style is also available with a tab that allows it to be secured with a padlock.

A hasp, or hasp and staple, latch is another basic style that can be locked with a padlock. Turning the swiveling eye to a horizontal position releases the latch; turning it to a vertical position closes it. A key-operated hasp eliminates the need for using a padlock. Hasps are effective, but I think there are more attractive options.

Automatic gate latches perform double duty, as both latch and stop. Their "automatic" feature refers

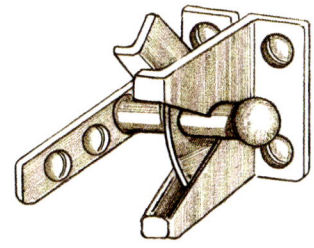

AUTOMATIC LATCH

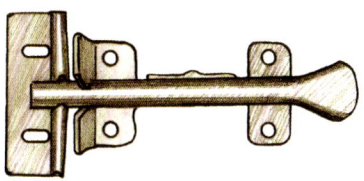

TOP-MOUNTED LATCH

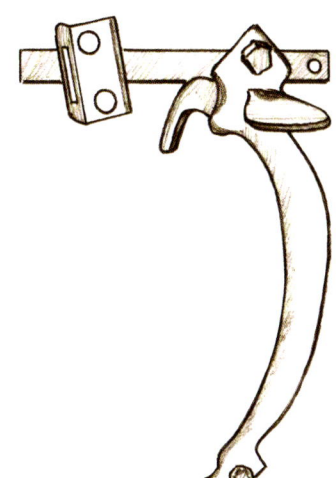

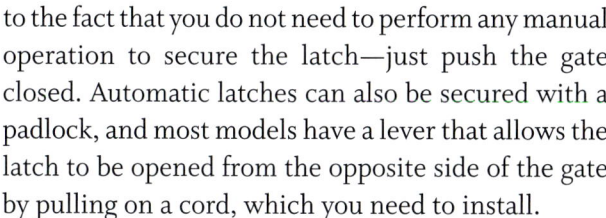

THUMB LATCH

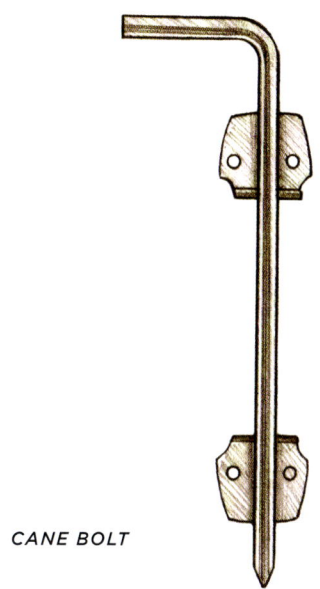

CANE BOLT

to the fact that you do not need to perform any manual operation to secure the latch—just push the gate closed. Automatic latches can also be secured with a padlock, and most models have a lever that allows the latch to be opened from the opposite side of the gate by pulling on a cord, which you need to install.

Thumb latches are probably the most common type of gate latch, in large part because they can be operated from both sides. They take a little more time to install than some of the other styles, as you must drill a hole through the gate to accommodate the thumb-activated bar, and the locking hardware on the back side needs to be carefully aligned. Nevertheless, a nice-looking thumb latch will complement any style of fence and gate nicely. One thing to watch out for, however, is the thickness of the gate. You may have trouble finding a thumb latch that can be used on a gate thicker than 2⅝ inches.

Top-mounted latches are attached to the horizontal top edge of the fence and gate. Because they are easy to see and can be operated with an elbow or even an otherwise full hand, they are convenient, but that is about all they have going for them. They tend to get banged around a lot and are completely exposed to the weather. I think they are an easy way to mess up the look of a nice gate.

Finally, a cane bolt, which drops down into the ground rather than across the gate to a catch on the fence, can be used to perform the same function as a gate latch. The bolt and hardware must be mounted on the gate. The bolt rests on the brackets while the gate is being used and is swiveled and dropped into a hole to latch the gate. The hole can be drilled in the concrete used to secure the adjacent gatepost, or it can be a pipe driven into the ground. You can also use a cane bolt to hold the gate open, if desired.

Building the Gate

REGARDLESS OF WHICH TYPE of frame or gate style you are building, take a few minutes to create a good work surface. As much as you want to build the entire fence plumb and level, this objective is doubly important with the gate. I like to work while standing up, so a piece of sturdy plywood across a couple of sawhorses is my work surface of choice for this type of carpentry chore. For those with healthier knees and stronger backs, though, a flat driveway or garage floor will certainly work better than bare earth.

It is also smart to work out every last detail and dimension of your gate on paper before you begin building it. At the very least, prepare an accurate scaled drawing. Even better is to draw a full-scale replica of the gate on a sheet of plywood, and then use the drawing as a template to align all of the pieces. This is a particularly good way to build gates with a Z-frame, since you can set the face of the gate down on the template with the boards aligned and spaced exactly as you want them and then attach the three framing boards. With a perimeter frame, it is usually easier to first construct the frame, and then attach the infill to it, as described below.

Measure the Opening

Measure the distance between gateposts at the top and bottom gate locations, which may or may not be the same as the top and bottom of the posts themselves. If the distances are the same, or very nearly the same, you're in luck. If not, then it probably means that one or both posts are either out of plumb or twisted. Fitting a square gate into a less-than-square opening may not be quite as fruitless as the legendary square-peg-in-round-hole effort, but it can be mighty frustrating. So, if necessary, take a little time to square the opening, either by repositioning, shaving a little wood from, or adding shims to one or both posts.

With the opening squared away, calculate the width of the necessary gate by subtracting ½ inch for clearance on the latch side and ¼ inch for the hinge side. This ¾-inch overall clearance is a standard dimension for gates utilizing normal T-hinges and routine

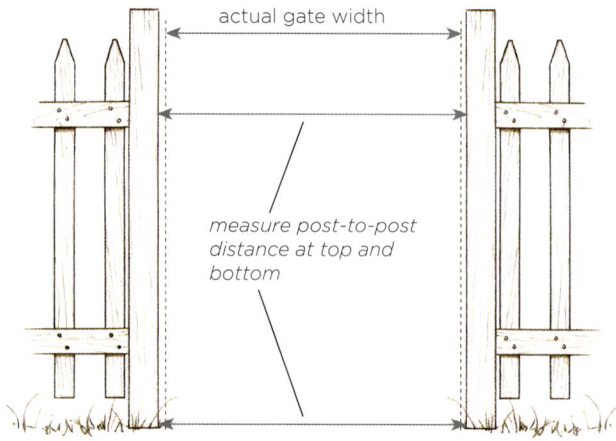

When measuring the gate opening to determine the width of the gate frame, remember to allow for clearance on both the latch and the hinge side.

Depending on the gate style, the height of the frame may be less than the height of the fence to permit infill to overhang the frame at the top and bottom.

latches. For other hardware, you may need to increase the clearance, especially on the hinge side (when using screw and bolt hinges, for example). Hold your hardware in place at the posts and check their clearance needs carefully. Some hardware comes with installation instructions that specify the necessary clearances.

Assemble the Frame

Cut the top and bottom frame members to the exact dimension of the gate's width, then cut the two side boards to fit in between. The length of these side (vertical) boards may depend on the design. You may want to make them short enough to allow infill boards to overlap at the bottom and, especially, at the top.

If you are building a perimeter frame with flat boards, cut and join the half-laps (see page 222). For an on-edge frame, set the four boards in position on the work surface. Use a framing square to make sure that all four corners are square. If you don't have a framing square, measure the diagonals from the outside corners. A rectangle will be truly square if the diagonals measure the same distance.

If you have a couple of pipe clamps, attach them to hold the boards tightly together while you drive screws or nails through the top and bottom boards and into the ends of the side boards. For a large, heavy gate, lag screws would provide a stronger connection.

With either type of frame, lay the uncut brace on the work surface and set the frame on top. Use the frame to mark the cut lines, then cut the brace to length and attach it to the frame.

The most foolproof way to make a perfectly sized brace is to position it beneath the gate frame and then mark the cut lines. Before doing so, however, make sure that the frame is square and supported at all four corners.

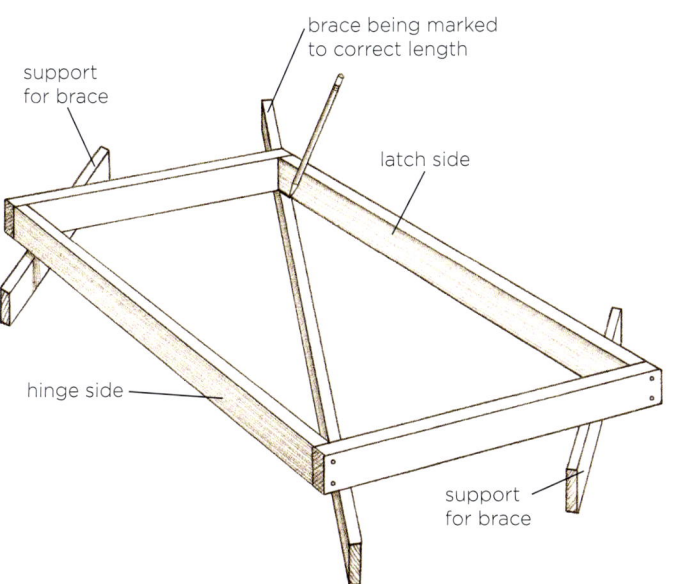

Attach the Infill

Set the frame flat on the work surface with the hinge side of the brace nearest to you. If you are using the same style of infill on the gate as you used on the fence, use the same fasteners as well. If your fence has spaced vertical boards or pickets, maintain consistent spacing from fence to gate.

You may want to set the infill in place loosely and check the fit before you start fastening. If you need to trim one board to fit, well, don't do it. Instead, trim a smaller amount off several boards so that the deviation is hard to notice.

It is usually best to install boards a bit too long on the top and bottom, with the intention of trimming them to length later. It is easier to lay out and cut a pattern on the top after the boards have been installed, and it is much less likely that you will split the board's ends with screws or nails, as can happen when they are installed too close to the board end. Set the gate in place and check the fit.

Install infill boards longer than their finished size and then mark and cut them to the correct length and shape. This technique is particularly useful for an arched design.

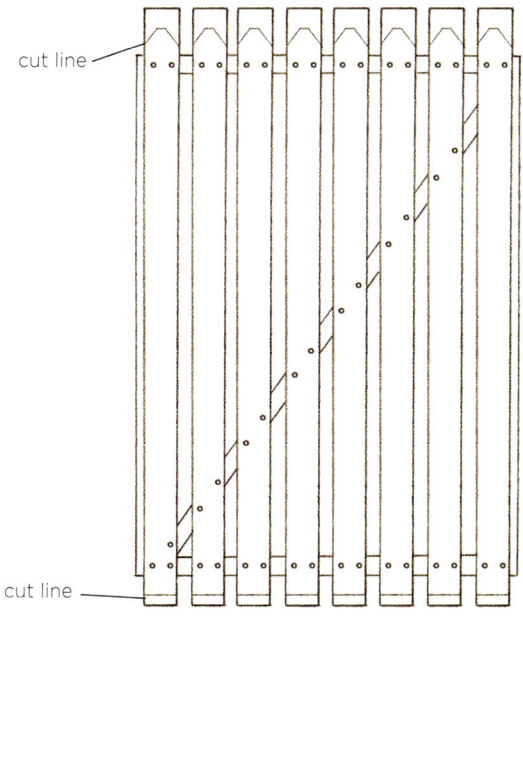

Install Hinges on the Gate

Hinges should always be attached to the frame, and not just to the infill. Two hinges are sufficient for most gates up to 4 feet tall, and three hinges are recommended for 6-foot-high gates. For those in between, make your decision based on the weight of the gate as well as the size and strength of the hinges themselves (two hefty hinges might well make for a stronger connection than three weak ones).

Use screws long enough to penetrate deep into the frame, but (obviously) not so long that they poke through the other side. If you are using large strap hinges on a heavy gate, use carriage bolts with washers and nuts rather than screws.

Hang the Gate

Set the gate back in the opening, resting it on wood blocks or bricks so that it is sitting at the exact installation height. Once the gate is level, I suggest slipping a small shim under the latch end. Once the hinge is installed and the gate is allowed to hang freely, gravity will usually bring the latch side down a bit into proper alignment.

With the gate sitting in place, drive a single screw into each hinge, then let the gate hang free while you check the alignment and make sure that the gate swings freely. Make any necessary adjustments, and then install the remaining screws.

If your gate design requires the use of a separate gate stop, install it right away. Be very careful not to let the gate swing beyond its planned arc, which can damage the hinge or pull out the screws as fast as you drove them in.

Install the Latch

New gates always seem to need a little time to settle comfortably into place, so you should not feel the need to rush into installing the latch. Instead, use the gate for a few days or even weeks, then add the latch.

Prop the gate in the opening so that it is level and at its intended height. Then slip a shim under the bottom of the gate on the latch side. Attach the hinges with the gate in this slightly tapered position. That way, once the props are removed and gravity takes over, the gate will settle into its anticipated alignment.

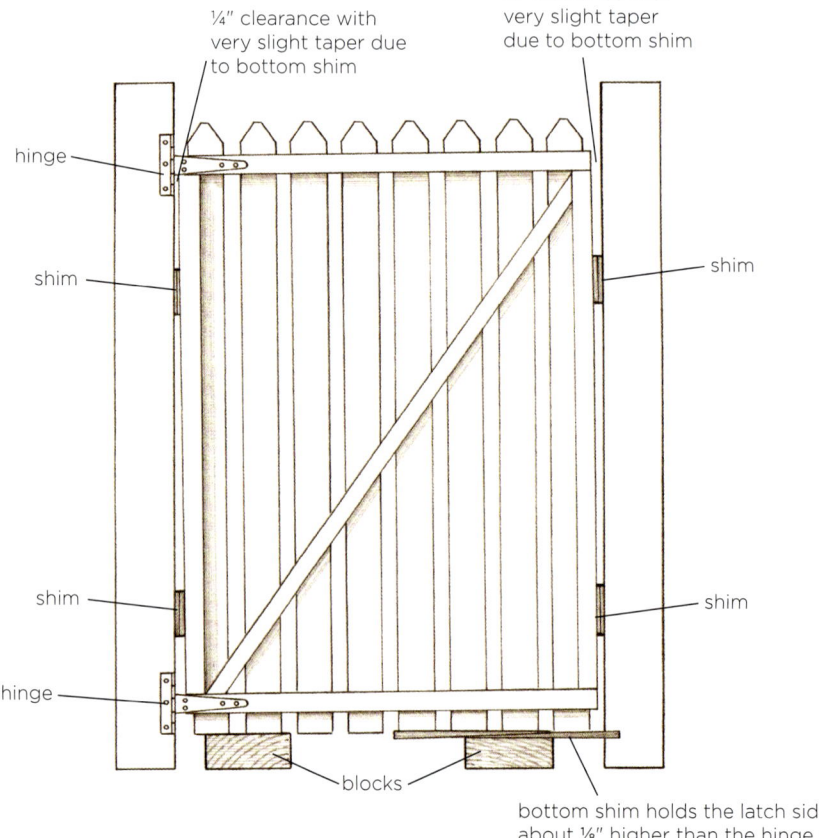

¼" clearance with very slight taper due to bottom shim

½" clearance with very slight taper due to bottom shim

hinge

shim

shim

shim

shim

hinge

blocks

bottom shim holds the latch side about ⅛" higher than the hinge side

Gateless Gateposts

If marking an entryway, such as a sidewalk or driveway, is all you need to do, you can apply the same techniques for fence and gate construction provided throughout this book to accomplish your goal. While any of these projects could serve as fully functional gateposts or terminal fence posts, they also can be used as freestanding gateways.

When marking either sidewalks or driveways, do not place the posts too close together. Leave plenty of room for traffic, both vehicular and pedestrian. And keep the posts set back far enough to be out of reach of snowplows.

Stone Pillars

Stacking stones into a pile is child's play, although many people find it a perfectly satisfactory method for creating driveway markers. Sometimes, however, what now looks like an intentionally created rubble pile was once a neat, dry-stacked pillar, which may help explain my preference for mortar to hold these small structures together.

The essential techniques for building with stone and mortar are covered in Chapter 4, and you should review that information before proceeding with this project. Those who enjoy a challenge may want to build pillars with round stones or irregular flagstone. The former require considerably more mortar than flat stones, and the latter require a good deal of cutting. For the best combination of strength, beauty, and ease of construction, use cut stone for your pillars. You can buy square and rectangular stones in various lengths and thicknesses if you want to give the pillars a complex design, but having the corners already squared off will make it much easier to fit the pieces together.

For pillars that will remain sound and solid for years to come, plan to set them on concrete footings and to top them with a solid cap. Good footings will resist damage from frost heave as well as from traffic-induced vibration. And a solid cap will not have any mortar joints that could weaken and allow water to seep inside the structure, where it could in turn freeze and cause serious damage.

Slightly oversized stone piers terminate this stone wall and help direct the visitor to the gateless entryway.

PREPARING THE FOOTING

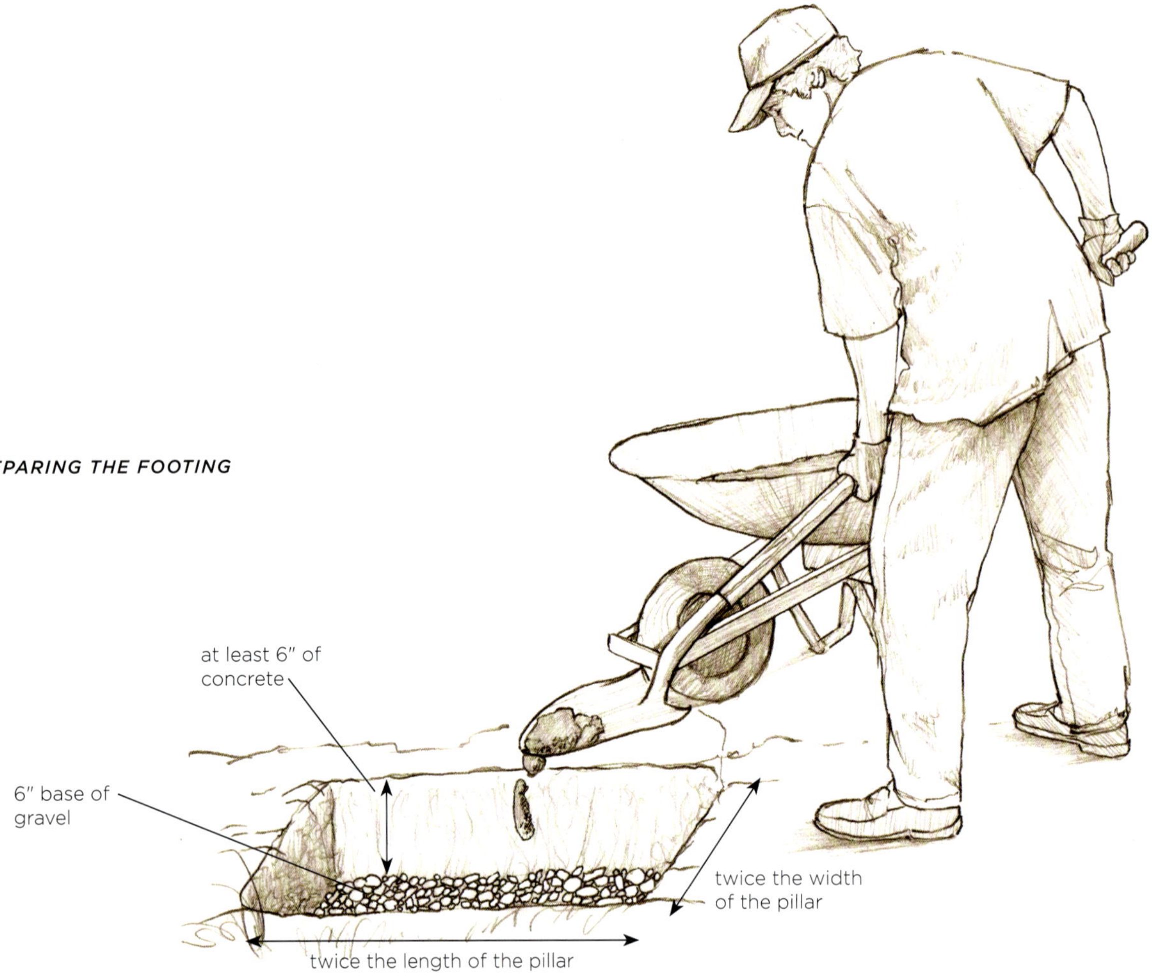

at least 6" of concrete

6" base of gravel

twice the width of the pillar

twice the length of the pillar

SETTING THE FIRST COURSE

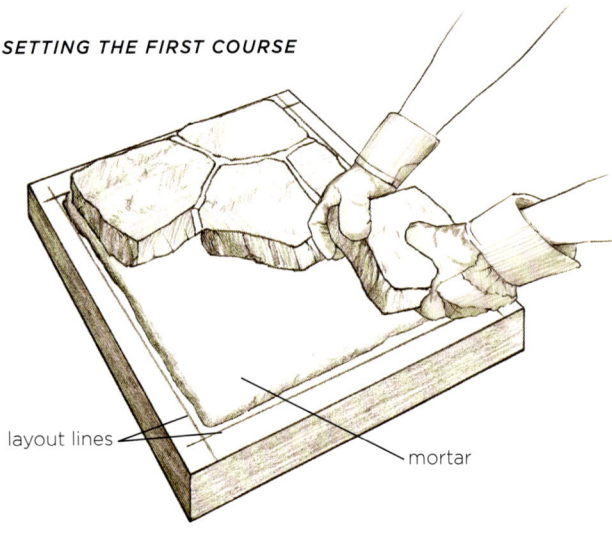

layout lines

mortar

I would not bother setting footings below the frost line for a project like this. I think it's just too much work, requiring too much concrete, for such a small-scale structure. If the posts heave a bit, it may not be visibly evident. Even in the event that they heave a great deal, the freestanding structure itself is all that will be damaged.

I would plan to dig at least 12 inches deep. If you want to hide the footings, dig a couple of inches deeper. Plan to make the footing twice the size of the footprint of the pillar; thus, if the pillar is going to be 2 feet square, make the footing 4 feet square. Dig the hole to the exact size of the intended footing, and use the earth itself as the form for the concrete. Make the sides straight and the edges neatly squared. Dump a 6-inch layer of gravel in the hole, and then compact and level it.

Mix concrete as described in Chapter 3, and then carefully pour it into the hole. Agitate the concrete a bit with a trowel or shovel to ensure that there are no voids. Fill to the top of the hole, or a bit below if you plan to keep the footing concealed. Just make sure that you have a 6-inch depth of concrete. Level and smooth the surface with a trowel. Let the concrete cure for a couple of days.

Mark an outline on the dry footing to ensure that your pillar is perfectly centered. Set the first course on a 1-inch bed of mortar, and use the same depth of mortar between subsequent courses. Keep stones on the same course an inch apart, and fill the gaps with mortar. Press each stone firmly into the mortar, and scrape away any that oozes out. Use a pointing tool to recess the mortar joint an inch or more so that it is not so visible.

The key to success is to select each stone for a course before you set any of them in the mortar. You might find that this is easiest to do on a scrap piece of plywood with an outline of the pillar drawn on it. Arrange stones on this template for a good fit, and then transfer the arrangement to the pillar. Any voids in the middle of the pillar should be filled with small stones and concrete. Keep a level close at hand to check that all four sides remain plumb and that each course is level.

The ideal capstone is wide enough to overhang the pillar on all four sides, but not so thick that it is hard to lift into place. Set the cap on a 1-inch layer of mortar. When you build the second pillar, make certain that the tops of both are at the same level.

SETTING THE CAPSTONE

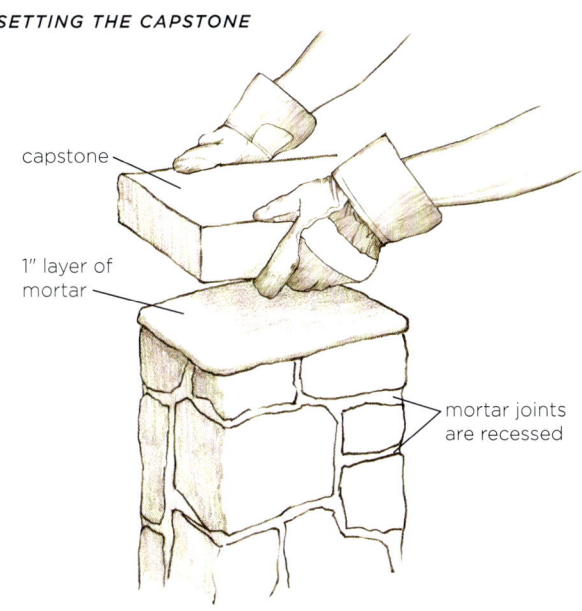

capstone

1" layer of mortar

mortar joints are recessed

FINISHED STONE PILLARS

capstones set at same height

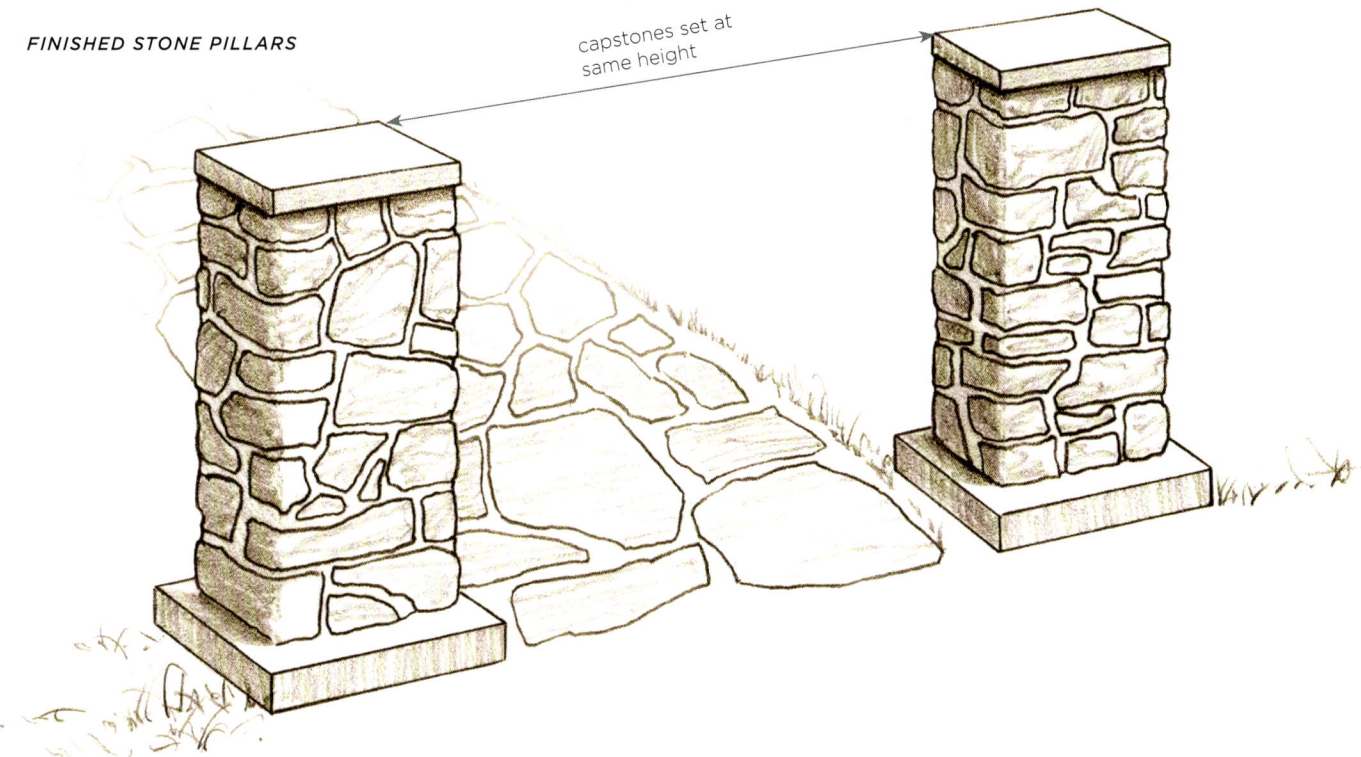

Brick Pillars

Pillars of brick can be built in much the same way as stone pillars. With bricks' size being so uniform, design and construction are also pretty uniform. From an aesthetic standpoint, I like the rectangularly shaped pillars shown in the illustrations on the facing page. Each course requires five standard-size bricks, and the mortar joints are pleasingly staggered on all four sides.

Since brick is much lighter than stone, you can set these pillars on smaller footings. Plan to make the footings extend 6 inches beyond each side of the pillar. Thus, these 16-inch by 12-inch pillars can be safely set on 22-inch by 18-inch footings. Prepare the footings as described for stone pillars on pages 234–235, and set the bricks as described in Chapter 4. Top the pillar with a large stone or a precast concrete cap. Another pleasing pattern is the square pillar shown in the photograph below. These are more substantial pillars, requiring eight standard-size bricks in each course. A course of offset bricks near the top of the pillars is a nice yet simple touch.

Solid brick pillars here separate and support a picket fence made with wood and an ornamental metal gate.

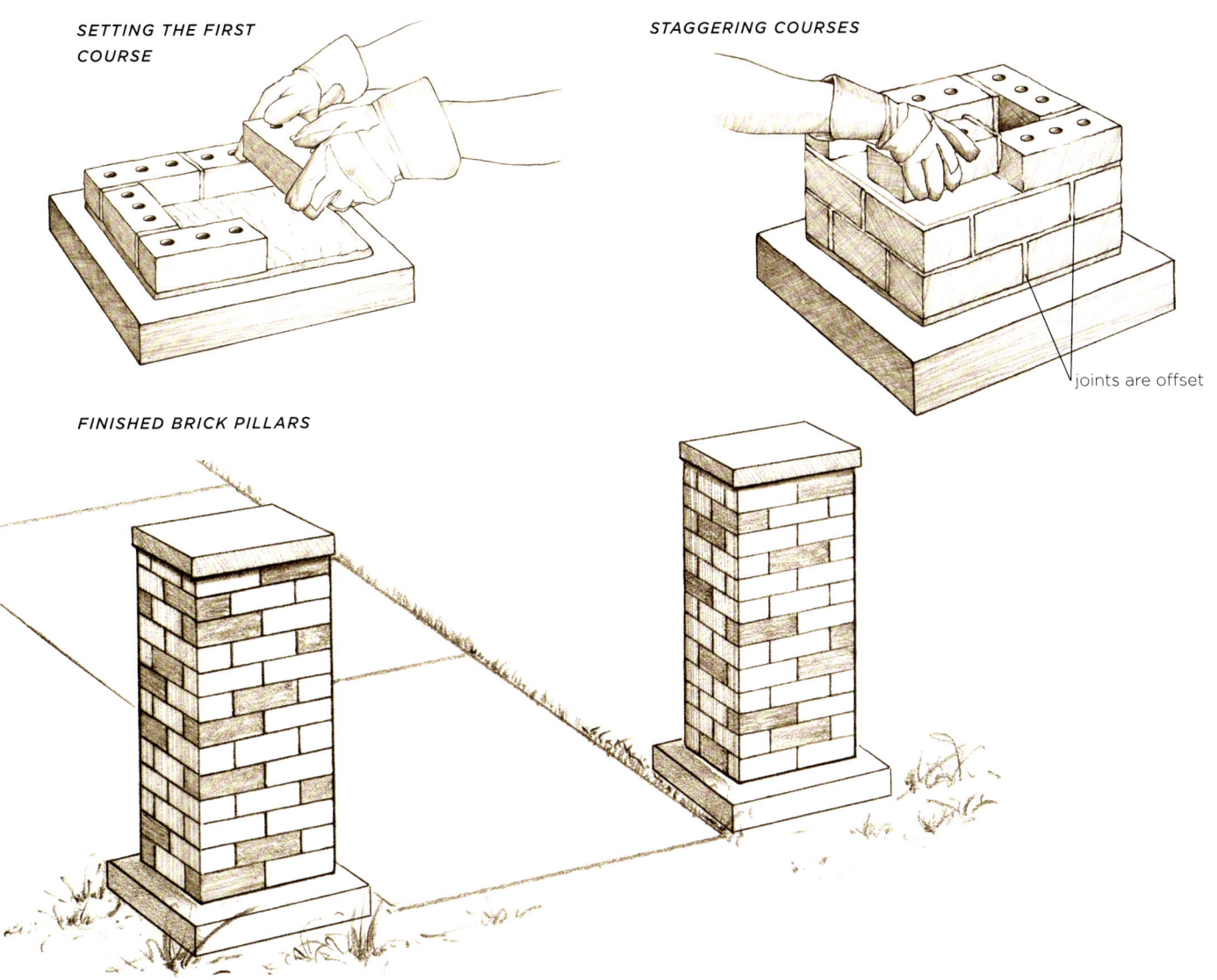

SETTING THE FIRST COURSE

STAGGERING COURSES

joints are offset

FINISHED BRICK PILLARS

Boulders

If you like the idea of freestanding markers but are not enthused about the work involved, take yourself and your checkbook to a nearby stone dealer or quarry. Find two large boulders that appeal to you and have them delivered and set in place. Just be sure that they are not going to have to be moved at any foreseeable point in time.

Wood-Framed Entry

Wood posts can also function as attractive freestanding entry markers, especially when converted into an arbor spanning a sidewalk. Use 4×4 or 6×6 posts, set in concrete as described in Chapter 3. Plan to have at least 7 feet of headroom beneath the arbor. The simple designs shown here utilize 2×4s and 2×6s, with their corners clipped or ends cut into a decorative shape, and 2×2s, all fastened with nails or screws. With some imagination, you should be able to come up with any number of designs from which to choose.

SINGLE-LAYER ARBOR

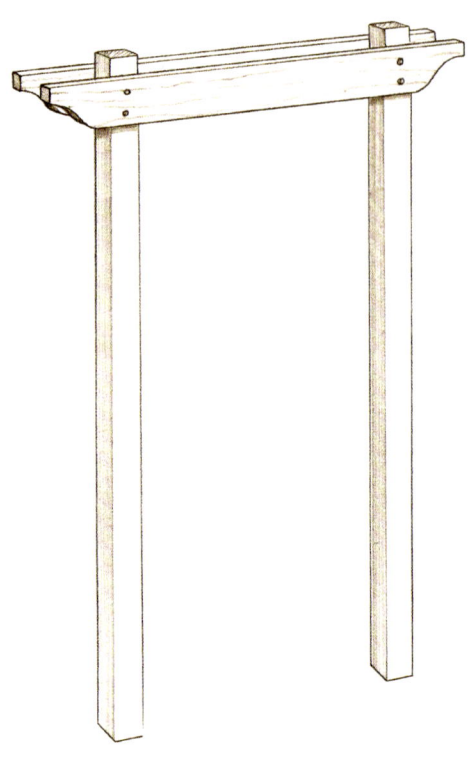

DOUBLE-LAYER ARBOR

TRIPLE-LAYER ARBOR

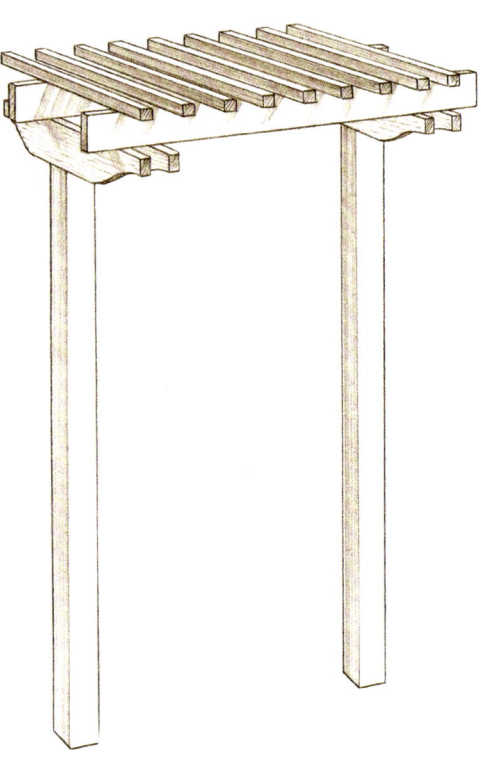

Gates of Simplicity Sometimes

the perfect gate is the one that simply and unobtrusively

performs its intended function, and nothing more.

Gates of Whimsy
Some fence builders see the gate as an opportunity to add some personal creativity to what may otherwise be a rather plain fence. Unusual shapes, different heights, and varying materials can all be used to fashion a one-of-a-kind gate.

Repairing Fences and Gates

THE BEST MAINTENANCE POLICY for a fence is to design it well, buy good-quality materials, and build it to last. But even the most carefully conceived and best-built fence is going to need repairs from time to time. Tend to small problems right away and you will have much less chance of having to deal with big problems later on. If you inherited an older fence when you bought your house, there's a good chance that this is the only chapter you will be needing for a while.

The weakest links in fences tend to be the posts, which is the primary reason I spent so much time on the subject in Chapter 3. The problems are twofold. First, because the posts are buried in the ground, they are most susceptible to rot and decay. Second, because the posts bear the weight of the fence and provide its lateral stability, when they start to fail, the whole fence follows suit. When parts or all of a fence line start sagging, the first place to look for a cause is at the posts. It is a good idea to catch any problems before they advance too far, and the best way to do that is to perform an annual inspection and keep up on the maintenance.

Key elements of a long-lasting wood fence are a good coating with paint or some other wood finish [ABOVE], renewed as needed, and solid construction combined with design features that encourage water to drain away from the wood [RIGHT].

Refinishing a Wood Fence

ALL WOOD FINISHES begin to deteriorate over time. The speed with which any given finish needs to be renewed really can depend on a range of factors. On some fences, certain sections may wear more quickly than others. Paint fades, cracks, and peels. When that happens, water can more easily get beneath the surface, which only speeds up the process of deterioration. Solid-color stains do not necessarily peel and crack like paint, but they do fade and wear away. With both of these finishes, the best time to refinish is when you see these early signs of failure starting to appear.

Surface preparation is particularly important for reapplying film-forming finishes. Scrape any loose paint or stain, sand the rough edges, and then scrub the fence with a stiff brush and water. If the surface remains chalky after this, clean it again with a mild detergent and rinse thoroughly. If there is any mildew on the fence, thoroughly remove it as explained on page 246.

If any bare wood has been exposed, coat it with a paintable water-repellent preservative. When dry, add primer and one or two coats of 100 percent acrylic latex paint.

Penetrating finishes are much easier to reapply. Usually the only surface preparation required is to scrub the surface with a stiff brush and sweep or vacuum off any loose dirt and debris. With that work done, apply the new finish.

Removing Mildew

MILDEW IS MORE OF A NUISANCE than a danger. It is a fungus that lives on the surface of wood, finished or not, but does not feed on wood or wood finishes. But, even though it does not damage a fence, it can discolor it. Mildew is most common in warm, humid climates, but it can show up on any fence behind some bushes or where water splashes. If you are uncertain whether or not a discolored section is due to mildew, put a few drops of household bleach on it. If the bleach removes the stain, it is probably mildew. If the stain remains, it is more likely dirt, rust, or extractives bleeding through the wood.

To remove mildew, you can buy a mildew cleaner at your paint store and use as directed. Alternatively, make your own solution by combining 1 quart of liquid bleach with 2 quarts of warm water and ¼ cup of household detergent. Use a stiff brush or sponge to scrub the mildew with the solution, and then rinse thoroughly. Once the fence surface is dry, apply a finish containing a mildewcide.

Tightening Loose Fasteners

NAILS AND SCREWS can loosen over time. Metal fasteners can also rust and break. Sometimes all you need to do is tighten a few screws or whack a few nails with a hammer. But if that does not seem to do the trick, replace the troublesome fasteners. You can either remove the old fasteners and drive larger ones into the existing holes, or leave the existing fasteners in place and drive new ones nearby. Be sure to use appropriate fasteners, as described in Chapters 3 and 7. If some nails persist in loosening, replace them with screws.

Checking for Rot

Wood that has begun to rot will feel soft and spongy. During your annual inspection, use a small screwdriver to poke around the base of any and all wood posts. If you can dig down 6 inches, check for rot beneath the surface as well. If rot has set in, the only reasonable cure is to replace the post, although there are some temporary fixes you can try that may buy you additional time. Rot also commonly appears at connections between posts and rails, and between rails and pickets or boards. These parts are easier to replace, and in the meantime you can strengthen a rotted rail end by nailing a 2×4 cleat to the post beneath it. ■

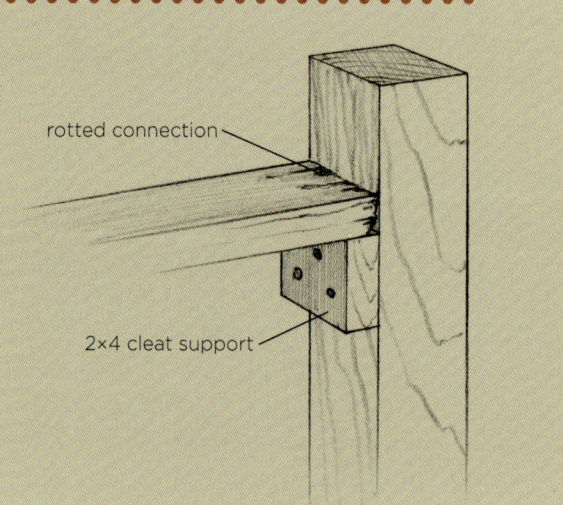

rotted connection

2×4 cleat support

Correcting Footing Failures

IF THE POSTS HAVE LOOSENED or fallen out of plumb, it is likely due to a problem with the footing or backfill. Concrete footings can crack or heave, and a tamped earth-and-gravel footing can soften over time. A failed concrete footing is hard to repair, and the best solution is to replace the post and footing. You can tighten the hold of an earth-and-gravel footing by digging out as much of the backfill as you can, replumbing the post, then backfilling and tamping as described in Chapter 3. You might also try fastening pressure-treated shims to all sides of the post before backfilling and tamping.

If a concrete footing seems secure but the post has shrunk away from the concrete and loosened, drive pressure-treated shims into the concrete on all four sides of the post, checking for plumb as you do so. When the shims are tight, trim them level with the concrete and run a bead of clear silicone caulk around the top to seal the joint. This may not provide a long-term solution, but it can buy a few more years.

SECURING A LOOSE POST

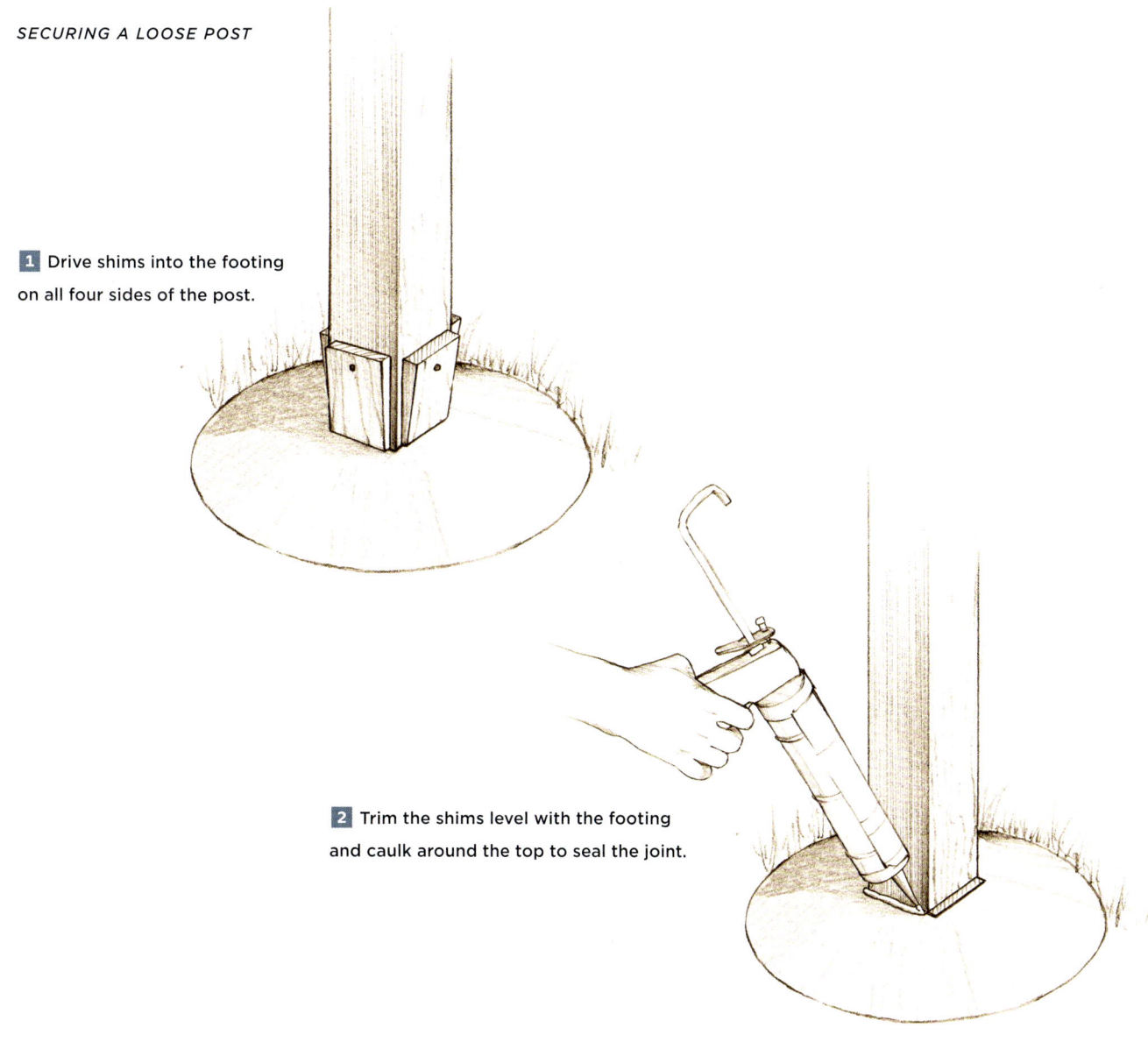

1 Drive shims into the footing on all four sides of the post.

2 Trim the shims level with the footing and caulk around the top to seal the joint.

Mending Decayed Posts

ROTTED POSTS CAN BE REPLACED, with new posts being set in new footings in the same location as the old ones. But on some fences, this can be a challenging job that requires a lot of work and threatens the strength and integrity of the rest of the fence. In that case, it often makes better sense to try and reinforce the old post with another, partial post.

Begin by shoring up the fence on both sides of the bad post. Place blocks beneath the rails at the very least, and for a high and heavy fence I suggest you brace the top rail and infill from both sides. Remove the backfill down to the bottom of the post or, with a concrete footing, dig a new posthole right next to the existing one. Cut the old post off an inch or two above ground level. Remove the post; if the post is set in concrete, you should be able to wiggle it loose and lift it out with a helper or two. Clean out the hole, and make sure the fence is plumb.

Now add about 4 inches of gravel to the bottom of the hole. Cut a new post so that it reaches from the bottom of the hole to about 3 feet aboveground. Cut the top at an angle to shed water, and set the uncut end of the post in the ground. Add a couple more inches of gravel to the hole, and tamp it firmly. Attach the reinforcement post to the old post with ½-inch carriage bolts spaced 8 inches apart. Fill the hole with concrete as described on pages 86–88. When the concrete has set, remove any bracing and blocks. Coat the bottom of the old post and all of the reinforcement post with a water-repellent preservative.

POST REPAIR

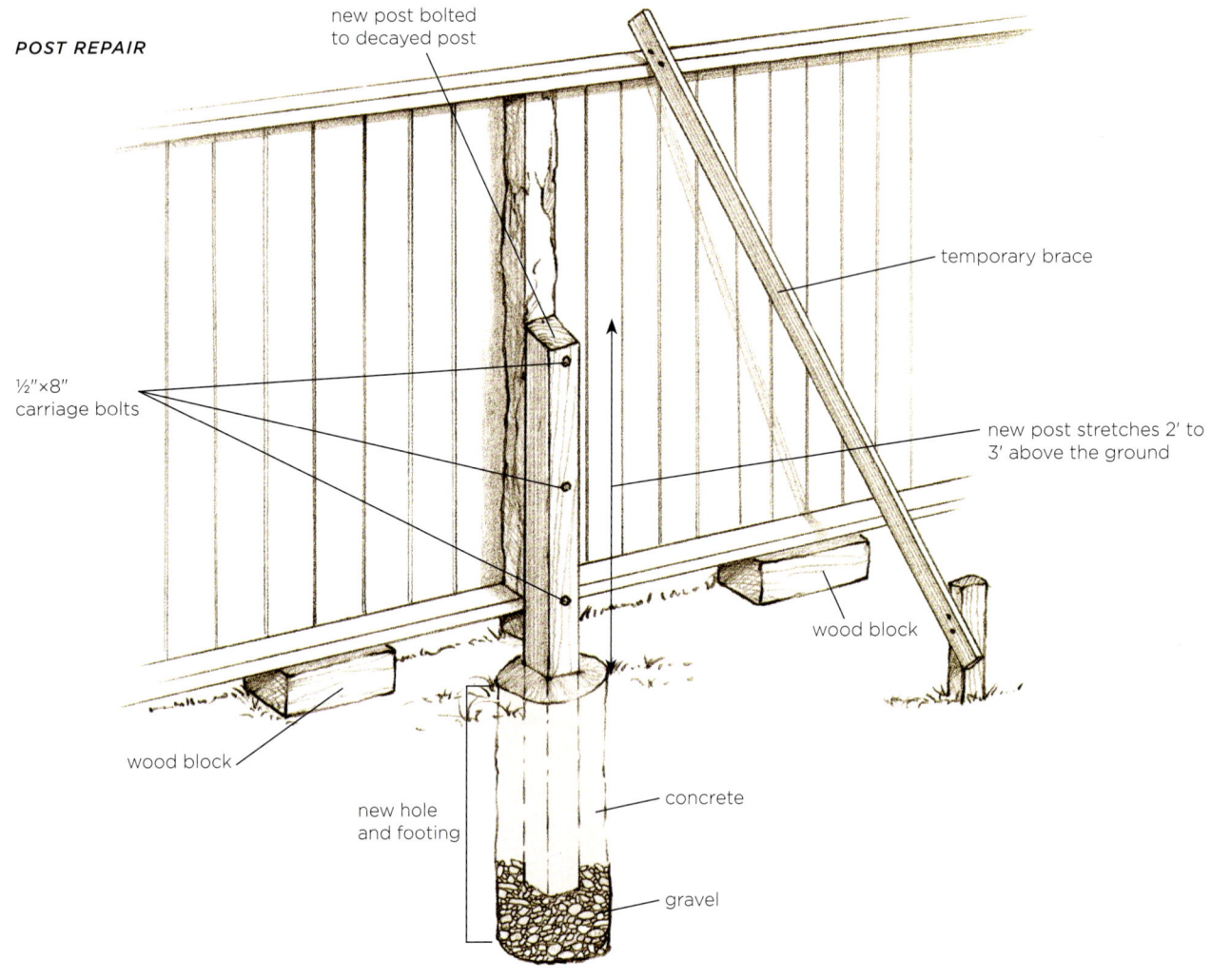

new post bolted to decayed post

temporary brace

½"×8" carriage bolts

new post stretches 2' to 3' above the ground

wood block

wood block

new hole and footing

concrete

gravel

Replacing Decayed Posts

IN MOST CASES, the best way to deal with rotted posts is to dig them out and replace them with new posts set in new footings. On many fences, this can be done fairly quickly, especially if you are able to remove the adjacent fence panels. Often this can be done by removing some screws or brackets, but if the rails are toenailed to the posts, I would use a reciprocating saw to cut through the nails and then carefully set the fence panels aside.

Now dig out the backfill around the post or loosen the earth around a concrete footing. Rocking the post back and forth can sometimes help loosen it up, although if it is badly rotted, this will just as likely cause it to break off. If you have several concrete footings to remove, it might make sense to rent an electric jackhammer for a half day to bust up the concrete into pieces that can be lifted out.

Remove any loose dirt and other debris from the hole, then proceed to set a new post with either a concrete or an earth-and-gravel footing, as described in Chapter 3. If you want to use concrete but find that removing the old footings created very large holes, you can set a 12-inch-diameter fiberboard form (commonly referred to as a Sonotube, after one of the brands) in the hole on top of the gravel, then position and brace the new post in the center of the form. Carefully backfill around the form, tamping the earth every 4 inches or so. Finally, pour concrete into the form. Once the concrete sets, you can cut away any of the form that remains aboveground.

With the new post or posts set and the concrete cured, reattach the fence sections. Handle the sections carefully. You may find that metal fence brackets make reattachment easier.

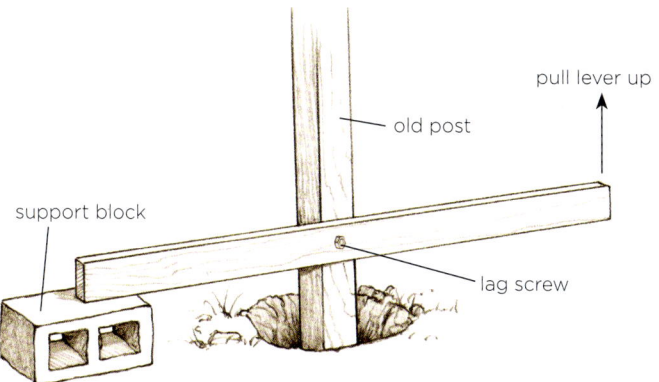

If a post resists being pulled out, attach a 2×4 lever to it with a lag screw. With one end supported by a wood or concrete block, lift the other end up to pull up the post.

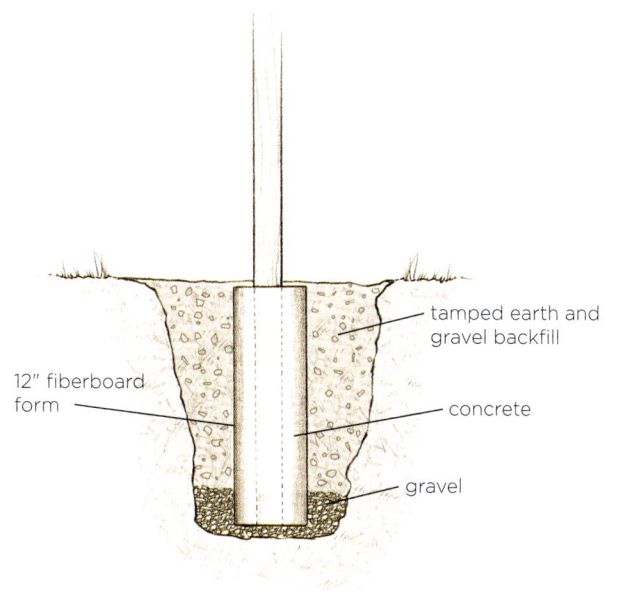

To set a new post in an oversize hole, set the post in a fiberboard form and then fill the form with concrete.

Repointing Mortar

THE WEAK LINK OF nearly all mortared stone and brick walls is the mortar itself. Over time, the mortar is almost guaranteed to form cracks and fall out in places. If and when this happens with your wall, tend to the repair as soon as possible. The longer you let failing mortar joints continue to fail, the more likely it is that the structural integrity of the wall will be compromised.

The practice of renewing mortar joints is known as repointing. It is not difficult, and it can be very effective. The illustrations show the work being done with brick, but the same tools and steps apply to stone. The first step is to remove all of the loose mortar. Use a hammer and small masonry chisel to clean out the joint to a depth of at least 1 inch (be sure to wear eye protection!). If you can get a hose to the site, clean out the joints with a good dose of water. Otherwise, use a brush to remove any dust and debris, and then use a spray bottle to moisten the joint.

Mix a batch of mortar. Scoop some mortar onto a trowel. Holding the trowel next to the joint, push mortar into the joint with a pointing tool. This is easier to do on horizontal than on vertical joints. With the latter, you have to fill the joint a bit at a time, working from bottom to top. Shave off any excess mortar from the joint, and then use the pointing tool to smooth and finish the joint.

If you need to repoint an old wall, you may want to approach it differently. The brick that was widely used prior to the 1930s was softer than the modern product, and the mortar mix used on these bricks contained lime rather than the Portland cement that is used in modern mixes. Portland cement mortar does not work too well with soft brick, and the colors of the two types of mortars can differ quite a bit. If you have doubts about the age of your wall, or concerns about color-matching the new and old mortars, I suggest that you talk to a mason with some experience in restoration work.

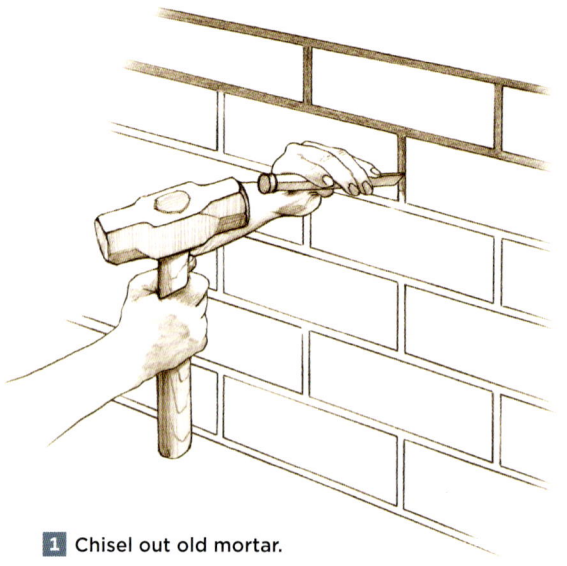

1 Chisel out old mortar.

2 Apply new mortar.

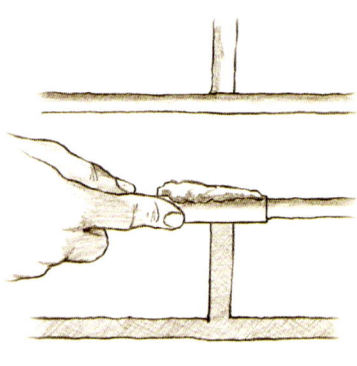

3 Finish joints with pointing tool.

Repairing Gates

GATES TAKE A LOT OF ABUSE. When they are in need of some repairs, they usually announce the fact by ceasing to work properly. Gate repairs, therefore, are a routine activity, which will be made all the easier if you tend to them sooner rather than later.

Most gate problems are related to the hardware, and most hardware problems are easy to fix. Sagging gates look bad and may not close or latch properly. The most common cause of sagging is a loose hinge. Tighten all of the hinge screws. If some screws still seem loose, try replacing them with larger screws. If that does not work or is not feasible, consider moving the hinge or hinges up or down a little.

When screw holes have become too big to do an adequate job of holding the threads, I reach for a ⅜-inch brad-point drill bit and a length of ⅜-inch wood dowel. I drill a clean hole in the old screw hole, then cut a piece of dowel a bit longer than the hole depth. Then I spread a little waterproof glue on the dowel and drive it into the hole with a rap or two of a hammer. After the glue has dried, cut the dowel flush with the post or board, and reattach the hinge with screws driven into the dowels.

An alternative to using dowels is to drill holes all the way through the post or board. Reattach the hinge using through bolts, washers, and nuts.

When the gate is sagging in a still-square gate opening, the problem can usually be easily fixed with a gate repair kit, which you should be able to find at most home improvement and hardware stores. Instructions will be included with the kit. The process involves attaching a couple of corner brackets on the gate, then running cable from each bracket to a turnbuckle. I find that it is easiest to prop the gate up so that it is level before attaching this hardware. Then, you need only turn the turnbuckle to pull the cables tight and bring the gate back into alignment. Not very attractive, but it is very effective. If only our sagging bodies could be fixed so easily.

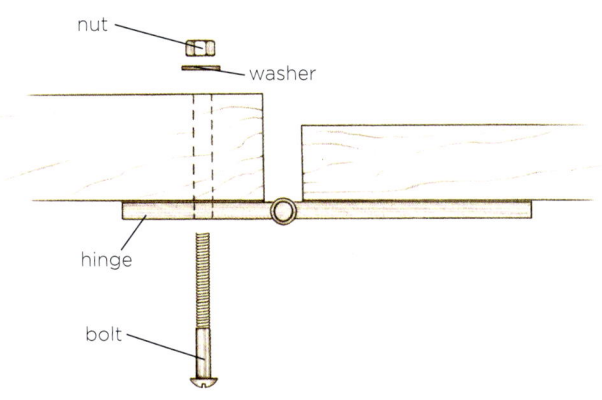

CHANGING THE FASTENER

nut
washer
hinge
bolt

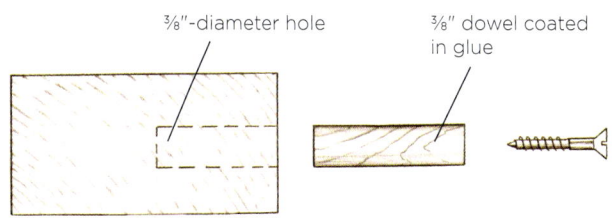

REPAIRING A SCREW HOLE

⅜"-diameter hole
⅜" dowel coated in glue

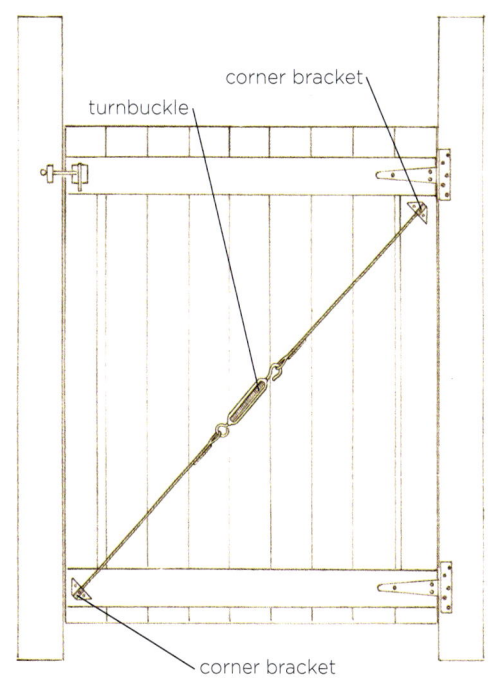

SAGGING GATE REPAIR

corner bracket
turnbuckle
corner bracket

Acknowledgments

BUILDING A FENCE can be, and often is, a solo endeavor. Creating a book about building fences, however, is a team sport, and I am fortunate and pleased to have had a terrific bunch of teammates in this effort.

Beginning with Deborah Balmuth, many staffers at Storey Publishing were involved in creating the finished product you are holding in your hands. My thanks to the editorial folks who helped shape and sell the idea for this book, and to the production and design gurus who carefully and skillfully fit the pieces together.

Two of my fellow freelancers deserve special thanks. My editor, Nancy Ringer, who kept me on message, and then helped make my message even clearer. And Melanie Powell, who took my illegible doodles and turned them into the splendid illustrations that fill these pages.

And to the folks upstairs at Storey, whose sales and publicity work gets into high gear only after I've finished writing these words, thanks in advance for what I am confident will be a fruitful effort. The frequency and quality of my vacations over the next few years rest squarely in your hands.

Interior Photography Credits

Photographs of hardware, materials, and tools © 2005 Storey Publishing

Additional photography by © Amelia/stock.adobe.com, 183; © Blake Gardner, 20; © Botanic World/Alamy Stock Photo, 200 b.; © Catriona Tudor Erler, ii 2nd fr.b. & b., vii r., 3 b., 16, 25, 29, 34–35, 42, 47, 50, 64, 93 t., 101 m.l., m.r., b.l. & b.m., 105 m.l. & b.l., 109 m.l., m.r. & b., 114 r., 121 t.l. & b., 125, 137 m.l., 138 b.l. & b.r., 145 l., 150, 153, 161 b.l., 166 b., 170, 171 t., 174 b.l., 175 b.l., 188 t.l., 192, 195, 204, 207, 215, 217 t.l. & t.r., 239, 240, 241 t.r., 242; © cpaulfell/Shutterstock.com, 201 b.l.; © David Cavagnaro, 28, 109 t.r., 137 t.l., t.r., m.r. & b.r., 145 r., 160 b., 173, 244; © Dolores M. Harvey/Shutterstock.com, 245; © Donna Boucher/Shutterstock.com, 137 b.l.; © Drepicter/Shutterstock.com, 138 t.l.; © Hilda DeSanctis/Alamy Stock Photo, x; © iofoto/Shutterstock.com, 2; © J Need/Shutterstock.com, 201 t.l.; © Joanne Dale/Shutterstock.com, 146 b.; © Jorge Salcedo/Shutterstock.com, 146 t.r.; © Joseph De Sciose, ii 2nd fr.t., vi r., 26, 56, 92 b., 101 t.r. & b.r., 105 t. & b.r., 108, 116, 161 t., 166 t.l., 171 b., 174 b.r., 182, 188 t.r., 189, 214, 236, 241 t.l.; © Keith Levit/Alamy Stock Photo, 3 t.; Kent Lew © Storey Publishing, 138 t.r.; © Kevin H Knuth/Shutterstock.com, 124; © Loong FeiLy/Shutterstock.com, 123 r.; © marina_lohrbach/Alamy Stock Photo, 196; © Marinodenisenko/Shutterstock.com, 201 m.l.; © Martin Hughes-Jones/Alamy Stock Photo, 200 t.; © MNStudio/Shutterstock.com, 27 t.; © Neil Lang/Shutterstock.com, 146 t.l.; © Ordasiphoto/Shutterstock.com, 212; © Rich Pomerantz, 92 t., 211; © Riepina Vladyslava/Shutterstock.com, 201 b.r.; © Roger Foley, ii t., vi l., vii l., 13, 32, 41, 66, 67 t., 90, 93 b., 114 l., 120, 135, 140, 158, 160 t.l. & t.r., 161 b.r., 168, 174 t., 175 t. & b.r., 194, 198, 209 t., 217 b.; © Rosalind Creasy, 121 t.r., 132, 134, 188 b., 216, 233, 241 b.; © Rosemary Kautzky, 67 b., 105 m.r., 123 l., 128, 130, 166 t.r.; © Tim Wright/Alamy Stock Photo, 209 b.; © Wiert nieuman/Shutterstock.com, 201 t.r.; © Wolfgang Volz/laif/Redux, 37

Index

Page numbers in *italic* indicate illustrations or photographs.

More Ways to Enhance Your Outdoor Living Space

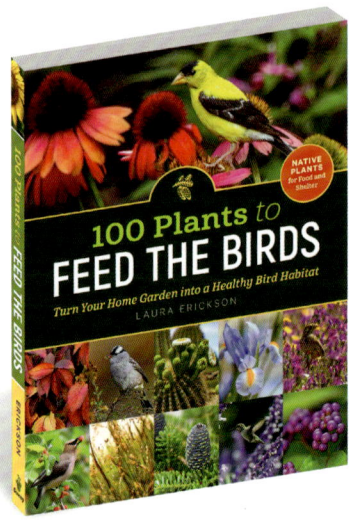

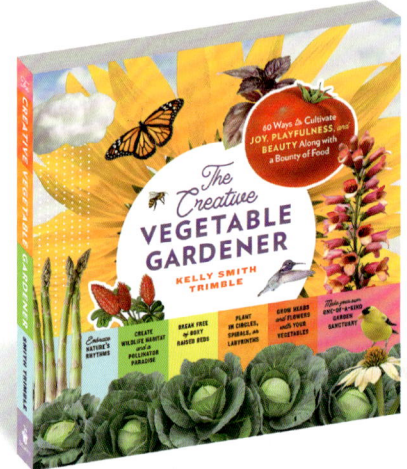

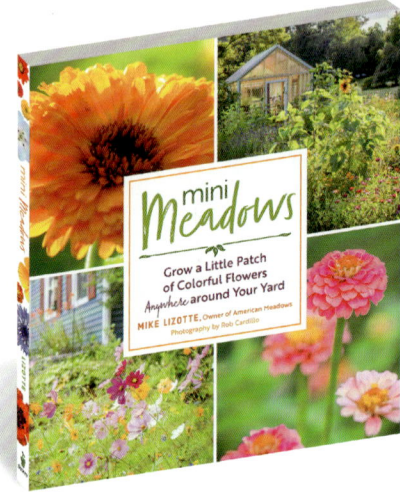

100 Plants to Feed the Birds

by Laura Erickson

Learn how to create a healthy landscape that will attract, feed, and shelter birds all year long, including during breeding and migrating periods. This expert guide offers in-depth planting and care information for 100 native plant species, plus color photographs and range maps.

The Creative Vegetable Gardener

by Kelly Smith Trimble

Imagine a different kind of vegetable garden—one that reflects your own unique aesthetic and offers an inspiring sanctuary as well as a source of homegrown food. Gardeners of all levels will love this book's advice on making the garden a place that nourishes the soul as well as the body.

Mini Meadows

by Mike Lizotte

With as little as 50 square feet and for less than $20, you can plant a colorful meadow that needs minimal maintenance, requires less water than a lawn, and provides a healthy habitat for pollinators. This guide walks you through the simple process of creating a mini meadow that suits your climate, soil, and desires.

Join the conversation. Share your experience with this book, learn more about Storey Publishing's authors, and read original essays and book excerpts at storey.com. Look for our books wherever quality books are sold or call 800-441-5700.